Progress in Plastic Deformation of Metals and Alloys (Second Volume)

Progress in Plastic Deformation of Metals and Alloys (Second Volume)

Guest Editor
Wojciech Borek

Basel • Beijing • Wuhan • Barcelona • Belgrade • Novi Sad • Cluj • Manchester

Guest Editor
Wojciech Borek
Department of Engineering
Materials and Biomaterials
Silesian University of Technology
Gliwice
Poland

Editorial Office
MDPI AG
Grosspeteranlage 5
4052 Basel, Switzerland

This is a reprint of the Special Issue, published open access by the journal *Materials* (ISSN 1996-1944), freely accessible at: www.mdpi.com/journal/materials/special_issues/plastic_deformation_metals_alloys_second_volume.

For citation purposes, cite each article independently as indicated on the article page online and using the guide below:

Lastname, A.A.; Lastname, B.B. Article Title. *Journal Name* **Year**, *Volume Number*, Page Range.

ISBN 978-3-7258-3396-2 (Hbk)
ISBN 978-3-7258-3395-5 (PDF)
https://doi.org/10.3390/books978-3-7258-3395-5

Cover image courtesy of Wojciech Borek

© 2025 by the authors. Articles in this book are Open Access and distributed under the Creative Commons Attribution (CC BY) license. The book as a whole is distributed by MDPI under the terms and conditions of the Creative Commons Attribution-NonCommercial-NoDerivs (CC BY-NC-ND) license (https://creativecommons.org/licenses/by-nc-nd/4.0/).

Contents

About the Editor vii

Preface ix

Khaled Elanany, Wojciech Borek and Saad Ebied
Constitutive Modelling Analysis and Hot Deformation Process of AISI 8822H Steel
Reprinted from: *Materials* 2024, 17, 5713, https://doi.org/10.3390/ma17235713 1

Martin Pitoňák, Anna Mičietová, Ján Moravec, Jiří Čapek, Miroslav Neslušan and Nikolaj Ganev
Influence of Strain Rate on Barkhausen Noise in Trip Steel
Reprinted from: *Materials* 2024, 17, 5330, https://doi.org/10.3390/ma17215330 25

Ndanduleni Lesley Lethole and Patrick Mukumba
Ab Initio Studies of Mechanical, Dynamical, and Thermodynamic Properties of Fe-Pt Alloys
Reprinted from: *Materials* 2024, 17, 3879, https://doi.org/10.3390/ma17153879 42

Tomasz Dyl, Dariusz Rydz, Arkadiusz Szarek, Grzegorz Stradomski, Joanna Fik and Michał Opydo
The Influence of Slide Burnishing on the Technological Quality of X2CrNiMo17-12-2 Steel
Reprinted from: *Materials* 2024, 17, 3403, https://doi.org/10.3390/ma17143403 56

Huan Qi, Qihang Pang, Weijuan Li and Shouyuan Bian
The Influence of the Second Phase on the Microstructure Evolution of the Welding Heat-Affected Zone of Q690 Steel with High Heat Input
Reprinted from: *Materials* 2024, 17, 613, https://doi.org/10.3390/ma17030613 72

Xingwen Yang, Jingtao Han and Ruilong Lu
Research on Cold Roll Forming Process of Strips for Truss Rods for Space Construction
Reprinted from: *Materials* 2023, 16, 7608, https://doi.org/10.3390/ma16247608 84

Chang-Feng Wan, Li-Gang Sun, Hai-Long Qin, Zhong-Nan Bi and Dong-Feng Li
A Molecular Dynamics Study on the Dislocation-Precipitate Interaction in a Nickel Based Superalloy During the Tensile Deformation
Reprinted from: *Materials* 2023, 16, 6140, https://doi.org/10.3390/ma16186140 102

Lovro Liverić, Tamara Holjevac Grgurić, Vilko Mandić and Robert Chulist
Influence of Manganese Content on Martensitic Transformation of Cu-Al-Mn-Ag Alloy
Reprinted from: *Materials* 2023, 16, 5782, https://doi.org/10.3390/ma16175782 115

Fei Wu, Yihao Hong, Zhengrong Zhang, Chun Huang and Zhenrong Huang
Effect of Lankford Coefficients on Springback Behavior during Deep Drawing of Stainless Steel Cylinders
Reprinted from: *Materials* 2023, 16, 4321, https://doi.org/10.3390/ma16124321 128

Błażej Tomiczek, Przemysław Snopiński, Wojciech Borek, Mariusz Król, Ana Romero Gutiérrez and Grzegorz Matula
Hot Deformation Behaviour of Additively Manufactured 18Ni-300 Maraging Steel
Reprinted from: *Materials* 2023, 16, 2412, https://doi.org/10.3390/ma16062412 147

Léo Thiercelin, Sophie Cazottes, Aurélien Saulot, Frédéric Lebon, Florian Mercier and Christophe Le Bourlot et al.
Development of Temperature-Controlled Shear Tests to Reproduce White-Etching-Layer Formation in Pearlitic Rail Steel
Reprinted from: *Materials* **2022**, *15*, 6590, https://doi.org/10.3390/ma15196590 . . . . **162**

About the Editor

Wojciech Borek

Dr. Wojciech Borek, PhD and MSc in Engineering, is an assistant professor in the Department of Engineering Materials and Biomaterials at the Silesian University of Technology in Gliwice, Poland. His scientific interests include materials science, heat treatment, thermomechanical treatment, plastic deformations, and Gleeble simulations; also, he is a specialist in steels, stainless steel, high-manganese austenitic steels, titanium alloys, and light metal alloys. He is an author and coauthor of, ca., 120 scientific publications worldwide including 10 chapters in books and more than 40 publications in the Web of Science database; he won 14 awards and honors, national and international; and he has served, or is currently serving, as a contractor of more than 12 research and didactic projects in Poland and abroad, a reviewer of numerous scientific publications, and co-promoter of two doctoral dissertations.

Preface

The plastic deformation of engineering materials involves changes to the geometrical shape of the investigated specimen and microstructures and affects how the deformed material reacts to the imposed stresses, and the value of strains depends primarily on the type of material, its chemical composition, and thus on its microstructure and texture. The Second Volume of this Special Issue focuses on new trends and progress in the hot and cold plastic deformation of metals and alloys and all new developments in the relationships between their structure and mechanical properties. All aspects related to plastic deformation from low to ultra-high strain, new methods, new technologies, and new applications in the broadly defined field of plastic deformation, as well as innovative approaches in this area, are welcomed. In addition, we cover thermomechanical processing, hot-rolling, heat treatment after plastic deformation, physical and numerical simulation of plastic deformation, and structural characterization. This Special Issue provides a multiscale approach to better understand the principal mechanisms of the plastic deformation of materials and their applications.

Wojciech Borek
Guest Editor

Article
Constitutive Modelling Analysis and Hot Deformation Process of AISI 8822H Steel

Khaled Elanany [1], Wojciech Borek [2,*] and Saad Ebied [3]

1. Silesian University of Technology, 2A Akademicka Str., 44-100 Gliwice, Poland; ke900594@student.polsl.pl
2. Department of Engineering Materials and Biomaterials, Silesian University of Technology, 18A Konarskiego Str., 44-100 Gliwice, Poland
3. Department of Production Engineering and Mechanical Design, Faculty of Engineering, Tanta University, Tanta 31527, Egypt; saad_ebied@f-eng.tanta.edu.eg
* Correspondence: wojciech.borek@polsl.pl

Abstract: This study used the Gleeble 3800 thermomechanical simulator to examine the hot deformation characteristics of AISI 8822H steel. The main goal was to understand the alloy's behaviour under various thermomechanical settings, emphasising temperature ranges between 1173 K and 1323 K and strain rates from 0.01 s^{-1} to 10 s^{-1}. This study aimed to enhance the alloy's manufacturing process by offering a thorough understanding of the material's response to these conditions. Four various constitutive models—Arrhenius-type, Johnson–Cook, modified Johnson–Cook, and Trimble—were used in a comprehensive technique to forecast flow stress values in order to meet the study's goals. The accuracy of each model in forecasting the behaviour of the material under the given circumstances was assessed. A thorough comparison investigation revealed that the Trimble model was the most accurate model allowing prediction of material behaviour, with the maximum correlation factor (R = 0.99) and at least average absolute relative error (1.7%). On the other hand, the Johnson–Cook model had the least correlation factor (R = 0.92) and the maximum average absolute relative error (32.2%), indicating that it was the least accurate because it could not account for all softening effects.

Keywords: AISI 8822H steel; hot deformation; strain rate; temperature; flow stress; Gleeble 3800; constitutive models

1. Introduction

Steel alloys are fundamental to technological and industrial advancements due to their versatility, strength, and customizable properties [1]. Composed primarily of iron and carbon, steel alloys can include additional elements to enhance specific characteristics [2]. For example, carbon increases hardness and strength, manganese improves wear resistance, nickel improves impact and corrosion resistance, and chromium boosts corrosion resistance [3]. To obtain the required qualities, iron ore must be melted, impurities must be eliminated, and alloying materials have to be added [4]. Tool steel, stainless steel, and carbon steel are the three types of steel that are suitable for different uses. Carbon steel is prevalent in structural and automotive uses; stainless steel is ideal for medical equipment and kitchenware due to its corrosion resistance; and tool steel's remarkable strength and heat endurance make it suitable for cutting tools. The adaptability and robust properties of steel alloys make them indispensable in industries like manufacturing, aerospace, construction, and medical equipment [5–9].

AISI 8822H is a nickel–chromium–molybdenum steel classified under H-steel grades, known for its high strength and toughness [10]. Its specific composition enhances its properties beyond standard carbon steels, making it ideal for mechanical applications [11]. It is commonly used for manufacturing crankshafts, gears, fasteners, axles, and shafts, all requiring high strength and hardness. Owing to its endurance and resistance to wear, this alloy is also appropriate for heavy-duty machinery and tools [12]. AISI 8822H is especially valuable in applications needing superior strength-to-weight ratios and outstanding mechanical performance, such as drilling tools and equipment used in extreme environments [13]. Forging and extrusion and other hot-forming processes can be used to further shape the structure and properties of this material and thus enable its use in new areas of industry.

Numerous constitutive models have been published to illustrate the flow stress behaviours of various alloys throughout temperature and strain rate ranges. The Arrhenius-type hyperbolic sine model, introduced by Sellars and Tegart [14], along with Zener and Hollomon [15], is one of the most common and earliest constitutive models. This model determines the relationship linking stress and strain rate, making it particularly suitable for elevated temperatures [16]. One of the main limitations of the Arrhenius model is its inability to consider strain. To take into account the strain effect and evaluate the flow stress behaviour in 42CrMO steel, Lin et al. developed an adjusted Arrhenius-type model incorporating strain compensation [17].

Another common constitutive model is the Johnson–Cook model. This model determines the alloy's flow stress behaviour during hot deformation testing using adiabatic and isothermal circumstances. Nevertheless, just one strain rate was used to study the impact of temperature [18]. To refine the initial Johnson–Cook model constants for the 2024Al-T351 alloy, Adibi-Sedeh et al. [19] carried out machining operations.

Based on actual compression data, Maheshwari et al. introduced an improved version of the Johnson–Cook model [20] to explain the flow stress characteristics of 2024Al alloy. The modified model was found to correlate better with experimental data in most scenarios compared to the original version. Maheshwari [21] developed a novel phenomenological constitutive model with a significantly higher correlation with data from experiments than the modified Johnson–Cook model previously provided. Khan and Liu [22] conducted compression tests on 2024Al-T351 alloy samples and, based on the results, proposed a novel phenomenological model describing the behaviour of flow stresses. Building on the original version of the Johnson–Cook model, LIN et al. [23] proposed an additional constitutive approach to forecast the deformation behaviour of Al-Zn-Mg-Cu, Al-Cu-Mg [24], and 7075Al [25] alloys during hot tensile tests. The authors observed higher prediction accuracy when compared to the Johnson–Cook model. To describe the flow stress behaviour of 7075Al, the researchers Trimble and O'Donnell [26] developed a newer model with a unique approach to constitutive modelling, described in detail in the research section of this article.

The main goal was to understand the alloy's behaviour under various thermomechanical settings, emphasising temperature ranges between 1173 K and 1323 K and strain rates from $0.01~\text{s}^{-1}$ to $10~\text{s}^{-1}$. This study aimed to enhance the alloy's manufacturing process by offering a thorough understanding of the material's response to these conditions. Four various constitutive models—Arrhenius-type, Johnson–Cook, modified Johnson–Cook, and Trimble—were used in a comprehensive technique to forecast flow stress values in order to meet the study's goals.

2. Experiments

The test samples were manufactured in a cylindrical shape, measuring 10 mm in diameter, 12 mm in length, and 12 mm in height. Table 1 shows the weight proportion of the AISI 8822H alloy's chemical composition [10]. The Gleeble 3800 thermomechanical simulator was used to conduct the hot uniaxial compression experiments at four distinct temperatures (1173, 1223, 1273, and 1323 K) and strain rates (0.01, 0.1, 1, and $10~\text{s}^{-1}$). To guarantee a consistent temperature distribution, the sample was heated to the deformation temperature with a heating rate of 3 °C/s and isothermal hold for one minute before being compressed, as illustrated in Figure 1. The Gleeble 3800 (Dynamic Systems Inc., Poughkeepsie, NY, USA) is equipped with direct resistance heating. This simulator is capable of heating samples at heating rates of over 10,000 °C/s and maintaining the temperature to within ±1 °C. The compressed air quenching procedure was used to maintain the deformed microstructure as soon as the samples were exposed to an actual true strain of 0.69. High-temperature nickel-based grease was applied to tantalum foils and on the contact surface of the sample to minimise friction between the sample tungsten carbide anvils. Type K thermocouple wires were used to monitor temperature throughout the test. For the hot compression test, ORIGIN PRO® 2024 was used to create true strain–true stress curves.

Table 1. AISI 8822H steel's chemical composition in weight percentage.

C	Mn	Si	Cr	Ni	Mo	Cu	Co	Fe
0.27	0.94	0.25	0.44	0.63	0.36	0.17	0.01	Bal.

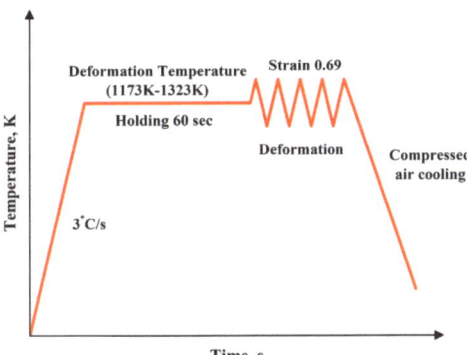

Figure 1. Schematic illustration representing the hot deformation process of AISI 8822H steel using the Gleeble simulator.

3. Results

Figure 2 displays typical true strain–true stress curves of the AISI 8822H alloy at various temperatures (1173, 1223, 1273, and 1323 K) and strain rates (0.01, 0.1, 1, and 10 s^{-1}) attained during the hot compression experiments. The curves in this figure were used to retrieve the flow stress values at different temperatures and strain rates within the true strain range of 0.2–0.6.

Due to work hardening, independent of the strain rate at initial strains, the flow stress increased with increasing strain under all test conditions and temperatures. The initial application of strain causes subgrain formation and an increase in dislocation density. The maximum stress and work hardening rate increased as expected with decreasing temperature and increasing strain rate.

The maximum stresses in the curves can be observed in the strain range of 0.1–0.3, with minor variations depending on strain rate and deformation temperature. The flow stress shows a characteristic maximum value of stress conservation at low strain rates of 0.01 and 0.1 s^{-1} for all temperatures. This is followed by flow softening with further straining. The flow stress curves displayed the usual behaviour of maximum stress peak and flow softening followed by a steady state, especially at lower strain rates (0.01–0.1 s^{-1}) and higher temperatures (1273–1323 K).

Shared softening features at low temperatures (usually 1173 K) and low strain rates (0.01 s^{-1}) could be explained by flow localisation. However, a subsequent increase in strain rate led to the flow localisation gradually disappearing as the flow stress behaviour attained stability, a feature of dynamic recovery (DRV). The flow curves should be able to show that they are approaching a steady state at 1 s^{-1}. Because of dynamic recovery, the steady-state flow imitates a dynamic equilibrium between flow softening and strain hardening. However, a decrease in the flow is observed around 10 s^{-1}, which may be caused by adiabatic heating. During the heated deformation phase, an adiabatic temperature rise is encouraged at a high strain rate of 10 s^{-1}. At the same strain rate, as a result, the flow stress curves decrease, particularly at low temperatures.

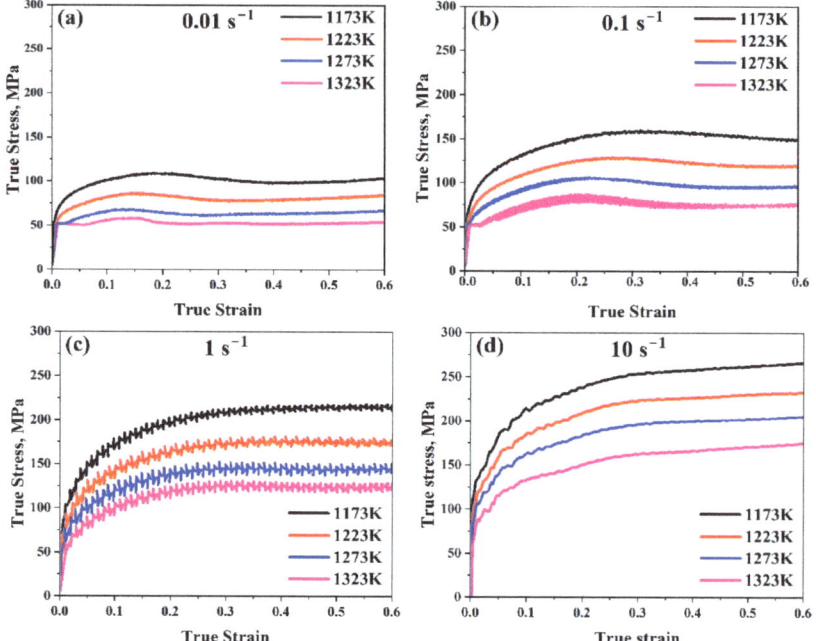

Figure 2. True strain–stress curves of AISI 8822H steel at strain rates (**a**) 0.01 s^{-1}, (**b**) 0.1 s^{-1}, (**c**) 1 s^{-1}, (**d**) 10 s^{-1}.

The flow stress behaviour of the AISI 8822H alloy was studied using four constitutive models at a strain step of 0.05. The models covered a temperature range of 1173–1323 K, a strain range of (0.2–0.6), and a strain rate range of 0.01–10 s^{-1}. The Arrhenius-based [14,15], original and modified Johnson–Cook [20,21], and Trimble models [26] were employed in this investigation. Following an assessment of each model's output, a comparison of experimental and anticipated flow stress levels was formed between the four models, determining which was better suited to assess the most accurate data.

3.1. Arrhenius-Type Model

The general formula of this model, as shown in Equation (1) [14], can be separated into three different forms depending on the level of stress [15], as shown in Equations (2)–(4).

$$\dot{\varepsilon} = AF(\sigma) exp\left(-\frac{Q_{def}}{RT}\right) \quad (1)$$

$$\dot{\varepsilon} = A_1 \sigma^{n'} exp\left(-\frac{Q_{def}}{RT}\right), \alpha\sigma < 0.8 \quad (2)$$

$$\dot{\varepsilon} = A_2 exp(\beta\sigma) exp\left(-\frac{Q_{def}}{RT}\right), \alpha\sigma > 1.2 \quad (3)$$

$$\dot{\varepsilon} = A[sinh(\alpha\sigma)]^n exp\left(-\frac{Q_{def}}{RT}\right), for\ all\ stresses \quad (4)$$

where σ is the flow stress (MPa); $\dot{\varepsilon}$ is the strain rate (s^{-1}); Q_{def} is the activation energy (J/mol); R is the universal gas constant (8.314 J/(mol·K)); T is the temperature (K); A, A_1, A_2, α, β are material constants; and n, n' are strain indices.

Equations (2) and (3) were used to evaluate n' and β, respectively, by taking natural logarithm in both sides, as shown in Equations (5) and (6). n' and β could be calculated as the mean slopes' values of the linear fits of the curves $(ln\dot{\varepsilon}$ vs. $ln\sigma)$ and $(ln\dot{\varepsilon}$ vs. $\sigma)$, as shown in Figure 3a and 3b, respectively. Therefore, α might be estimated using Equation (7).

$$ln\dot{\varepsilon} = lnA_1 + n'ln\sigma - \frac{Q_{def}}{RT} \tag{5}$$

$$ln\dot{\varepsilon} = lnA_2 + \beta\sigma - \frac{Q_{def}}{RT} \tag{6}$$

$$\alpha = \beta/n' \tag{7}$$

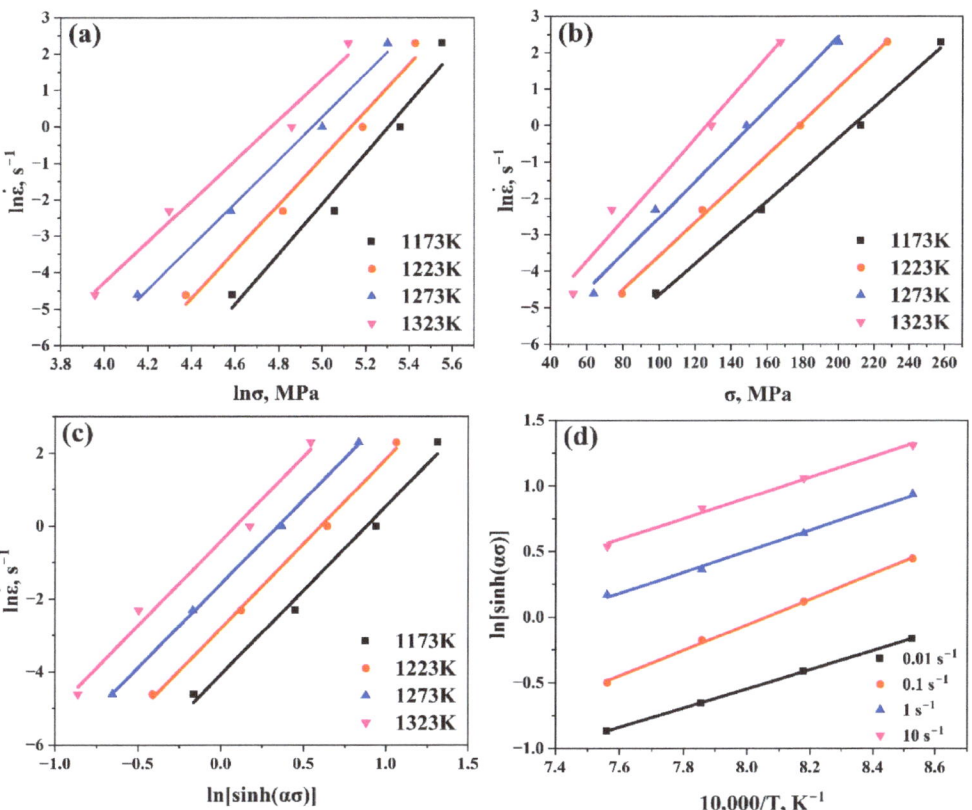

Figure 3. Plots of (**a**) $(ln\dot{\varepsilon}$ vs. $ln\sigma)$, (**b**) $(ln\dot{\varepsilon}$ vs. $\sigma)$, (**c**) $(ln\dot{\varepsilon}$ vs. $ln[sinh(\alpha\sigma)])$, (**d**) $(ln[sinh(\alpha\sigma)]$ vs. $10000/T)$ to evaluate n', β, n, and s, respectively, at 0.4 ε.

After that, n was estimated using Equation (8) as the mean slopes' values of the linear fits of the curve $(ln\dot{\varepsilon}$ vs. $ln[sinh(\alpha\sigma)])$, as shown in Figure 3c.

$$ln\dot{\varepsilon} = lnA + n\,ln[sinh(\alpha\sigma)] - \frac{Q_{def}}{RT} \tag{8}$$

In the next step, a secondary constant s was estimated using Equation (9) as the mean slopes' values of the linear fits of the curve ($ln[sinh(\alpha\sigma)]$ vs. $1/T$), as shown in Figure 3d. And, therefore, given n and R, Q could be estimated using Equation (10).

$$\frac{Q_{def}}{Rn} = \frac{\partial ln[sinh(\alpha\sigma)]}{\partial(1/T)} = s \tag{9}$$

$$Q_{def} = Rns \tag{10}$$

Equation (11) illustrates the temperature-compensated strain rate Z that Zener and Hollomon established. To obtain A, the intercept of the linear fit of the curve (lnZ vs. $ln[sinh(\alpha\sigma)]$) was found to provide lnA by taking the natural logarithm for each side, as seen in Figure 4, and consequently, A could be estimated.

$$Z = \dot{\varepsilon}\exp\left(\frac{Q_{def}}{RT}\right) = A[sinh(\alpha\sigma)]^n \tag{11}$$

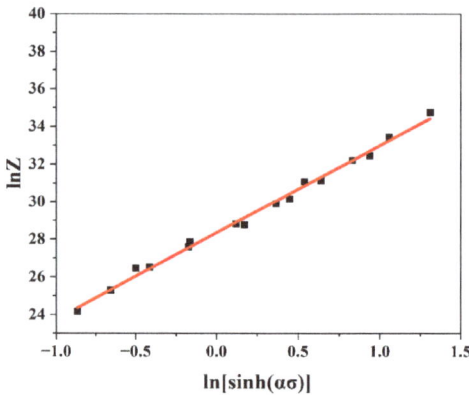

Figure 4. Plots of (lnZ vs. $ln[sinh(\alpha\sigma)]$) to evaluate lnA at 0.4 ε.

Using Figures 3 and 4, the values of n', β, α, n, Q_{def} (kJ/mol), and A at 0.4 ε could be evaluated as listed in Table 2.

Table 2. Values of the constants at 0.4 ε for the Arrhenius-type model.

n'	β	α	n	Q_{def}	A
6.2	0.05	0.008	4.6	316.5	2×10^{12}

Using Equation (11), the Arrhenius-type constitutive formula can be recast as follows:

$$\sigma = \frac{1}{\alpha}ln\left\{\left(\frac{Z}{A}\right)^{\frac{1}{n}} + \left[\left(\frac{Z}{A}\right)^{\frac{2}{n}} + 1\right]^{\frac{1}{2}}\right\} \tag{12}$$

Using Equation (12), the constitutive equations for all strains can be stated as follows:

$$\sigma_{0.4} = \frac{1}{0.008}ln\left\{\left(\frac{Z_{0.4}}{2*10^{12}}\right)^{\frac{1}{4.6}} + \left[\left(\frac{Z_{0.2}}{2*10^{12}}\right)^{\frac{2}{4.6}} + 1\right]^{\frac{1}{2}}\right\} \tag{13}$$

Repeating all the previous steps at a strain range of (0.2–0.6), five values of n', $β$, $α$, n, Q_{def}, and A could then be obtained. These values could be used to define all constitutive equations at the strain range. By substituting Z values into the constitutive equations, the predicted flow stress values at all temperature, strain, and strain rate ranges could be calculated.

A polynomial fit was produced for each constant evaluated under different strains, and it was demonstrated that the best match was achieved when the fourth order of polynomials was used, as illustrated in Figure 5.

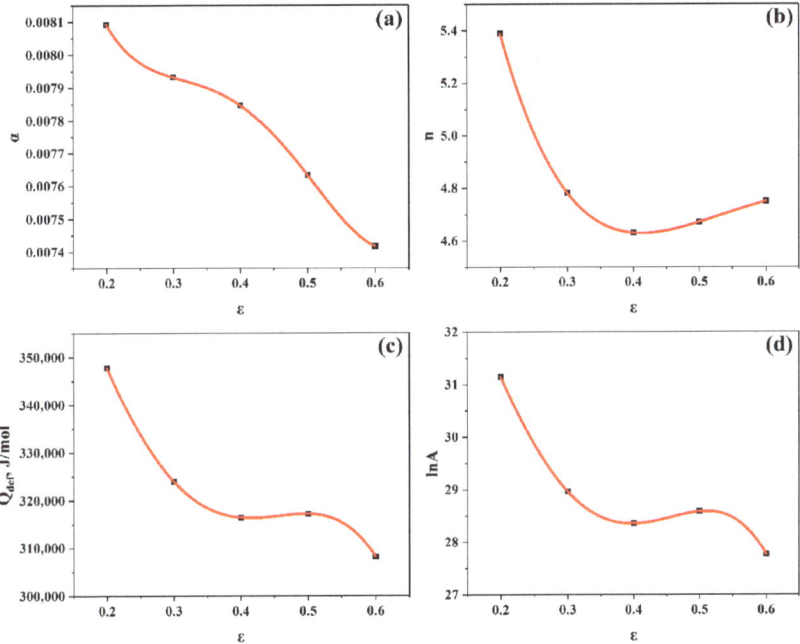

Figure 5. Polynomial fitting of (**a**) $α$ vs. $ε$, (**b**) n vs. $ε$, (**c**) Q_{def} vs. $ε$, (**d**) lnA vs. $ε$.

As a result, the regression in the equations of the relevant material constants $α$, n, Q_{def}, and lnA as a function of strain was reasonably described by Equations (14)–(17), respectively.

$$α = 0.0111 - 0.03348ε + 0.13187ε^2 - 0.22624ε^3 + 0.13731ε^4 \tag{14}$$

$$n = 9.55976 - 35.76636ε + 94.07557ε^2 - 106.60521ε^3 + 44.83021ε^4 \tag{15}$$

$$Q_{def} = 4.26 * 10^5 - 3.47 * 10^5 ε - 9.6 * 10^5 ε^2 + 4.53 * 10^6 ε^3 - 4.19 * 10^6 ε^4 \tag{16}$$

$$lnA = 37.54237 - 20.44644ε - 146.89034ε^2 + 541.78232ε^3 - 475.74717ε^4 \tag{17}$$

Using the correlation coefficient (R) and average absolute relative error ($AARE$), the predicted flow stress deviation was assessed to compare the predictability of such a constitutive model.

The correlation coefficient illustrates the significance of the linear correlation of the experimental and predicted values. It should be noted that an elevated (R) value does not always suggest better performance because the model tends to be biased toward higher or lower values. However, the average absolute relative error ($AARE$) is a statistical measure that may be used to objectively assess a model's predictability because it is computed by comparing the relative deviations term by term [15,27,28].

Equations (18) and (19) can be used to express (*AARE*) and (*R*). For the Arrhenius-type model, the values of (*AARE*) and (*R*) are 2.59% and 0.99, respectively, as shown in Figure 6.

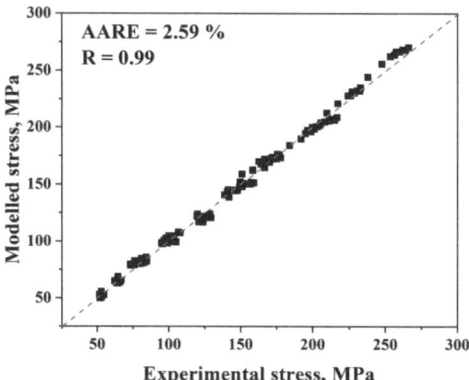

Figure 6. Comparison of the Arrhenius-type model's anticipated and experimental stress data.

$$AARE = \frac{1}{N}\sum_{i=1}^{N}\left|\frac{E_i - P_i}{E_i}\right| \times 100\% \qquad (18)$$

$$R = \frac{\sum_{i=1}^{N}(E_i - \overline{E})(P_i - \overline{P})}{\sqrt{\sum_{i=1}^{N}(E_i - \overline{E})^2 \sum_{i=1}^{N}(P_i - \overline{P})^2}} \qquad (19)$$

As illustrated in Figure 7, the strain–stress experimental curves could be compared with the predicted flow stress values after evaluation.

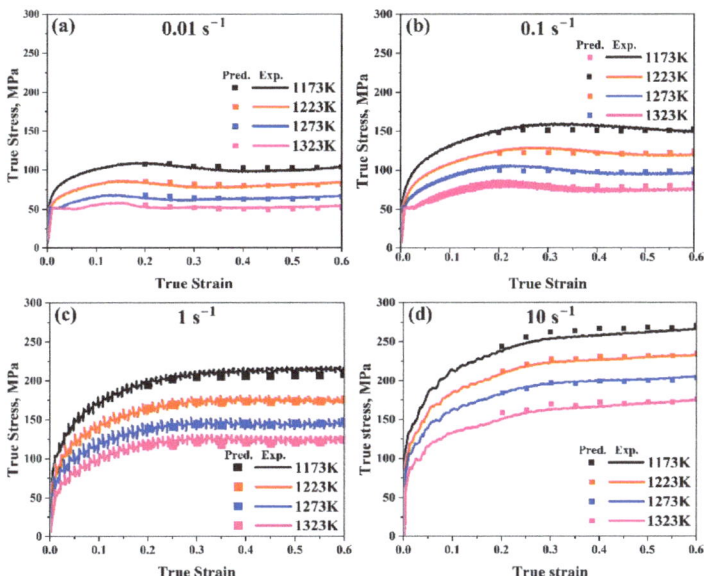

Figure 7. Comparison of the Arrhenius-type model's true strain–true stress anticipated and experimental flow stress data at temperature range (1173–1323 K) at (**a**) 0.01 s^{-1}, (**b**) 0.1 s^{-1}, (**c**) 1 s^{-1}, and (**d**) 10 s^{-1}, respectively.

3.2. Johnson–Cook Model (J–C)

An illustration of the Johnson–Cook model is as follows [16] in Equation (20):

$$\sigma = (A_j + B_j \varepsilon^{n_j})(1 + C_j ln\dot{\varepsilon}^*)(1 - (T_j^*)^{m_j}) \quad (20)$$

where ε is the strain; σ is the flow stress (MPa); $\dot{\varepsilon}^*$ is the dimensionless strain rate ($\dot{\varepsilon}^* = \dot{\varepsilon}/\dot{\varepsilon}_0$); $\dot{\varepsilon}$ is the strain rate (s^{-1}); $\dot{\varepsilon}_0$ is the reference strain rate (s^{-1}); T_j^* equals to $(T - T_r)/(T_m - T_r)$; T is the current temperature (K); T_m is the melting temperature (K); T_r is the reference temperature (K); A_j is the yield strength at the reference strain rate and temperature; and B_j, C_j, n_j, m_j are material constants.

Since the temperature range is (1173–1323 K), the reference temperature, 1173 K, was assumed to be the lowest value of this range. The reference strain rate was assumed to be 1 s^{-1}. Moreover, as shown in Figure 2c, the AISI 8822H alloy yield stress at reference conditions is approximately 60 MPa. It was estimated that our alloy melts around 1743 K. Yield stress, melting temperature, and reference values are exceptionally important for solving this model's general formula, as mentioned in Equation (20).

Equation (20) can be rewritten as follows by multiplying both sides by the natural logarithm at the reference conditions:

$$ln(\sigma - A_j) = lnB_j + n_j ln\varepsilon \quad (21)$$

Equation (21) was used to evaluate n_j and lnB_j by taking the slope and intercept of the linear fit of the curve $ln(\sigma - A_j)$ vs. $ln\varepsilon$, as shown in Figure 8. Hence, it was simple to evaluate B_j.

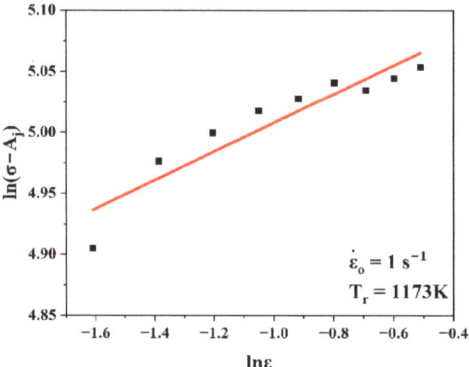

Figure 8. Plots of $ln(\sigma - A_j)$ vs. $ln\varepsilon$ to evaluate n_j and lnB_j.

Equation (20) can be modified to the following at reference temperature:

$$\frac{\sigma}{(A_j + B_j\varepsilon^{n_j})} - 1 = C_j ln\dot{\varepsilon}^* \tag{22}$$

After that, at the reference temperature, C_j could be estimated using Equation (22) by taking the mean slope value of the linear fit of the curve $\frac{\sigma}{(A_j+B\varepsilon^{n_j})} - 1$ vs. $ln\dot{\varepsilon}^*$, as shown in Figure 9a.

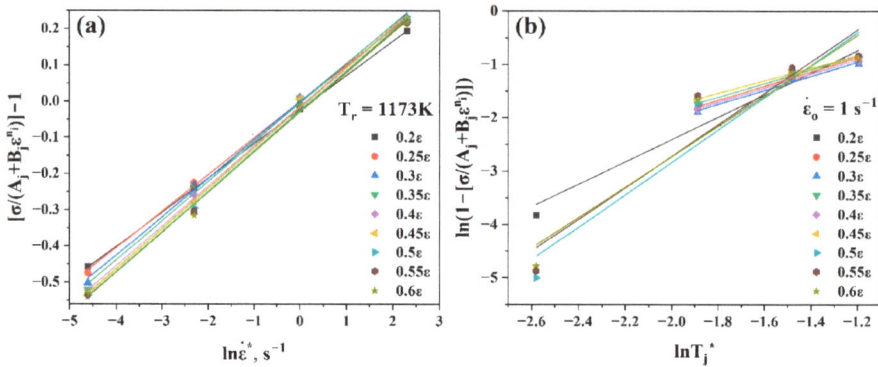

Figure 9. Plots of (a) $\frac{\sigma}{(A_j+B_j\varepsilon^{n_j})} - 1$ vs. $ln\dot{\varepsilon}^*$ and (b) $ln\left(1 - \frac{\sigma}{(A_j+B_j\varepsilon^{n_j})}\right)$ vs. lnT_j^* at strain range (0.2–0.6) to evaluate C_j and m_j, respectively.

Equation (20) can be rewritten as follows by multiplying both sides by the natural logarithm at the reference strain rate:

$$ln\left(1 - \frac{\sigma}{(A_j + B_j\varepsilon^{n_j})}\right) = m_j lnT_j^* \tag{23}$$

In the next step, at the reference strain rate, m_j was estimated using Equation (23) by taking the mean slope value of the linear fit of the curve $ln\left(1 - \frac{\sigma}{(A_j+B_j\varepsilon^{n_j})}\right)$ vs. lnT_j^*, as shown in Figure 9b. Since there are not available data at 1173 K for this curve, it was assumed that the reference temperature would be changed from 1173 K to 1123 K.

Using Figures 8 and 9, the values of A_j (MPa), n_j, B_j, C_j, and m_j at the strain range (0.2–0.6) could be evaluated as listed in Table 3. By substituting all the values of the constants, A_j, n_j, B_j, C_j, corresponding strain and strain rate values, and T_j^* into Equation (20), the anticipated flow stress values for all strain rates, temperatures, and strain ranges may be calculated.

Table 3. Values of the constants at strain range (0.2–0.6) for the Johnson–Cook model.

A_j	n_j	B_j	C_j	m_j
60	0.117	168.188	0.107	1.935

Therefore, it is possible to verify the validity of the Johnson–Cook model by comparing the experimental and anticipated flow stress levels using Equations (18) and (19), as shown in Figure 10. For the Johnson–Cook model, ($AARE$) and (R) are equal to 32.2% and 0.92, respectively.

As illustrated in Figure 11, the strain–stress experimental curves could be compared with the predicted flow stress values after evaluation.

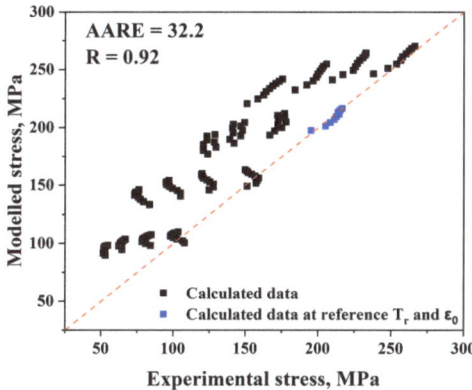

Figure 10. Comparison of the Johnson–Cook model's anticipated and experimental stress data.

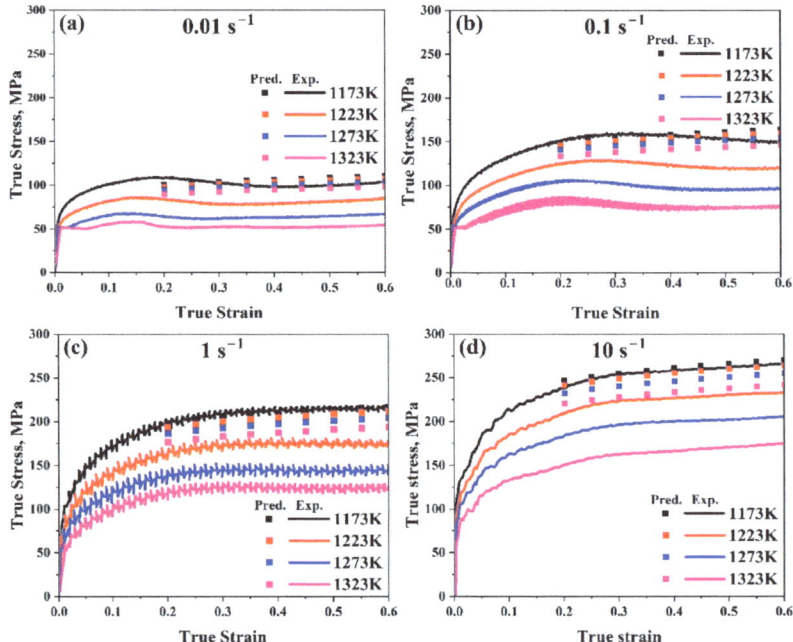

Figure 11. Comparison of the Johnson–Cook model's true strain–true stress anticipated and experimental flow stress data at temperature range (1173–1323 K) at (**a**) 0.01 s^{-1}, (**b**) 0.1 s^{-1}, (**c**) 1 s^{-1}, and (**d**) 10 s^{-1}, respectively.

3.3. Modified Johnson–Cook Model (Modified J–C)

The following formula can be used to depict the modified version of the Johnson–Cook model [17]:

$$\sigma = \left(P_j + Q_j \varepsilon^{n'_j}\right) \dot{\varepsilon}^{*r} \left[1 + \left(\frac{\sigma_m}{\sigma_y} - 1\right) exp\left(-\alpha_j T_j^{*/\beta_j}\right)\right] \quad (24)$$

where ε is the strain; σ is the flow stress (MPa); $\dot{\varepsilon}^*$ is the dimensionless strain rate ($\dot{\varepsilon}^* = \dot{\varepsilon}/\dot{\varepsilon}_o$); $\dot{\varepsilon}$ is the strain rate (s^{-1}); $\dot{\varepsilon}_o$ is the reference strain rate (s^{-1}); $T_j^{*/}$ equals to $0(T_m - T)/(T - T_r)$; T is the current temperature (K); T_m is the melting temperature (K); T_r is the reference temperature (K); P_j is the yield strength at the reference strain rate and temperature; and $Q_j, n'_j, r, \alpha_j, \beta_j$ are material constants. In order to solve this model, similarly to the original one, it was necessary to identify the reference strain rate, temperature, yield stress at these points, and melting point for AISI 8822H steel. These values were found to be 1 s^{-1}, 1173 K, 60 MPa, and 1743 K, respectively.

Equation (24) can be rewritten as follows by multiplying both sides by the natural logarithm at the reference conditions:

$$ln\left(\sigma - P_j\right) = lnQ_j + n'_j ln\varepsilon \quad (25)$$

Equation (25) was used to evaluate n'_j and lnQ_j by taking the slope and intercept of the linear fit of the curve $ln(\sigma - P_j)$ vs. $ln\varepsilon$, as shown in Figure 12. Hence, it was simple to evaluate Q_j.

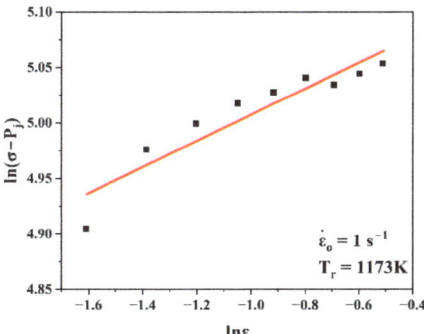

Figure 12. Plots of $ln(\sigma - P_j)$ vs. $ln\varepsilon$ to evaluate n'_j and lnQ_j.

Equation (24) can be modified to the following at reference temperature:

$$ln\left(\frac{\sigma}{P_j + Q_j\varepsilon^{n'_j}}\right) = rln\dot{\varepsilon}^* \qquad (26)$$

After that, at the reference temperature, r could be estimated using Equation (26) by taking the mean slope value of the linear fit of the curve $ln\left(\frac{\sigma}{P_j+Q_j\varepsilon^{n'_j}}\right)$ vs. $ln\dot{\varepsilon}^*$, as shown in Figure 13a.

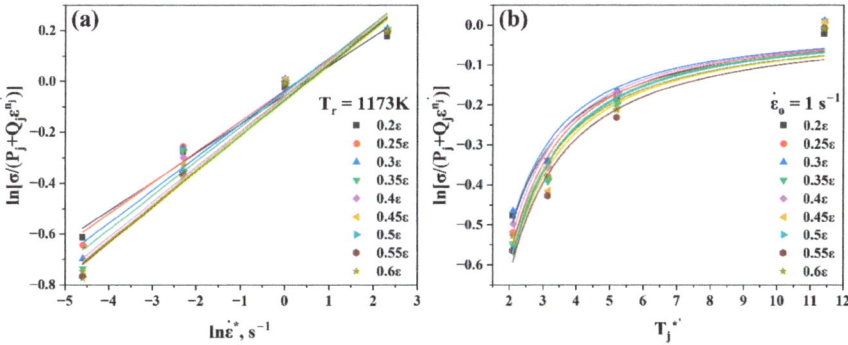

Figure 13. Plots of (a) $ln\left(\frac{\sigma}{P_j+Q_j\varepsilon^{n'_j}}\right)$ vs. $ln\dot{\varepsilon}^*$ and (b) $ln\left(\frac{\sigma}{P_j+Q_j\varepsilon^{n'_j}}\right)$ vs. $T_j^{*\prime}$ at strain range (0.2–0.6) to evaluate r, α_j, and β_j, respectively.

Equation (24) can be rewritten as follows by multiplying both sides by the natural logarithm at the reference strain rate:

$$ln\left(\frac{\sigma}{P_j + Q_j\varepsilon^{n'_j}\left(\dot{\varepsilon}^{*r}\right)}\right) = -\alpha_j T_j^{*\prime \beta_j} \qquad (27)$$

In the next step, at the reference strain rate, α_j and β were estimated using Equation (27) by taking the first-order power function fit of the curve $ln\left(\frac{\sigma}{P_j+Q_j\varepsilon^{n'_j}}\right)$ vs. $T_j^{*\prime}$, as shown in Figure 13b.

Similarly to the Johnson–Cook model, the reference temperature was changed from 1173 K to 1123 K. Using Figures 8 and 9, the values of P_j (MPa), n'_j, Q_j, r, α_j, and β_j at the strain range (0.2–0.6) could be evaluated as listed in Table 4. By substituting all the values of the constants, P_j, n'_j, Q_j, r, α_j, β_j, corresponding strain and strain rate values, and T_j^* into Equation (24), the anticipated flow stress values for all strain rates, temperatures, and strain ranges may be calculated.

Table 4. Values of the constants at strain range (0.2–0.6) for the modified Johnson–Cook model.

P_j	n'_j	Q_j	r	α_j	β_j
60	0.12	168.19	0.13	1.36	−1.23

It is therefore possible to verify the validity of the Johnson–Cook model by comparing the experimental and anticipated flow stress levels using Equations (18) and (19), as shown in Figure 14. For the modified Johnson–Cook model, ($AARE$) and (R) are equal to 9.2% and 0.98, respectively.

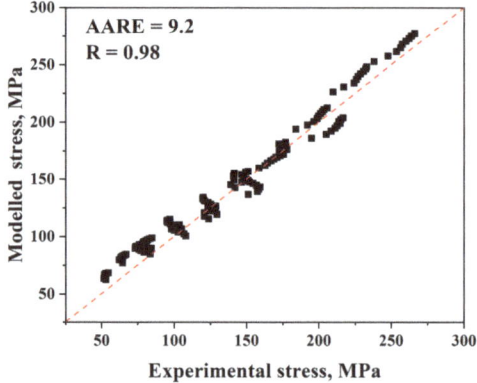

Figure 14. Comparison of the modified Johnson–Cook model's anticipated and experimental stress data.

As illustrated in Figure 15, the true strain–true stress experimental curves could be compared with the predicted flow stress values after evaluation.

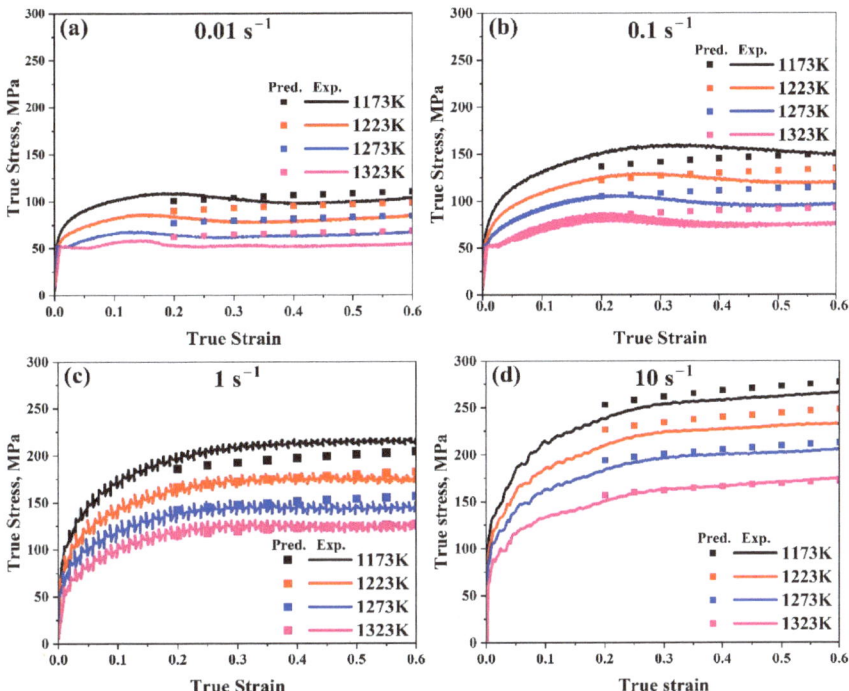

Figure 15. Comparison of the modified Johnson–Cook model's true strain–true stress anticipated and experimental flow stress data at temperature range (1173–1323 K) at (**a**) 0.01 s^{-1}, (**b**) 0.1 s^{-1}, (**c**) 1 s^{-1}, and (**d**) 10 s^{-1}, respectively.

3.4. Trimble Model

The Trimble model is defined by the following formula [18]:

$$\sigma = A_t \varepsilon^{n_t} exp(B_t \varepsilon + C_t) T_t^* \tag{28}$$

where σ is the flow stress (MPa); ε is the strain; T_t^* equals to $(T - T_r)$; T is the current temperature (K); T_r is the reference temperature (K); and A_t, n_t, B_t, C_t are material constants. In contrast to the Johnson–Cook models, this model started with a reference temperature of 1123 K.

Equation (24) can be rewritten as follows by multiplying both sides by the natural logarithm:

$$ln\sigma = lnA_t + n_t ln\varepsilon + (B_t \varepsilon + C_t) T_t^* \tag{29}$$

For each strain value at one particular strain rate, to be able to solve this model, two additional parameters were assumed, as mentioned in Equations (30) and (31):

$$S_t = B_t \varepsilon + C_t \tag{30}$$

$$I_t = lnA_t + n_t ln\varepsilon \tag{31}$$

Equation (29) was used to evaluate S_t and I_t by taking the slope and intercept of the linear fit of the curve $ln\sigma$ vs. T_t^* at the strain range (0.2–0.6), as shown in Figure 16.

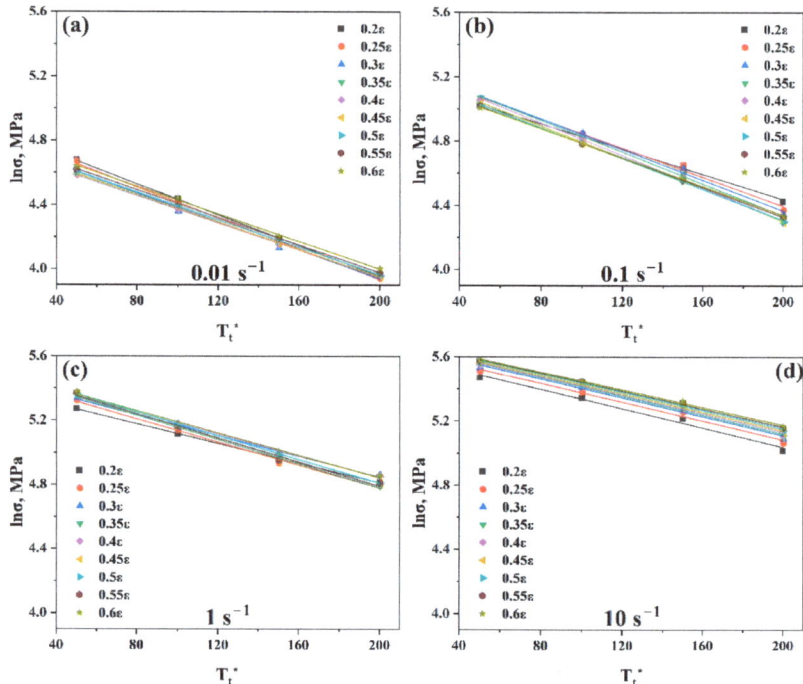

Figure 16. Plots of ($ln\sigma$ vs. T_t^*) at strain range (0.2–0.6) at (**a**) 0.01 s^{-1}, (**b**) 0.1 s^{-1}, (**c**) 1 s^{-1}, and (**d**) 10 s^{-1} to evaluate S_t and I_t, respectively.

Equation (30) was used to evaluate B_t and C_t by taking the slope and intercept of the linear fit of the curve (S_t vs. ε), as shown in Figure 17a.

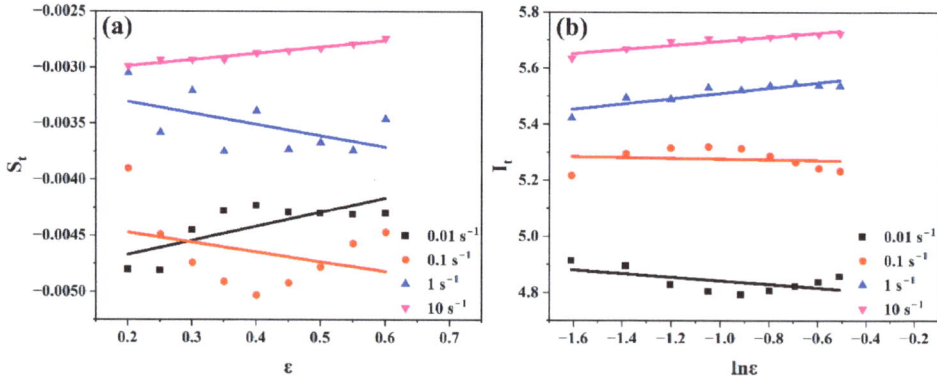

Figure 17. Plots of (**a**) (S_t vs. ε) to evaluate B_t and C_t and (**b**) (I_t vs. $ln\varepsilon$) to evaluate n_t and lnA_t at strain rate range (0.01–10 s^{-1}).

Equation (30) was used to evaluate n_t and lnA_t by taking the slope and intercept of the linear fit of the curve I_t vs. $ln\varepsilon$, as shown in Figure 17b. Hence, it is simple to evaluate A_t.

Using Figure 17, the values of B_t, C_t, n_t, and A_t at the strain rate range (0.01–10 s^{-1}) could be evaluated as listed in Table 5.

Table 5. Values of the constants at strain rate range (0.01–10 s^{-1}) for the Trimble model.

$\dot{\varepsilon}$ (s^{-1})	Constants			
	B_t	C_t	n_t	A_t
0.01	0.0013	−0.0049	−0.0633	118.96
0.1	−0.0009	−0.0043	−0.0127	193.46
1	−0.001	−0.00311	0.0947	271.8
10	−0.0006	−0.0031	0.0739	320.91

By substituting all the values of the constants, B_t, C_t, n_t, A_t corresponding strain values, and T_t^* into Equation (29), the anticipated flow stress values for all strain rates, temperatures, and strain ranges may be calculated.

A polynomial fit was produced for each constant evaluated under different strain rates, and it was demonstrated that the best match was achieved when the third order of polynomials was used, as illustrated in Figure 18.

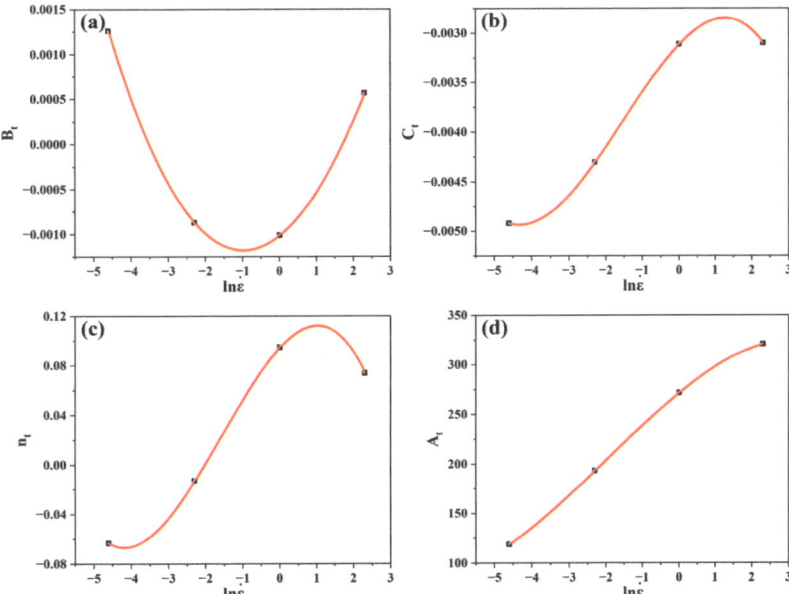

Figure 18. Polynomial fitting of (**a**) B_t vs. $ln\dot{\varepsilon}$, (**b**) C_t vs. $ln\dot{\varepsilon}$, (**c**) n_t vs. $ln\dot{\varepsilon}$, and (**d**) A_t vs. $ln\dot{\varepsilon}$.

As a result, the regression in the equations of the relevant material constants B_t, C_t, n_t, and A_t as a function of strain rate was reasonably described by Equations (32)–(35), respectively.

$$B_t = -0.001 + 3.32*10^{-4} ln\dot{\varepsilon} + 1.62*10^{-4} ln\dot{\varepsilon}^2 - 3.73*10^{-6} ln\dot{\varepsilon}^3 \tag{32}$$

$$C_t = -0.0031 + 3.87*10^{-4} ln\dot{\varepsilon} - 1.11*10^{-4} ln\dot{\varepsilon}^2 - 2.39*10^{-5} ln\dot{\varepsilon}^3 \tag{33}$$

$$n_t = 0.0947 + 0.0322 ln\dot{\varepsilon} - 0.0121 ln\dot{\varepsilon}^2 - 0.0025 ln\dot{\varepsilon}^3 \tag{34}$$

$$A_t = 271.8 + 30.07 ln\dot{\varepsilon} - 2.76 ln\dot{\varepsilon}^2 - 0.4516 ln\dot{\varepsilon}^3 \tag{35}$$

It is therefore possible to verify the validity of the Trimble model by comparing the experimental and anticipated flow stress levels using Equations (18) and (19), as shown in Figure 19. For the Trimble model, (*AARE*) and (*R*) are equal to 1.7% and 0.99, respectively.

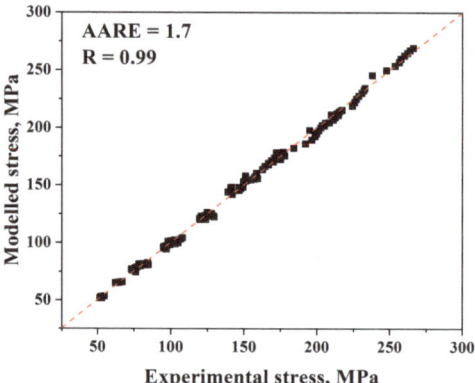

Figure 19. Comparison of the Trimble model's anticipated and experimental stress data.

As illustrated in Figure 20, the true strain–true stress experimental curves could be compared with the predicted flow stress values after evaluation.

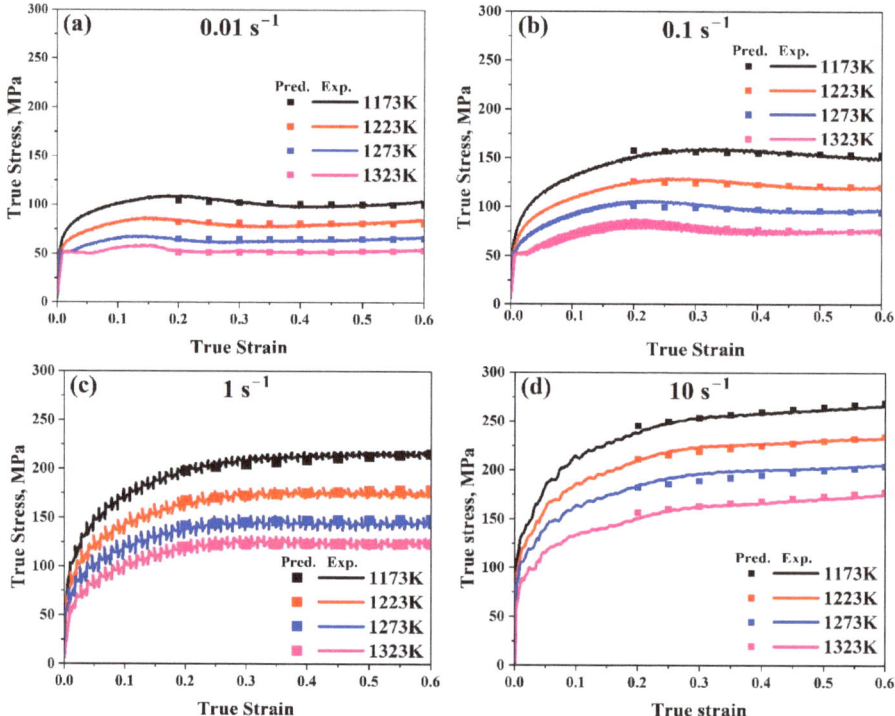

Figure 20. Comparison of the Trimble model's true strain–true stress anticipated and experimental stress data at temperature range (1173–1323 K) at (**a**) 0.01 s^{-1}, (**b**) 0.1 s^{-1}, (**c**) 1 s^{-1}, and (**d**) 10 s^{-1}, respectively.

4. Discussion

The Arrhenius-type model is one of the most widely used constitutive models for predicting flow stress values in hot deformation testing. After comparing our experimental and anticipated data, it was observed from Figure 6 that the results are very close, which confirms that this model is quite accurate at elevated temperatures. Furthermore, by contrasting the measured curves for a true strain–true stress with the anticipated flow stress values, as illustrated in Figure 7, it is observed that for all ranges, predicted data are very close to the experimental ones, proving that this model is accurate to predict our data. Furthermore, it can be found that ($AARE$) is too low (2.6%) and (R) is too high (0.99). Taking into account that the lower the ($AARE$) and the higher the R, the more accurate the model, it can be determined that, in general, among the models that forecast flow stress values, one of the most accurate is the Arrhenius-type.

For the Johnson–Cook model, it is evident from Figure 10 that experimental and predicted data only exhibit good consistency when the strain rate and reference temperature are met; under all other circumstances, there are significant deviations. Moreover, by making a comparison between true strain–true stress experimental and predicted flow stress values, it was observed from Figure 11 that there is a significant variation in the data as the temperature rises, even though the general pattern in the experimental data curve and the predicted value match. Additionally, the more the temperature, strain rate, and reference value vary, the more significant the variation between the experimental and anticipated data.

The main reason for this is that an accurate and successful model of metals should take into consideration the impacts of the three distinct forms of softening, in contrast to the Johnson–Cook constitutive model, which treats them as different elements [16]. Researchers who have examined this topic have discovered that these three impacts are coupled. The coupling effect, which can result in nonlinear changes in the rheological behaviour of metals depending on the situation, is caused by the intricate interaction between dislocation build-up and recovery mechanisms throughout plastic deformation.

Specifically, the temperature softening effect and transient strain rate may interact positively or negatively, while the material's various microstructural properties and processing conditions cause the strain hardening effect. Thus, compounding the impact of several components cannot adequately capture the complicated plastic deformation behaviour. However, some metals are more vulnerable to the coupling impact of many factors. The study results show that this coupling impact is evident in the plastic deformation of AISI 8822H steel at high temperatures. Therefore, to construct a constitutive model of AISI 8822H steel with more accuracy throughout the plastic deformation, the above model must be altered.

Furthermore, it can be found that ($AARE$) is too high (32.2%) and (R) is too low (0.92). Therefore, it may be concluded that the Johnson–Cook model is generally not a reliable model for predicting flow stress values.

More recent constitutive models have focused on developing or modifying new models. According to experimental hot compression values, a modified model explains the flow stress behaviour of the AISI 8822H alloy. In most cases, the updated model was found to have a higher correlation with experimental data than the original model, as illustrated in Figure 21.

As illustrated in Figure 14, the updated model's anticipated flow stress values are excessively near the experimental values, increasing the model's accuracy.

Furthermore, as illustrated in Figure 15, contrasting the measured curves for a true strain–true stress with the anticipated flow stress values reveals that for all ranges, the predicted data for the modified model are significantly closer to the experimental ones than the original one. This demonstrates how well the updated model predicts our data compared to the original.

Additionally, it is discovered that the (*AARE*) and (*R*) values are equal to 9.2% and 0.98, respectively. Hence, it can be determined that, in general, the modified Johnson–Cook model is clearly more accurate than the original one for predicting flow stress values.

The Trimble model is the newest model used to forecast flow stress data. After comparing our experimental and anticipated data in the Trimble model, it was observed from Figure 19 that the results are very close, which confirms how accurate it is. Furthermore, by contrasting the measured curves for a true strain–true stress with the anticipated flow stress values, as illustrated in Figure 20, it is observed that for all ranges, predicted data are very close to the experimental ones, which again proves that this model is accurate for predicting our data. Furthermore, it can be found that (*AARE*) is too low (1.7%) and (*R*) is too high (0.99). Therefore, the Trimble model is one of the best models for predicting flow stress levels.

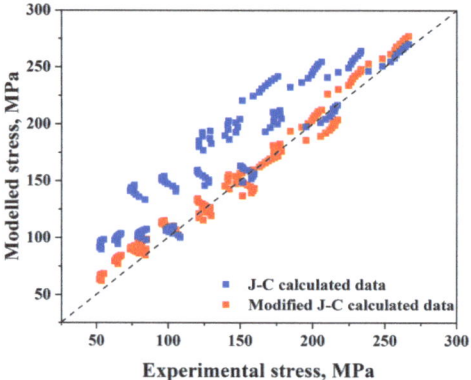

Figure 21. Comparison of the experimental and anticipated stress data for the original and modified Johnson–Cook models.

After evaluating all constitutive equations and flow stress values of the four models, it is now time to compare each one to identify the most accurate and appropriate model for assessing the anticipated flow stress values.

First, the experimental and expected flow stress values for each of these models were compared, as illustrated in Figure 22. This figure shows that the Arrhenius-type and Trimble models are the most constitutive accurate ones, while the Johnson–Cook model is the least accurate one for forecasting the flow stress values.

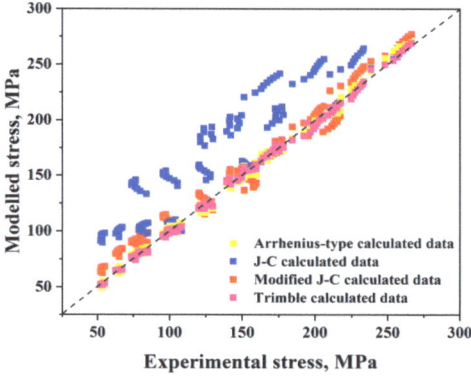

Figure 22. Comparison of the experimental and anticipated stress data for the four models.

Additionally, the modified Johnson–Cook model is much better than the original but less accurate than the Arrhenius-type and Trimble models.

Furthermore, by making a comparison between true strain–true stress experimental data and predicted flow stress values for these models, in our study, the behaviour of flow stress of AISI 8822H steel was checked concerning two cases. The first case was applied at the range (1173–1323 K) with a particular strain rate value of $1~\text{s}^{-1}$, for example, as illustrated in Figure 23. The second case was applied at the range (0.01–10 s^{-1}) with a specific temperature value of 1173 K, for example, as shown in Figure 24. The Arrhenius-type and Trimble models are the most accurate models since the variation between the anticipated and experimental values in both models is always too low for all ranges. The Johnson–Cook model is accurate at all strain rate values, especially at 1173 K, since the variation between the anticipated and experimental data is low. The higher the temperature, the lower the accuracy since the variation between the anticipated and experimental data becomes higher and higher for all strain rate values. The weak accuracy of the Johnson–Cook model is improved by using the modified model, particularly at temperatures over 1173 K. Consequently, it is evident that the variation between the anticipated and experimental data in this model is always too low for all strain rates and temperature values higher than 1173 K.

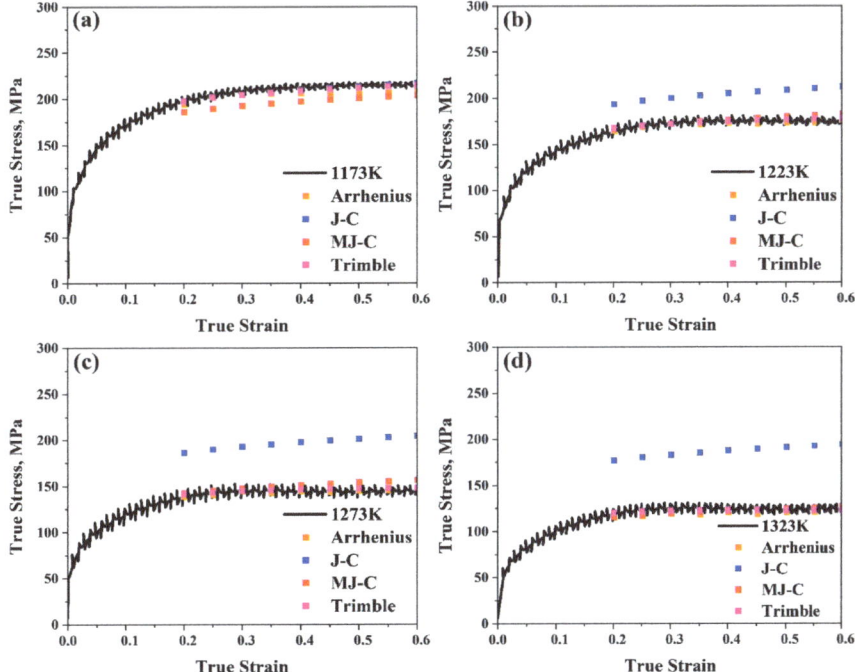

Figure 23. Comparison between true strain–true stress experimental curves and predicted flow stress values for the Arrhenius-type, Johnson–Cook, modified Johnson–Cook, and Trimble models at $1~\text{s}^{-1}$ at (**a**) 1173 K, (**b**) 1223 K, (**c**) 1273 K, and (**d**) 1323 K, respectively.

One more approach to comparing the four models is to compare the values of ($AARE$) and (R), as represented in Table 6. This table clearly observes that the smallest value of ($AARE$) and the largest value of (R) exist in the Trimble model, which emphasises that, for AISI 8822H steel, the most accurate model to assess the anticipated flow stress values is the Trimble model.

Table 6. Comparison between the Arrhenius-type, Johnson–Cook, modified Johnson–Cook, and Trimble models with respect to ARRE % and R.

	Constitutive Models			
	Arrhenius-Type	Johnson–Cook	Modified Johnson–Cook	Trimble
ARRE %	2.6	32.2	9.2	1.7
R	0.99	0.92	0.98	0.99

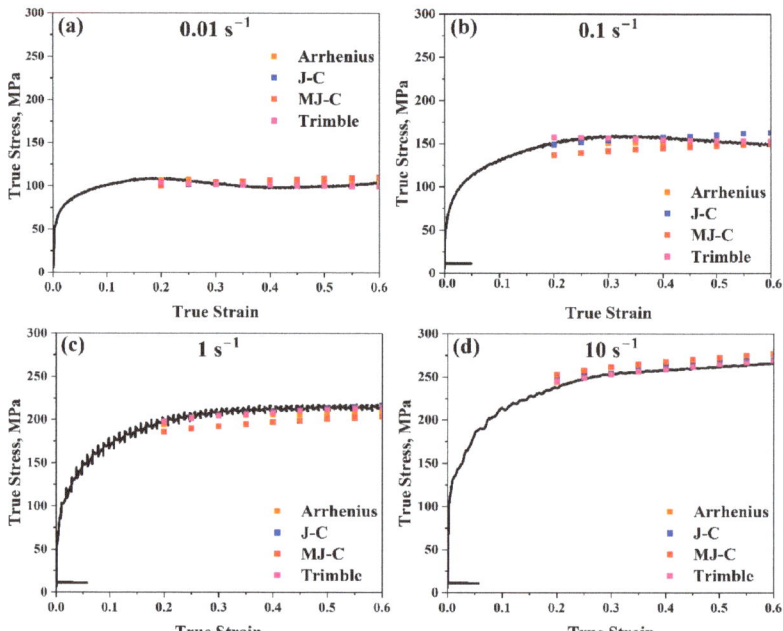

Figure 24. Comparison between true strain–true stress experimental curves and modelled flow stress values for the Arrhenius-type, Johnson–Cook, modified Johnson–Cook, and Trimble models at 1173 K at (a) 0.01 s^{-1}, (b) 0.1 s^{-1}, (c) 1 s^{-1}, and (d) 10 s^{-1}, respectively.

5. Conclusions

The following observations were developed after consideration of the compression tests performed on axisymmetric samples and an investigation of the test data that were obtained:

(1) The flow stress values for AISI 8822H steel increased when the strain rate increased at a constant temperature and when the deformation temperature decreased at a constant strain rate.

(2) Based on the results of (AARE) and (R) for the four constitutive models, the Trimble model was found to have the highest (R) value, which equals 0.99, and the lowest AARE value, which equals 1.7%. Consequently, the Trimble model is the most suitable for predicting the hot deformation behaviour of AISI 8822H steel over the processing range investigated in this study.

(3) The Johnson–Cook model was found to have the lowest (R) value, 0.92, and the highest AARE value, 32.2%. Consequently, the Johnson–Cook model is the least suitable for predicting the hot deformation behaviour of AISI 8822H steel over the processing range investigated in this study.

Author Contributions: Conceptualisation, W.B. and K.E.; methodology, W.B. and K.E.; validation, W.B. and S.E.; formal analysis, W.B. and S.E.; investigation, K.E.; resources, W.B. and K.E.; data curation, W.B. and K.E.; writing—original draft preparation, K.E.; writing—review and editing, W.B. and K.E.; visualisation, K.E.; supervision, W.B. and S.E.; project administration W.B.; funding acquisition, W.B. All authors have read and agreed to the published version of the manuscript.

Funding: This research received no external funding.

Institutional Review Board Statement: Not applicable.

Informed Consent Statement: Not applicable.

Data Availability Statement: The generated data can be obtained from wojciech.borek@polsl.pl or ke900594@student.polsl.pl.

Acknowledgments: This work was performed at the Silesian University of Technology, Gliwice, Poland in cooperation with researcher from Tanta University, Egypt.

Conflicts of Interest: The authors declare no conflicts of interest.

References

1. El-Bitar, T.; El-Meligy, M.; Borek, W.; Ebied, S. Characterization of Hot Deformation Behavior for Ultra-High Strength Steel Containing Tungsten. *Acta Metall. Slovaca* **2023**, *29*, 144–147. [CrossRef]
2. Liu, X.; Bao, Y.; Zhao, L.; Gu, C. Establishment and Application of Steel Composition Prediction Model Based on t-Distributed Stochastic Neighbor Embedding (t-SNE) Dimensionality Reduction Algorithm. *J. Sustain. Metall.* **2024**, *10*, 509–524. [CrossRef]
3. Wang, Y.; Oh, M.K.; Kim, T.S.; Karasev, A.; Mu, W.; Park, J.H.; Jönsson, P.G. Effect of LCFeCr alloy additions on the non-metallic inclusion characteristics in Ti-containing ferritic stainless steel. *Metall. Mater. Trans. B* **2021**, *52*, 3815–3832. [CrossRef]
4. Tomiczek, B.; Snopiński, P.; Borek, W.; Król, M.; Gutiérrez, A.R.; Matula, G. Hot deformation behaviour of additively manufactured 18Ni-300 maraging steel. *Materials* **2023**, *16*, 2412. [CrossRef]
5. Mohrbacher, H.; Kern, A. Nickel alloying in carbon steel: Fundamentals and applications. *Alloys* **2023**, *2*, 1–28. [CrossRef]
6. Xia, T.; Ma, Y.; Zhang, Y.; Li, J.; Xu, H. Effect of Mo and Cr on the Microstructure and Properties of Low-Alloy Wear-Resistant Steels. *Materials* **2024**, *17*, 2408. [CrossRef]
7. Backhouse, A.; Baddoo, N. Recent developments of stainless steels in structural applications. *ce/papers* **2021**, *4*, 2349–2355. [CrossRef]
8. Reséndiz-Calderón, C.D.; Cao-Romero-Gallegos, J.A.; Farfan-Cabrera, L.I.; Campos-Silva, I.; Soriano-Vargas, O. Influence of boriding on the tribological behavior of AISI D2 tool steel for dry deep drawing of stainless steel and aluminum. *Surf. Coat. Technol.* **2024**, *484*, 130832. [CrossRef]
9. dos Santos, F.A.M.; Martorano, M.A.; Padilha, A.F. Delta ferrite formation and evolution during slab processing from an 80-ton industrial heat of AISI 304 austenitic stainless steel. *REM-Int. Eng. J.* **2023**, *76*, 47–54. [CrossRef]
10. Lopez-Hirata, V.M.; Saucedo-Muñoz, M.L.; Rodríguez-Rodríguez, K.; Dorantes-Rosales, H.J. Homogenizing Treatment of AISI 420 Stainless and AISI 8620 Steels. In Proceedings of the TMS Annual Meeting & Exhibition, Orlando, FL, USA, 3–7 March 2024; Springer: Berlin/Heidelberg, Germany, 2024; pp. 497–505.
11. Alshareef, A.J.; Marinescu, I.D.; Basudan, I.M.; Alqahtani, B.M.; Tharwan, M.Y. Ball-burnishing factors affecting residual stress of AISI 8620 steel. *Int. J. Adv. Manuf. Technol.* **2020**, *107*, 1387–1397. [CrossRef]
12. Liu, Z.; Cui, T.; Chen, Y.; Dong, Z. Effect of Cu addition to AISI 8630 steel on the resistance to microbial corrosion. *Bioelectrochemistry* **2023**, *152*, 108412. [CrossRef] [PubMed]
13. Banerjee, A.; Ntovas, M.; Da Silva, L.; O'Neill, R.; Rahimi, S. Continuous drive friction welding of AISI 8630 low-alloy steel: Experimental investigations on microstructure evolution and mechanical properties. *J. Manuf. Sci. Eng.* **2022**, *144*, 71001. [CrossRef]
14. Sellars, C.M.; Tegart, W.J.M. Hot workability. *Int. Metall. Rev.* **1972**, *17*, 1–24. [CrossRef]
15. Zener, C.; Hollomon, J.H. Effect of strain rate upon plastic flow of steel. *J. Appl. Phys.* **1944**, *15*, 22–32. [CrossRef]
16. Zhou, Y.; Yang, G.; He, X.; Zhou, L.; Zhai, Y. Constitutive Model of TNM Alloy Using Arrhenius-Type Model and Artificial Neural Network Model. *J. Phys. Conf. Ser.* **2023**, *2437*, 12062. [CrossRef]
17. Lin, Y.C.; Chen, M.-S.; Zhong, J. Constitutive modeling for elevated temperature flow behavior of 42CrMo steel. *Comput. Mater. Sci.* **2008**, *42*, 470–477. [CrossRef]
18. Wang, Z.; Jiang, C.; Wei, B.; Wang, Y. Analysis of the High Temperature Plastic Deformation Characteristics of 18CrNi4A Steel and Establishment of a Modified Johnson–Cook Constitutive Model. *Coatings* **2023**, *13*, 1697. [CrossRef]
19. Adibi-Sedeh, A.H.; Madhavan, V.; Bahr, B. Extension of Oxley's analysis of machining to use different material models. *J. Manuf. Sci. Eng.* **2003**, *125*, 656–666. [CrossRef]
20. Maheshwari, A.K.; Pathak, K.K.; Ramakrishnan, N.; Narayan, S.P. Modified Johnson-Cook material flow model for hot deformation processing. *J. Mater. Sci.* **2010**, *45*, 859–864. [CrossRef]
21. Maheshwari, A.K. Prediction of flow stress for hot deformation processing. *Comput. Mater. Sci.* **2013**, *69*, 350–358. [CrossRef]

22. Khan, A.S.; Liu, H. Variable strain rate sensitivity in an aluminum alloy: Response and constitutive modeling. *Int. J. Plast.* **2012**, *36*, 1–14. [CrossRef]
23. Lin, Y.C.; Ding, Y.; Chen, M.-S.; Deng, J. A new phenomenological constitutive model for hot tensile deformation behaviors of a typical Al–Cu–Mg alloy. *Mater. Des.* **2013**, *52*, 118–127. [CrossRef]
24. Lin, Y.C.; Li, L.-T.; Jiang, Y.-Q. A phenomenological constitutive model for describing thermo-viscoplastic behavior of Al-Zn-Mg-Cu alloy under hot working condition. *Exp. Mech.* **2012**, *52*, 993–1002. [CrossRef]
25. Lin, Y.C.; Li, L.-T.; Fu, Y.-X.; Jiang, Y.-Q. Hot compressive deformation behavior of 7075 Al alloy under elevated temperature. *J. Mater. Sci.* **2012**, *47*, 1306–1318. [CrossRef]
26. Trimble, D.; O'Donnell, G.E. Constitutive modelling for elevated temperature flow behaviour of AA7075. *Mater. Des.* **2015**, *76*, 150–168. [CrossRef]
27. Ghosh, S.; Hamada, A.; Patnamsetty, M.; Borek, W.; Gouda, M.; Chiba, A.; Ebied, S. Constitutive modeling and hot deformation processing map of a new biomaterial Ti–14Cr alloy. *J. Mater. Res. Technol.* **2022**, *20*, 4097–4113. [CrossRef]
28. Żukowska, L.W.; Śliwa, A.; Mikuła, J.; Bonek, M.; Kwaśny, W.; Sroka, M.; Pakuła, D. Finite element prediction for the internal stresses of (Ti,Al)N coatings. *Arch. Metall. Mater.* **2016**, *61*, 149–152. [CrossRef]

Disclaimer/Publisher's Note: The statements, opinions and data contained in all publications are solely those of the individual author(s) and contributor(s) and not of MDPI and/or the editor(s). MDPI and/or the editor(s) disclaim responsibility for any injury to people or property resulting from any ideas, methods, instructions or products referred to in the content.

Article

Influence of Strain Rate on Barkhausen Noise in Trip Steel

Martin Pitoňák [1], Anna Mičietová [2], Ján Moravec [2], Jiří Čapek [3], Miroslav Neslušan [1,*] and Nikolaj Ganev [3]

1. Faculty of Civil Engineering, University of Žilina, Univerzitná 1, 010 26 Žilina, Slovakia; martin.pitonak@uniza.sk
2. Faculty of Mechanical Engineering, University of Žilina, Univerzitná 1, 010 26 Žilina, Slovakia; anna.micietova@fstroj.uniza.sk (A.M.); jan.moravec@fstroj.uniza.sk (J.M.)
3. Faculty of Nuclear Sciences and Physical Engineering, Czech Technical University in Prague, Trojanova 13, 120 00 Prague, Czech Republic; jiri.capek@fjfi.cvut.cz (J.Č.); nikolaj.ganev@fjfi.cvut.cz (N.G.)
* Correspondence: miroslav.neslusan@fstroj.uniza.sk; Tel.: +421-908-811-973

Abstract: This paper deals with Barkhausen noise in Trip steel RAK 40/70+Z1000MBO subjected to uniaxial plastic straining under variable strain rates. Barkhausen noise is investigated especially with respect to microstructure alterations expressed in terms of phase composition and dislocation density. The effects of sample heating and the corresponding Taylor–Quinney coefficient are considered as well. Barkhausen noise of the tensile test is measured in situ as well as after unloading of the samples. In this way, the contribution of external and residual stresses on Barkhausen noise can be distinguished in the direction of tensile loading, as well as in the transversal direction. It was found that the in situ-measured Barkhausen noise grows in both directions as a result of tensile stresses and the realignment of domain walls. The post situ-measured Barkhausen noise drops down in the direction of tensile load due to the high opposition of dislocation density at the expense of the growing transversal direction due to the prevailing effect of the realignment of domain walls. The temperature of the sample remarkably grows along with the increasing strain rate which corresponds with the increasing Taylor–Quinney coefficient. However, this effect plays only a minor role, and the density of the lattice imperfection expressed especially in terms of dislocation density prevails.

Keywords: trip steel; Barkhausen noise; tensile test; strain rate; plastic straining; self-heating

Citation: Pitoňák, M.; Mičietová, A.; Moravec, J.; Čapek, J.; Neslušan, M.; Ganev, N. Influence of Strain Rate on Barkhausen Noise in Trip Steel. *Materials* **2024**, *17*, 5330. https://doi.org/10.3390/ma17215330

Academic Editor: Wojciech Borek

Received: 2 October 2024
Revised: 16 October 2024
Accepted: 22 October 2024
Published: 31 October 2024

Copyright: © 2024 by the authors. Licensee MDPI, Basel, Switzerland. This article is an open access article distributed under the terms and conditions of the Creative Commons Attribution (CC BY) license (https://creativecommons.org/licenses/by/4.0/).

1. Introduction

The motion of dislocations, their interference, and their increasing density are the prevailing mechanisms of the strain hardening of ferritic steels beyond their yielding [1,2]. Prolonged plasticity (elongation at break) is usually compensated by their lower yield as well as ultimate strengths. On the other hand, the strain-induced transformation of austenite to martensite can be found as the predominating mechanism of strain hardening for some austenitic steels [3–5]. This phenomenon is well known as the transformation-induced plasticity (Trip) effect [6]. This Trip effect delays the phase of strain instability as well as the corresponding elongation at break. Moreover, the phase transformation contributes to the increasing heating especially at higher strain rates when the adiabatic conditions are met [7,8]. Trip steels are considered bodies with complex microstructure containing a mixture of ferrite, bainite, or primary martensite and austenite [4,9]. The increased strength, delayed plastic instability, and prolonged elongation at break of these bodies are due to the superimposing contribution of two main strain hardening mechanisms such as dislocation slip and the strain-induced transformation of austenite to martensite [10–12]. Trip steels are produced in a diversity of microstructures in which the chemistry and/or fraction of the aforementioned phases can vary in order to customise the final mechanical properties and the corresponding formability. These materials are very often employed in the automotive industry for body-in-white parts requiring high crashworthiness, reinforced components, bumpers, etc. [13]. Moreover, these steels are employed in civil engineering as well for components that transform the energy of impacts into the energy consumed during their plastic deformation.

Alterations in microstructure during the plastic deformation of Trip steels are very complex. The mechanism of dislocations is directly stored in the matrix with respect to its increasing density, whereas the valuable sample heating can be obtained at higher strain rates mostly [14]. However, the strain-induced transformation of austenite to martensite contributes to the more intensive sample heating as well [8,15]. Moreover, the rate of austenite decomposition decelerates along with the increasing strain hardening in the Trip steels containing lower C content [16–19].

An increasing degree of matrix alteration expressed in terms of an increasing density of lattice imperfection can be analysed using the Taylor–Quinney coefficient (β_{INT}) [12,13]. The values of this coefficient near zero indicate that nearly whole plastic work is stored in the matrix in the form of lattice imperfections, whereas a coefficient near 1 indicates a nearly unaffected matrix and the transformation of plastic work into sample heating. The Taylor–Quinney coefficient can be obtained from the first law of thermodynamics when the adiabatic conditions are met [14,20]:

$$\beta_{INT} = \frac{\rho C_p \Delta T}{\int dW_p} \quad (1)$$

where W_p is the incremental plastic work, ρ is the matrix density, C_p refers to the heat capacity, and ΔT is the temperature growth. β_{INT} can exceed 1 when another source such as phase transformation is involved [8]. Rittel et al. [20] investigated β_{INT} in Ti, Al, and Fe alloys and found that β_{INT} is higher in materials with poor thermal conductivity. Zaera et al. [8] found that β_{INT} is higher for austenitic steels due to the Trip effect. Smith [21] studied the Taylor–Quinney coefficient as a function of strain rate. Soares et al. [14] and Rittel [22] also studied the influence of strain rate on β_{INT} in high-entropy alloys and glassy polymers.

Magnetic Barkhausen noise (MBN) is affected by lattice imperfections due to their encountering by domain walls (DWs) in motion [23,24]. The origin of MBN is directly connected to the presence of pinning sites blocking DWs in their positions and their sudden irreversible and discontinuous motion as soon as the magnetic field attains the critical threshold [24,25]. The superimposing contribution of stresses should be considered, which also reorganise (realign) DWs [26,27]. The MBN technique can be used for monitoring the components in real industry (grinding, heat treatment, etc. [28–30]) or/and as a reliable and very fast tool for the characterisation of materials with respect to their stressing or/and microstructure alterations [31,32].

The MBN technique has already been employed for the investigation of Trip steels. Lindgren and Lepistö [33] investigated MBN anisotropy during uniaxial plastic straining. Neslušan et al. [34] reported the deceleration of austenite transformation in Trip steel beyond the yielding and the decreasing MBN after a tensile test due to the increased dislocation density. Similar conclusions were reported by Vértesy et al. [35]. Tavares et al. [36] correlated MBN with martensite/austenite partitioning in stainless steel. However, the contribution of increasing ferrite phase on MBN in Trip steels is only minor [34].

MBN has been reported as a suitable technique for monitoring Trip steels, especially with respect to their microstructure. However, the degree of these alterations can be affected by strain rates as well as the developed strains [14,21] due to altering the ratio between the plastic work consumed for heating and the increasing density of lattice imperfection. As mentioned above, this behaviour is quite complex in Trip. Furthermore, Trip steels are usually employed for components consuming the energy under random impacts when high strain rates are developed and kept within the whole deformation process (for example, under traffic collisions). The evolution of component deformation, microstructure alterations, phase stability, and their residual stress state is driven by the variety of mechanisms whose contribution is mixed and difficult to unwrap. Strain rates very often play a significant role in the aforementioned behaviour and might therefore affect Barkhausen noise emission as well. For these reasons, this study investigates this topic by employing the Taylor–Quinney coefficient. Furthermore, this study investigates the influence of the variable strain rate on Barkhausen noise emission when the contribution of the residual stresses (post situ measurements of MBN), the external stressing (in situ measurements of the tensile test), and the microstructure alterations are investigated as well.

2. Materials and Methods

2.1. Materials

Experiments were carried out on galvanised Trip steel RAK 40/70+Z1000MBO. The microstructure of the investigated steel (as-received) is illustrated in Figure 1. The as-received microstructure of this Trip steel is composed of austenite (13.6%), ferrite (47.6%), bainitic ferrite (28.3%), and martensite (10.5%) [34]. The way in which the fraction of the particular phases was obtained was reported earlier [34]. The samples for the uniaxial tensile test as shown in Figure 2 were cut from a sheet of thickness 0.75 mm. Each test was repeated 3 times. The surface of the sheet was subjected to galvanising and the thickness of the Zn-galvanised layer ranged from 6 to 7 μm. The chemical composition can be found in Table 1 and the mechanical properties in Table 2.

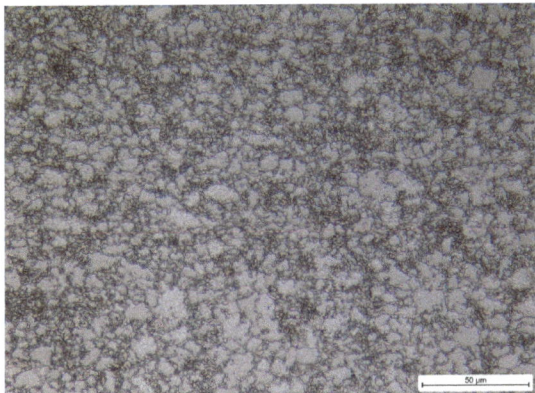

Figure 1. Metallographic image of the investigated Trip steel, 3%Nital.

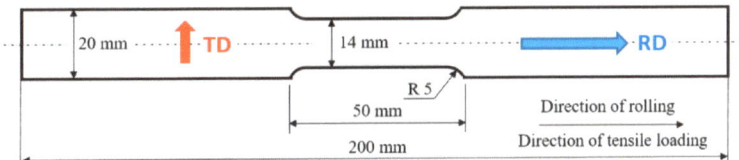

Figure 2. Gauged sample for the uniaxial test.

Table 1. Chemical composition of the employed Trip steel (wt. %).

Fe	C	Mn	Cr	Si	Al	P	Ni	Ti	Cu
Bal.	0.20	1.68	0.06	0.20	1.73	0.02	0.02	0.01	0.03

Table 2. Mechanical properties of the employed Trip steel.

Yield Strength	Ultimate Strength	Elongation at Break	Strain Hardening Coefficient	Young's Modulus
440 MPa	770 MPa	28%	0.29	210 GPa

2.2. MBN Measurements

MBN measurements were carried out in situ as well as after the unloading of samples in the predefined plastic strains. The in situ measurements were performed using the software ViewScan 4.0.0 CZ through the whole stress–strain curves in the frequency range of MBN from 70 to 200 kHz. The changing magnetic field was in the direction of the Trip rolling direction (RD) as well as in the transversal direction (TD) (see Figure 2). In order the obtain a richer MBN signal (MBN signal in the frequency range from 10 to 1000 kHz), MBN was also measured in situ using software MicroScan 5.4.1 in the predefined plastic strains (6%, 10%, 20%, and 27.5%) for four strain rates ($1.67 \times 10^{-3} \cdot s^{-1}$, $8.33 \times 10^{-3} \cdot s^{-1}$, $25.00 \times 10^{-3} \cdot s^{-1}$, and $41.67 \times 10^{-3} \cdot s^{-1}$). The uniaxial tensile test was performed using Instron 5985 (Instron, Norwood, MA, USA) true strains were checked by the 2620-602 extensometer). Apart from the conventional effective (rms) value of the signals referred to as MBN, the MBN envelopes and the corresponding PP value (refers to the position of the magnetic field in which MBN attains a maximum) were also analysed as well. MBN was also measured after the unloading of samples (post situ) when the tensile test was stopped at the aforementioned plastic strains. A magnetising voltage of 5 V and magnetising frequency of 125 Hz were obtained using the voltage and frequency sweeps. The MBN signal was sampled at a frequency of 6.7 MHz. MBN measurements were carried out through the galvanised layer using sensor S1-18-12-01. The Barkhausen noise sensor (Stresstech Oy, Jyväskylä, Finland) was calibrated by the use of the Barkhausen noise reference sample Emuge 2267523/20-11 of hardness 60 HRC made of Vanadis 4 sintered powder (Stresstech Oy, Jyväskylä, Finland). Calibration was carried out before MBN measurements and verified after measurements as well. This sintered sample provides long-term and very stable MBN emission and enables stable Barkhausen noise emission of the employed sensor to be obtained.

2.3. Microhardness and Temperature Measurements

All measurements were carried in the centre of the gauged part with respect to its width as well as length. Microhardness ($HV1$) was measured using an Innova Test 400™ (INNOVATEST Europe BV, Maastricht, The Netherlands) tester (five repetitive measurements for each plastic strain and strain rate). Information about W_p was calculated on the basis of the loading force exported by the Instron software. $\rho = 7850$ kg·m^{-3} and $C_p = 450$ J·kg^{-1}K^{-1} were employed for the β_{INT} calculations. Temperature T was measured during the tensile test using sensor PPG101A6 (software Tera Term 5.3, sampling frequency 100 Hz).

2.4. XRD Measurements

Finally, X-ray diffraction (XRD) measurements were carried out as well in order to measure residual stresses, dislocation density, phase composition, and texture (crystallographically preferred orientation) after unloading using the X'Pert PRO MPD (Malvern Panalytical, Malvern, UK) diffractometer with chromium and cobalt radiation. The galvanised Zn layer was etched before the XRD measurements. Diffraction angles 2θ were determined using the Pearson VII function and Rachinger method from the diffraction lines Cr$K\alpha_1$ of the {211} planes of the ferrite phase. The Winholtz and Cohen method and the X-ray elastic constants $\frac{1}{2}s_2 = 5.75$ TPa^{-1} and $s_1 = -1.25$ TPa^{-1} were used to determine residual stresses. Texture analysis was based on the orientation distribution function calculation from experimental pole figures (three diffraction lines {200},{211},{220}, and the MATLAB™ toolbox MTEX software 5.11.2 was used for data post-processing [37]). The dislocation density was calculated on the basis of the Williamson and Smallman method [38]. It is necessary to mention that XRD is not able to directly distinguish ferrite, bainite, and martensite in low-carbon steels, because of the small tetragonality of martensite, i.e., the observed diffraction maxima overlap.

2.5. Experimental Plan

The experiments comprised the following:
1. In situ measurements during the tensile test such as
   (a) Stress–strain evolution;
   (b) MBN measurements in the RD and TD;
   (c) Temperature measurements.
2. Post situ measurements after the tensile test such as the following:
   (a) Non-destructive measurements such as
      - MBN measurements in the RD and TD;
      - XRD measurements in the RD and TD.
   (b) Destructive measurements such as
      - Microhardness measurements.

The time sequence of the consecutive steps was directly linked with the items listed above in the same order.

3. Results of Experiments and Their Discussion

3.1. Mechanical Properties and the Taylor–Quinney Coefficient

Figure 3 illustrates the stress–strain curves for the different strain rates. Yield and ultimate strengths as well as the elongation at break were extracted from these curves (three repetitive measurements for each strain rate). Figure 4 demonstrates that the yield as well as the ultimate strength are nearly unaffected (together with their ratio), whereas the elongation at break gently drops down when the plastic deformation is accelerated. Also, the evolutions of incremental work plotted along with strain hardening overlap each other (see Figure 5a). On the other hand, the evolution of temperature is remarkably different at the higher strain rates as compared with the lower ones. Temperature growth becomes higher and the increase in temperature is accelerated at the higher strain rates (see Figure 5b). Two different aspects should be noted with respect to the different temperatures. The temperature is a quantity attributed to the motion of atoms in the lattice. It should be considered that the speed of dislocations is accelerated at higher strain rates. Therefore, the energy stored in the atoms when the dislocation encounters the lattice is higher. The second aspect is associated with the thermal conductivity during the tensile test. Conditions for the adiabatic process can be met at the higher strain rates mainly. However, certain heat transfers from the gauged to the neighbouring regions take place, which contributes to the lower temperatures at the lower strain rates as illustrated in Figure 5b. The increase in temperature is higher than that reported before [14] due to the contribution of an additional heat source when the strain-induced transformation of austenite to martensite as the typical behaviour of the improved formability of Trip steels takes place [11].

With the remarkably different evolution of the temperature T for the same W_p, the different evolution of β_{INT} can be calculated and plotted as depicted in Figure 6. A more remarkable temperature growth can easily be linked with the higher β_{INT} at the higher strain rates. β_{INT} evolutions exhibit local minima at lower strain rates followed by local maxima at higher strains, with a final progressive decrease afterward. The local minima are lower and shifted to lower strains at the lower strain rates. The rate of the progressive decrease drops down at the higher strain rates. The lower and higher evolutions of β_{INT} at the lower strains are due to less developed lattice alterations (early beyond the yielding), as well as due to the only-minor temperature growth. For this reason, the calculation of β_{INT} early beyond the yielding is quite tricky. As soon as the strain hardening as well as the temperature increase are more developed, $β_{INT}$ stabilises and can be considered the parameter linked with the degree of the aforementioned aspects.

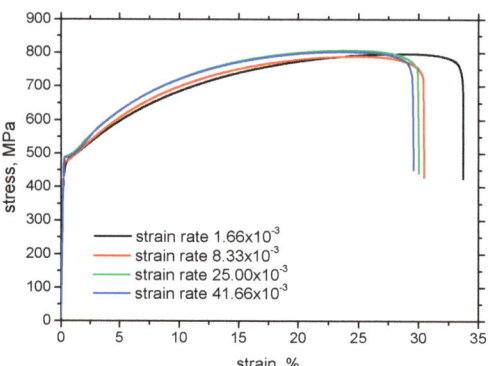

Figure 3. Stress–strain curves for the different strain rates.

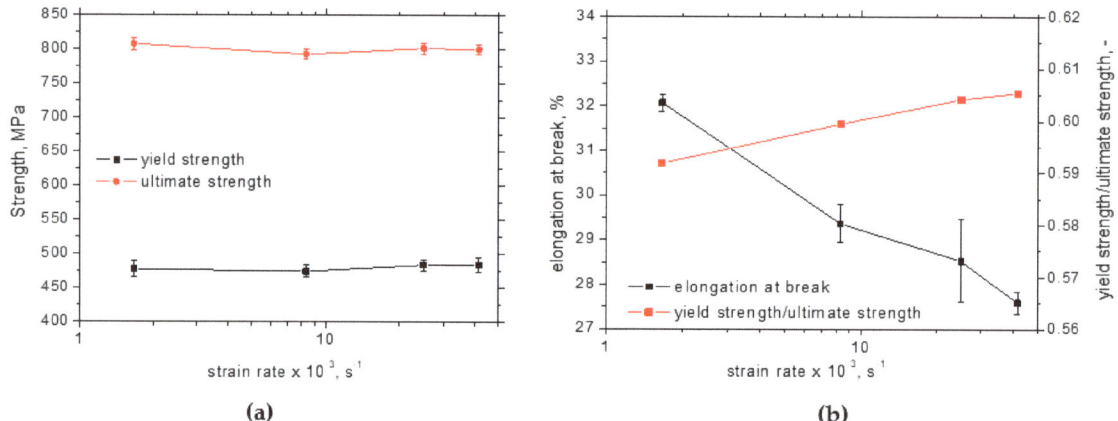

Figure 4. Mechanical properties as a function of strain rate: (**a**) yield and ultimate strength, (**b**) elongation at break.

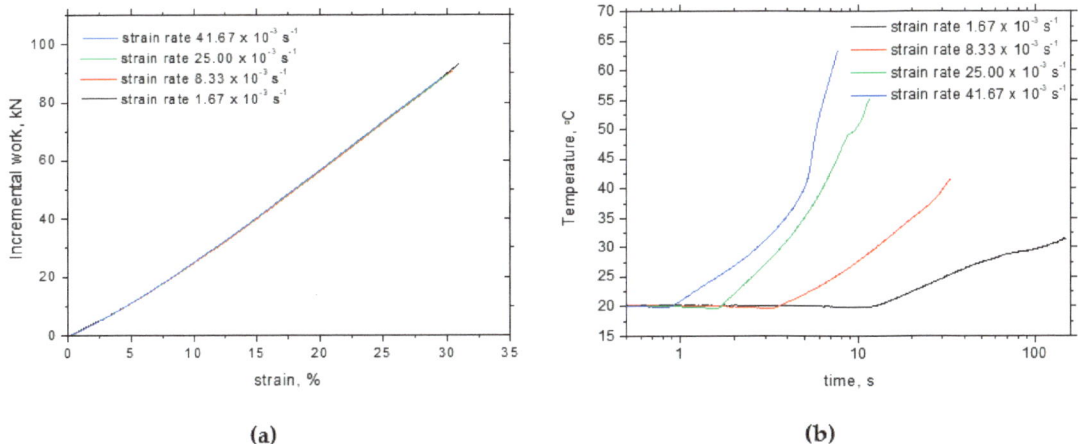

Figure 5. Evolution of incremental work and temperature during tensile test: (**a**) incremental work, (**b**) temperature.

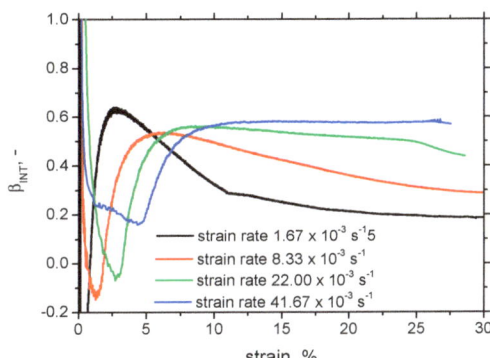

Figure 6. Evolution of βINT during tensile test.

3.2. XRD and Hardness Measurements

Figure 7b illustrates that the residual stresses in the TD are unaffected with respect to strain hardening, and the strain rate more or less corresponds to the bulk compressive stresses after rolling. However, the values of compressive stresses progressively grow in the RD along the strain hardening (see Figure 7a). The differences in residual stresses in RD among the strain rates are negligible beyond the yielding, followed by lower compressive residual stresses for the higher strain rates before the break.

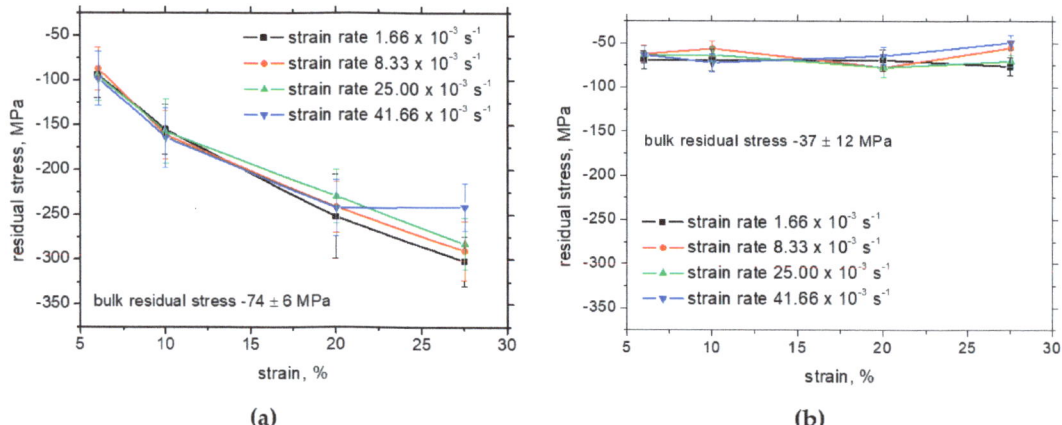

Figure 7. Evolution of residual stresses in ferrite along ε: (**a**) RD, (**b**) TD.

The strain hardening mechanism is based on austenite decomposition (phase transformation of austenite to strain-induced martensite—Trip effect) and the superimposing contribution of the interaction of dislocations (their increasing density). Figure 8a demonstrates that the rate of austenite decomposition decelerates at the higher strains [34] and the lower strain rates. On the other hand, austenite decomposition is delayed at the higher strain rates and the lower strains compensated by the lower austenite fraction at the higher strains. Figure 8 shows that the degree of strain-induced transformation of austenite to martensite plays a strong role with respect to *HV1*, since the higher fraction of strain-induced martensite can be linked with the higher dislocation density and the corresponding higher hardness *HV1* at the higher strain and strain rates (also see Figure 9).

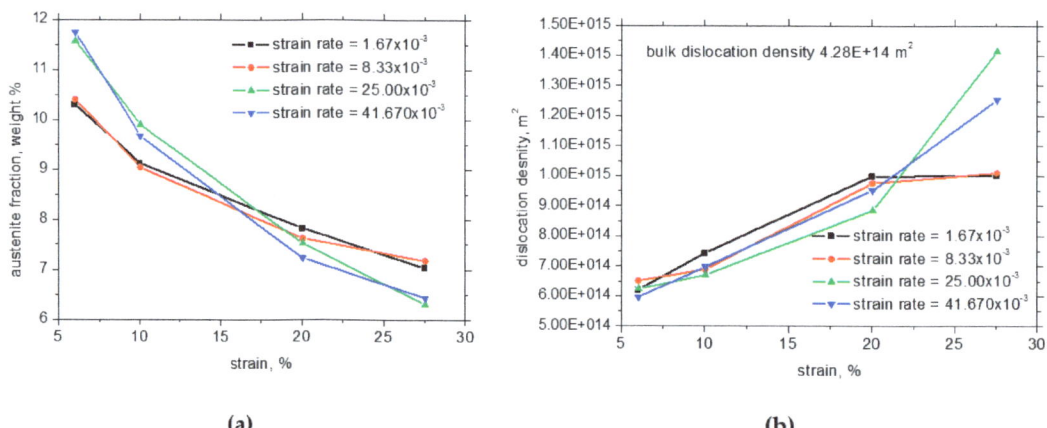

Figure 8. Evolution of austenite fraction and dislocation density in ferrite along ε: (**a**) austenite fraction, (**b**) dislocation density in ferrite.

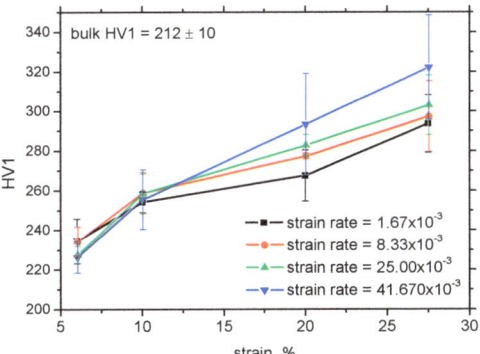

Figure 9. Evolution of *HV1* along ε.

It should also be noted that the higher temperatures measured at the higher strain rates should contribute to stabilisation of the austenite phase [4,5]. However, the findings associated with Figures 8 and 9 indicate that this effect is only minor and process dynamics prevails. Furthermore, the higher $β_{INT}$ at higher strain rates also oppose the higher dislocation density and *HV1* with contrast to $β_{INT}$ for the lower strain rates. Only a certain release of residual stresses in the RD due to the higher sample heating can be considered when the lower amplitude of compressive stresses for the higher strain rates and strains can be reported, as depicted in Figure 7a.

Finally, it should also be mentioned that the sample breaking is delayed, especially at the lowest strain rate, and the sample breaking is attained earlier for the higher strain rates. For this reason, the true strains and the corresponding density of lattice imperfections are more developed at the higher strain rates and engineering strains. This effect contributes to the higher dislocation density and the corresponding *HV1* value.

3.3. Barkhausen Noise Measurements

Two different phases of the MBN in situ evolution can be reported for the RD, TD, and all strain rates. The remarkable descending phase early beyond the yielding is followed by the progressive growth until the breakage. The descending phase can be found in some cases of low-alloyed steels when the Luders region can be found [39,40]. MBN drops down due to the missing interaction among the dislocations (dislocation density is kept nearly unaffected in this region [40]).

Trip steels do not exhibit this region in the engineering stress–strain curve (see Figure 3; the limited delayed region can be found for the higher strain rates). The in situ MBN evolution beyond the yielding indicates that this phenomenon takes place in the case of Trip steel as well. Moreover, it should be considered that the austenite decomposition is the major mechanism of strain hardening in this phase. The progressive increase in MBN in the RD following the MBN drop is due to the predominating effect of tensile stresses which tends to align DWs along the direction of the stress. Figure 10a also depicts that the rate of MBN growth in the RD decreases since the effect of tensile stress is compensated by the increasing opposition of dislocation density. Furthermore, higher MBNs in the RD for the higher strain rates are due to the higher stresses for the same strains, as shown in Figure 3. It has already been reported that the uniaxial tension tends to align MBN along a direction which is perpendicular to the load when the matrix is yielded [34,39,41]. The moderate growth as depicted in Figure 10 indicates that this effect prevails over the effect of external stresses (tends to align the DWs along the RD) and the effect of growing opposition of the increasing dislocation density. The differences among the in situ MBN in the TD are only minor.

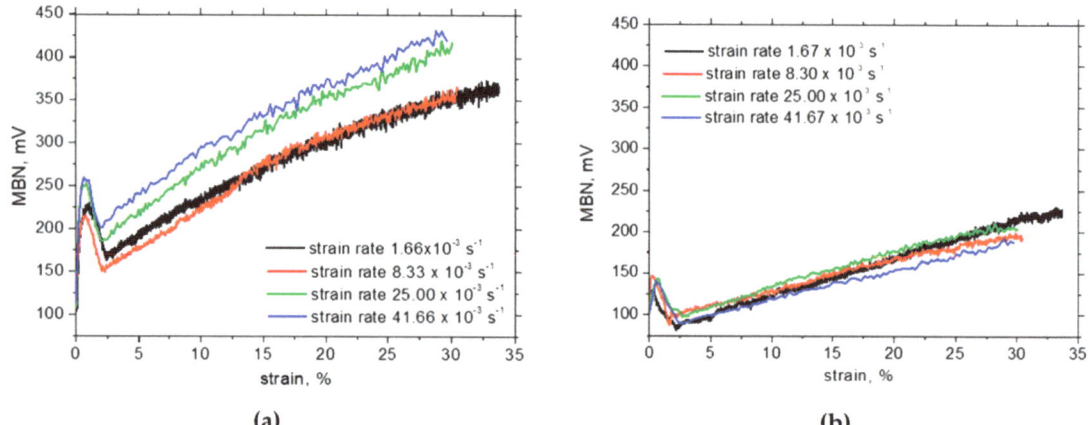

Figure 10. In situ MBN measurements—ViewScan: (**a**) RD, (**b**) TD.

The in situ MBNs measured using MicroScan software are approx. twice higher than those measured using ViewScan software due to the richer MBN signal with respect to the frequency of the obtained MBN signal (compare MBN in Figures 10 and 11). However, the evolutions and the differences among the strain rates and strains are quite similar but only shifted to the higher MBN using MicroScan software.

As soon as the external load is released, MBN evolutions with strain in the RD are remarkably different (see Figure 12). MBN in the RD remarkably decreases due to the increasing opposition of dislocation density and the realignment of DWs along the TD. The evolutions in Figure 12a indicate that the effect of an increasing fraction of the ferromagnetic phase plays no role since MBN drops down despite the paramagnetic austenite being replaced by the ferromagnetic martensite. On the other hand, the MBN measured in situ and post situ in the TD for the higher strain rates is very similar due to the realignment of DWs along the TD [34,39,41]. It should also be mentioned that the initially equiaxed grains become remarkably strained along the direction of tension and the matrix is also gently fragmented [34]. This behaviour remarkably contributes to the developed magnetic anisotropy as mentioned earlier. The realignment of crystals also contributes to the realignment of DWs and explains the increasing MBN in the TD at the expense of the RD [33] (see Figure 12). The influence of ascending temperature especially under the higher strain rates is only minor since the temperature growth is too low with respect to the matrix tempering as well as phase composition.

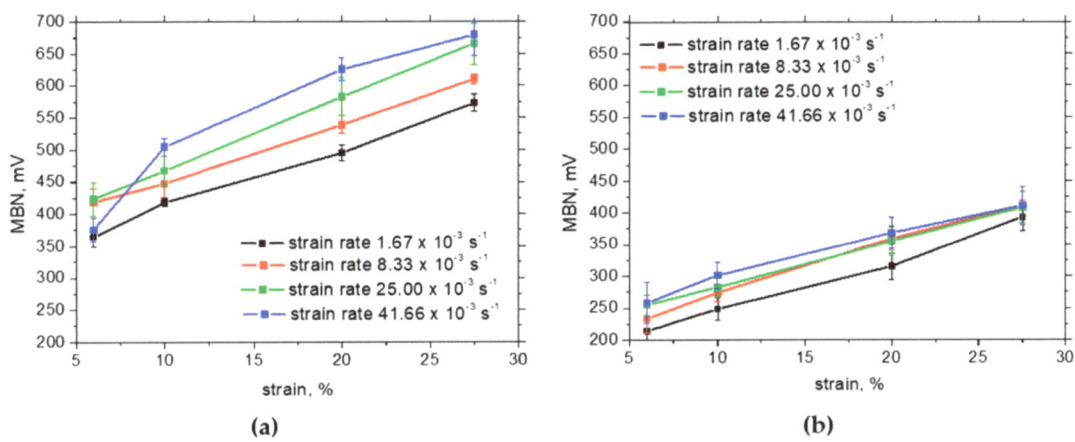

Figure 11. In situ MBN measurements—MicroScan: (**a**) RD, (**b**) TD.

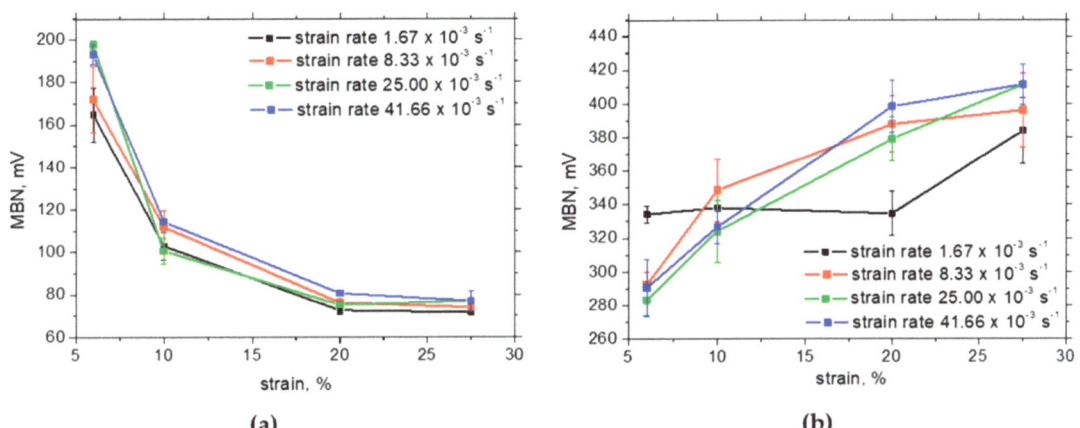

Figure 12. MBN measurements after unloading—MicroScan: (**a**) RD, (**b**) TD.

Fe alloys are ferromagnetic bodies of biaxial anisotropy with an easy *[001]* direction of magnetisation in a *bcc* lattice [23–25]. Figure 13 shows that after unloading, the easy axis of magnetisation has an inconclusive direction; nevertheless, the TD becomes the easy axis of magnetisation for higher strains. Moreover, Figure 13 clearly proves that the preferential crystallographic orientation of α in the matrix is strongly altered, which in turn can be linked to the realignment of domains and the corresponding DWs. A typical deformation texture was observed for all the measured samples. All typical texture components, typical for rolled steel, were observed: the rotated cubic *{100}<011>*, *{112}<011>*, and *{111}<011>*, which are part of the α_1-fibre *<110>*∥RD; in addition, the γ-fibre *<111>*∥ND is also very intensive. These two texture fibres prove the TD as the easy axis of magnetisation. The strength of all texture components grows with strain.

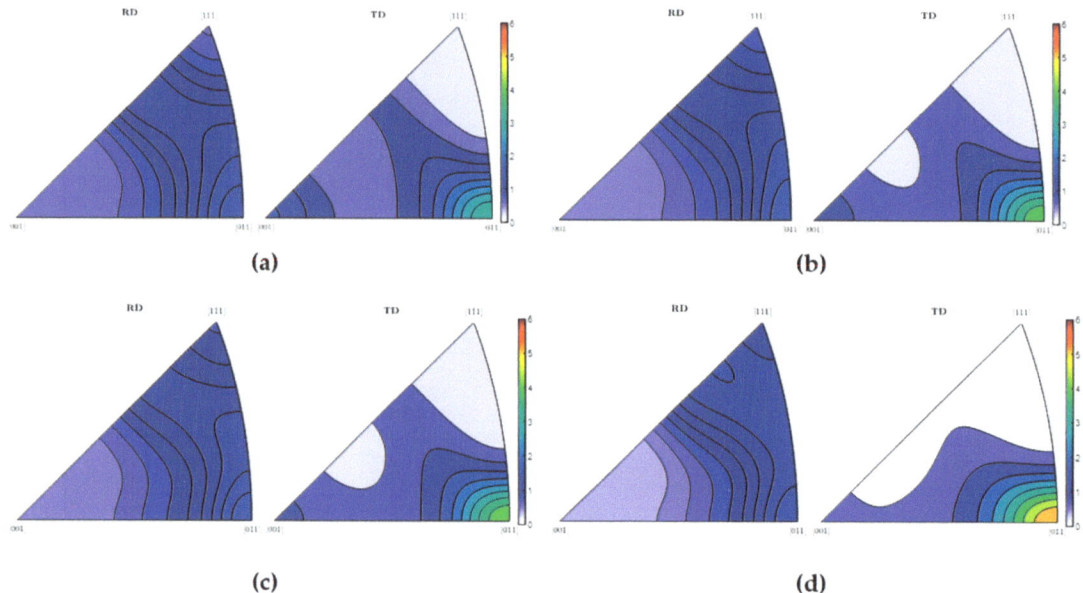

Figure 13. XRD inverse pole figures for strain rate of 8.33×10^{-3} s^{-1}: (**a**) bulk, (**b**) 6%, (**c**) 10%, (**d**) 27.5%.

The influence of external tensile stresses on MBN can be demonstrated when the MBNs measured in situ and after unloading are subtracted (see Figure 14).

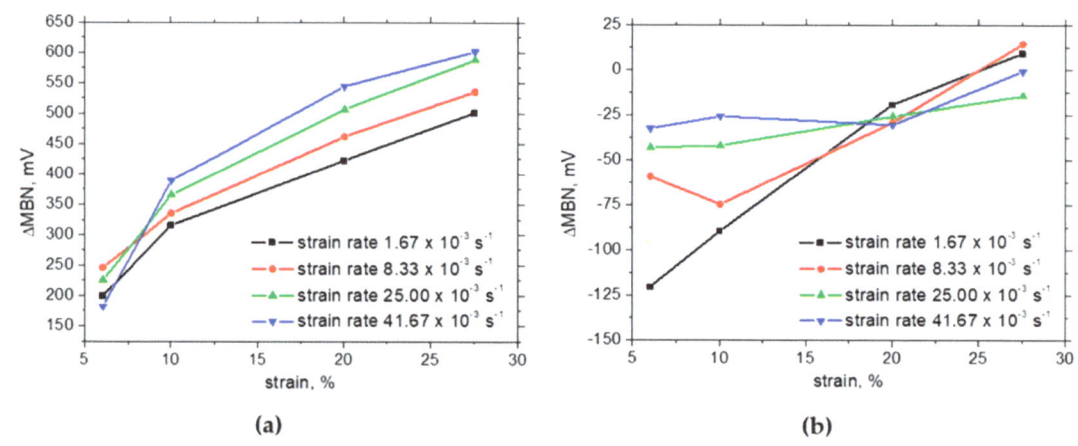

Figure 14. Differences between MBN measured in situ and after unloading—MicroScan: (**a**) RD, (**b**) TD.

Increasing external stresses are the main reason for differences between MBN measured in situ and after unloading in the RD. However, this effect is strongly compensated by the alignment of DWs in the TD. Therefore, the differences for the TD are much lower and drop down at the higher strains as contrasted against the RD. For easier navigation with respect to the different aspects affecting MBN evolution (apart from the martensite fraction) and their predomination, please check Figure 15.

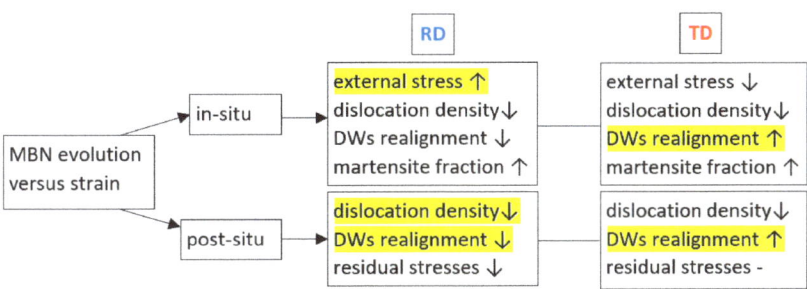

Figure 15. Aspects affecting MBN evolution along with the strain and their predomination (highlighted in yellow).

The evolutions of *PP* in situ are very flat and quite similar for all strain rates. The measured in situ *PP* values for the RD and TD can be found in the range from 0.85 to 1.05 kA.m^{-1}. Also, the evolution of *PP* in the TD after unloading is flat and unaffected by strain rate (see Figure 16b). *PP* values in the TD after unloading are lower by about 0.2 kA.m^{-1} as compared with those measured in situ due to the missing contribution of external stress, which tends to align DWs along the RD. On the other hand, the *PP* values in the RD after unloading exhibit a continuous and remarkable increase (see Figure 16a) due to the increasing opposition of pinning sites (expressed in terms of increasing dislocation density) and the synergistic effect of DWs aligned along the TD [34,39].

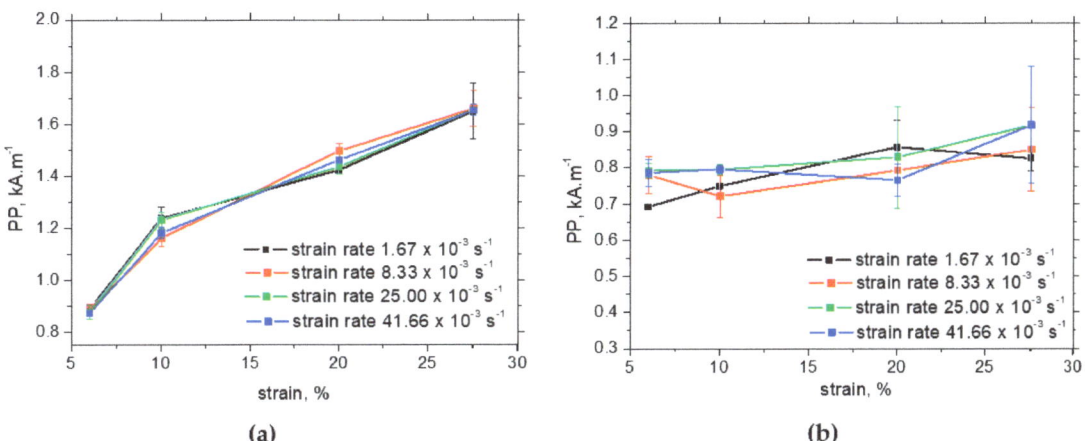

Figure 16. *PP* measurements after unloading—MicroScan: (**a**) RD, (**b**) TD.

One might consider that the increasing magnitude of compressive stresses also contributes to the decreasing MBN in the RD, as depicted in Figure 17 (also see correlation coefficients in Table 3 and Figure 15). However, this effect should be at the expense of decreasing MBN in the TD but is not (see Figure 17) due to the compensation contribution of the realignment of DWs along the TD. Similar conclusions can be reported with respect to the evolution of *PP* versus *HV1*. The correlation between *PP* and dislocation density expressed in terms of *HV1* is very strong (see Figure 18a), whereas the *PP* values grow only a little with *HV1* in the TD (see Figure 18b). Table 1 also proves the close correlation between the dislocation density and *HV1*.

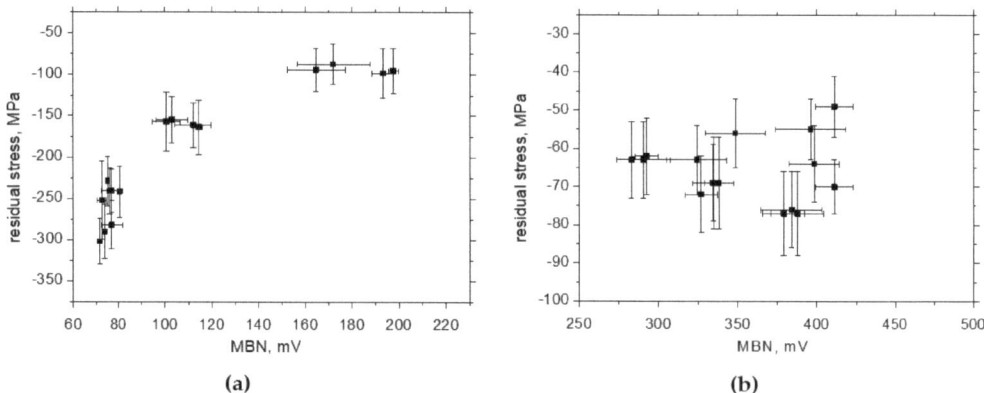

Figure 17. MBN versus residual stresses: (**a**) RD, (**b**) TD.

Table 3. Correlation coefficients ρ_p.

		Residual Stress	HV1	Dislocation Density
MBN	RD	0.91	−0.85	−0.73
	TD	−0.02	0.9	0.85
PP	RD		0.94	0.88
	TD		0.73	0.82

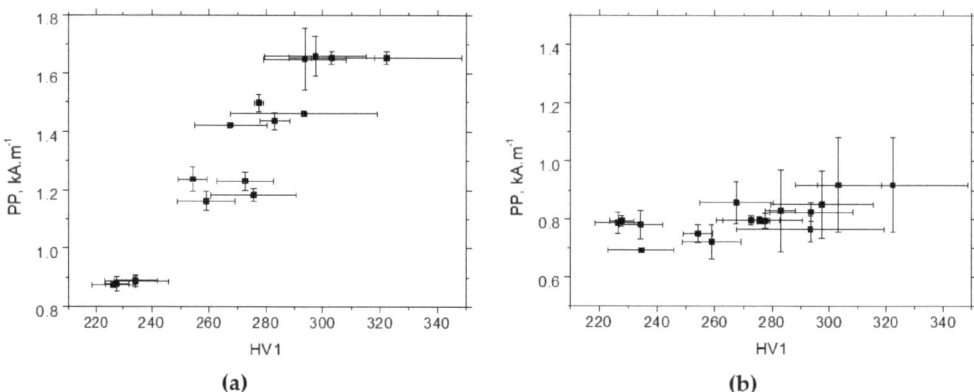

Figure 18. *HV1* versus *PP*: (**a**) RD, (**b**) TD.

4. Conclusions

It can be concluded that the strain rate has no valuable influence on the yield and ultimate strength, whereas the elongation at break drops gently down. The increasing strain rates result in increasing sample heating when the temperature at the low strain rate is about 32 °C near sample breakage as contrasted against 64 °C at the highest strain rate. With almost the same evolution of the plastic work, the Taylor–Quinney coefficient of 0.2 increases with the strain rate up to 0.55. However, the higher dislocation density (1.3 × 10^{15} m^2) at the higher strains and the strain rates indicates that the correlation between β_{INT} and the energy consumed by the matrix is low since the effect of delayed sample breakage at the lower strain rate dominates. The residual stresses in the TD are nearly unaffected along the employed strains and strain rates (about −60 MPa), whereas the increasing magnitude of the compressive stresses in the RD can be found especially along with the ascending plastic strains when the residual stress of −100 MPa beyond the yielding is replaced by -300 MPa at the end of the tensile test. Austenite decomposition at the higher strain rates is delayed as contrasted against the lower strain rates, but this evolution is reversed near the sample breakage. The increasing dislocation density strongly correlates with the increasing microhardness $HV1$ when the initial $HV1$ of 230 beyond the yielding increases up to 300 $HV1$.

In situ MBN in the RD ascends due to the presence of tensile stresses, whereas the realignment of DWs with respect to crystallographic alterations prevails in the TD. The steep increase in MBN in situ in the RD (MBN is growing in the range from 200 mV up to 400 mV) is replaced by the remarkable descending evolution after unloading due to the predominating influence of dislocation density and the superimposing crystallographic realignment of DWs (MBN is dropping down in the range from 180 mV down to 75 mV). On the other hand, MBN in the TD after the unloading is ascending due to the aforementioned realignment of DWs (MBN in TD is growing in the range from 280 mV up to 400 mV). The matrix after plastic straining becomes harder not only from the mechanical but also from the magnetic point of view due to the increased opposition of dislocation density.

The results of this study can be used for many real applications in which Trip steels are employed such as automotive parts that require very high strain hardening during crashes and a large amount of energy absorption. Moreover, information about deformation behaviour under the variable strain rates can also be used in components of high considered formability for the complex reinforcement of systems in order to improve safety in their operation.

Author Contributions: Conceptualization, M.P.; methodology, M.P.; software, J.M.; validation, A.M.; formal analysis, A.M.; investigation, M.P., A.M., J.M., J.Č., M.N. and N.G.; resources, M.P.; data curation, J.M.; writing—original draft preparation, M.P., A.M., J.Č., M.N. and N.G.; writing—review and editing, M.P. and M.N.; visualisation, J.Č. and N.G.; supervision, A.M.; project administration, M.N.; funding acquisition, M.N. All authors have read and agreed to the published version of the manuscript.

Funding: This work was realised within the frame of the project VEGA project 1/0052/22. This work was also supported by the Slovak Research and Development Agency under contract no. APVV-23-0665. This work was created within the project APVV-23-0665: Methods of exact verification of selected parameters for the road safety management.

Institutional Review Board Statement: Not applicable.

Informed Consent Statement: Not applicable.

Data Availability Statement: The original contributions presented in the study are included in the article, further inquiries can be directed to the corresponding author/s.

Conflicts of Interest: The authors declare no conflicts of interest.

References

1. Liang, L.W.; Wang, Y.J.; Chen, Y.; Wang, H.Y.; Dai, L.H. Dislocation nucleation and evolution at the ferrite-cementite interface under cyclic loadings. *Acta Mater.* **2020**, *186*, 267–277. [CrossRef]
2. Yaddanapudi, K.; Knezevic, M.; Mahajan, S.; Beyerlein, I.J. Plasticity and structure evolution of ferrite and martensite in DP 1180 during tension and cyclic bending under tension to large strains. *Mater. Sci. Eng. A* **2021**, *820*, 141536. [CrossRef]
3. Saberipour, S.; Hanzaki, A.Z.; Abedi, H.R.; Moallemi, M. Interplay of austenite and ferrite deformation mechanisms to enhance the strength and ductility of a duplex low-density steel. *J. Mater. Res. Technol.* **2022**, *18*, 755–768. [CrossRef]
4. Zackey, V.; Parker, E.; Fahr, D.; Busch, R. The enhancement of ductility in high strength steels. *Trans. ASME* **1967**, *60*, 252–259.
5. Fonstein, N. *Advanced High Strength Sheet Steels*, 1st ed.; Springer International Publishing: Cham, Switzerland, 2015. [CrossRef]
6. Mangonon, P.L.; Thomas, G. The martensite phases in 304 stainless steel. *Metall. Trans.* **1970**, *1*, 1577–1578. [CrossRef]
7. Tilak Kumar, J.V.; Sudha, J.; Padmanabhan, K.A.; Frolova, A.V.; Stolyarov, V.V. Influence of strain rate and strain at temperature on TRIP effect in metastable austenitic stainless steel. *Mater. Sci. Eng. A* **2020**, *777*, 139046. [CrossRef]
8. Zaera, R.; Rodríguez-Martínez, J.A.; Rittel, D. On the Taylor-Quinney coefficient in dynamically phase transforming materials application to 304 stainless steel. *Int. J. Plast.* **2013**, *40*, 185–201. [CrossRef]

9. Tan, X.; He, H.; Lu, W.; Yang, L.; Tang, B.; Yan, J.; Xu, Y.; Wu, D. Effect of matrix structures on TRIP effect and mechanical properties of low-C low-Si Al-added hot-rolled TRIP steels. *Mater. Sci. Eng. A* **2020**, *771*, 138629. [CrossRef]
10. Wang, M.; Huang, M.X. Abnormal TRIP effect on the work hardening behaviour of a quenching and partitioning steel at high strain rate. *Acta Mater.* **2020**, *188*, 551–559. [CrossRef]
11. Haušild, P.; Kolařík, K.; Karlík, M. Characterization of strain-induced martensitic transformation in A301 stainless steel by Barkhausen noise measurement. *Mater. Des.* **2013**, *44*, 548–554. [CrossRef]
12. Tan, X.; Ponge, D.; Lu, W.; Xu, Y.; He, H.; Yan, J.; Wu, D.; Raabe, D. Joint investigation of strain partitioning and chemical partitioning in ferrite-containing TRIP-assisted steels. *Acta Mater.* **2020**, *186*, 374–388. [CrossRef]
13. Kuziak, R.; Kawalla, R.; Waengler, S. Advanced high strength steels for automotive industry. *Archiv. Civ. Mech. Eng.* **2008**, *8*, 103–117. [CrossRef]
14. Soares, G.C.; Patnamsetty, M.; Peura, P.; Hokka, M. Effect of adiabatic heating and strain rate on the dynamic response of a CoCrFeMnNi high entropy alloy. *J. Dyn. Behav.* **2019**, *5*, 320–330. [CrossRef]
15. Haušild, P.; Davydov, V.; Drahokoupil, J.; Landa, M.; Pilvin, P. Characterization of strain-induced martensitic transformation in a metastable stainless steel. *Mater. Des.* **2010**, *31*, 1821–1827. [CrossRef]
16. Chiang, J.; Lawrence, B.; Boyd, J.D.; Pilkey, A.K. Effect of microstructure on retained austenite stability and work hardening of TRIP steels. *Mater. Sci. Eng. A* **2011**, *528*, 4516–4521. [CrossRef]
17. Sakuma, Y.; Itami, A.; Kawano, O.; Kimura, N. Next generation high strength sheet steel utilizing transformation induced plasticity. *Nippon. Steel Tech. Rep.* **1995**, *64*, 20–25.
18. Lloyd, J.T.; Field, D.M.; Limmer, K.R. A four parameter hardening model for TWIP and TRIP steels. *Mater. Des.* **2020**, *194*, 108878. [CrossRef]
19. Kong, H.; Chao, Q.; Rolfe, B.; Beladi, H. One-step quenching and partitioning treatment of a tailor welded blank of boron and TRIP steels for automotive applications. *Mater. Des.* **2019**, *174*, 107799. [CrossRef]
20. Rittel, D.; Zhang, L.H.; Osovski, S. The dependence of the Taylor-Quinney coefficient on the dynamic loading mode. *J. Mech. Phys. Solids* **2017**, *107*, 96–114. [CrossRef]
21. Smith, J.L. Full-Field Determination of the Taylor-Quinney Coefficient in Tension Tests of Ti-6Al-4V Aluminum 2024-T351, and Inconel 718 at Various Strain Rates. Ph.D. Dissertation, The Ohio State University, Columbus, OH, USA, 2019.
22. Rittel, D. On the conversion of plastic work to heat during high strain rate deformation of glassy polymers. *Mech. Mater.* **1999**, *31*, 131–139. [CrossRef]
23. Cullity, B.D.; Graham, C.D. *Introduction to the Magnetic Materials*, 2nd ed.; IEEE Press: New Jersey, NJ, USA, 2009.
24. Jiles, D. *Introduction to Magnetizm and Magnetic Materials*, 3rd ed.; Taylor & Francis Group: New York, NY, USA, 2016.
25. Chikazumi, S. *Physics of Ferromagnetizm*, 2nd ed.; Oxford University Press: Oxford, UK, 2005.
26. Liu, J.; Tian, G.Y.; Gao, B.; Zeng, K.; Zheng, Y.; Chen, J. Micro-macro characteristics between domain wall motion and magnetic Barkhausen noise under tensile stress. *J. Magn. Magn. Mater.* **2020**, *493*, 165719. [CrossRef]
27. Górka, J.; Przybyła, M. Research on the Influence of HMFI and PWHT Treatments on the Properties and Stress States of MAG-Welded S690QL Steel Joints. *Materials* **2024**, *17*, 3560. [CrossRef]
28. Neslušan, M.; Čížek, J.; Kolařík, K.; Minárik, P.; Čilliková, M.; Melikhová, O. Monitoring of grinding burn via Barkhausen noise emission in case-hardened steel in large-bearing production. *J. Mater. Process. Technol.* **2017**, *240*, 104–117. [CrossRef]
29. Ghanei, S.; Kashefi, M.; Mazinani, M. Comparative study of eddy current and Barkhausen noise non destructive testing methods in microstructural examination of ferrite-martensite dual-phase steel. *J. Magn. Magn. Mater.* **2014**, *356*, 103–110. [CrossRef]
30. Jarrahi, F.; Kashefi, M.; Ahmadzade-Beiraki, E. An investigation into the applicability of Barkhausen noise technique in evaluation of machining properties of high carbon steel parts with different degrees of spheroidization. *J. Magn. Magn. Mater.* **2015**, *385*, 107–111. [CrossRef]
31. Ktena, A.; Hristoforou, E.; Gerhardt, G.J.L.; Missell, F.P.; Landgraf, F.J.G.; Rodrigues, D.L.; Albertis-Campos, M. Barkhausen noise as a microstructure characterization tool. *Physica B Condensed Matter.* **2014**, *435*, 109–112. [CrossRef]
32. Neslušan, M.; Minárik, P.; Čilliková, M.; Kolařík, K.; Rubešová, K. Barkhausen noise emission in tool steel X210Cr12 after semi-solid processing. *Mater. Charact.* **2019**, *157*, 109891. [CrossRef]
33. Lindgren, M.; Lepistö, T. Effect of prestraining on Barkhausen noise vs. stress relation. *NDT&E Int.* **2001**, *34*, 337–344. [CrossRef]
34. Neslušan, M.; Pitoňák, M.; Čapek, J.; Kejzlar, P.; Trško, L.; Zgútová, K.; Slota, J. Measurement of the rate of transformation induced plasticity in TRIP steel by the use of Barkhausen noise emission as a function of plastic straining. *ISA Trans.* **2022**, *125*, 318–329. [CrossRef]
35. Vértesy, G.; Mészáros, I.; Tomáš, I. Nondestructive magnetic characterization of TRIP steels. *NDT&E Int.* **2013**, *54*, 107–114. [CrossRef]
36. Tavares, S.S.M.; Noris, L.F.; Pardal, J.M.; da Silva, M.R. Temper embrittlement of super martensitic stainless steel and non-destructive inspection by magnetic Barkhausen noise. *Eng. Fail. Anal.* **2019**, *100*, 322–328. [CrossRef]
37. Matěj, Z.; Kužel, R.; Nichtová, L. XRD total pattern fitting applied to study of microstructure of TiO_2 films. *Powder Diff.* **2010**, *25*, 125–131. [CrossRef]
38. Williamson, G.K.; Smallman, R.E. Dislocation densities in some annealed and cold-worked metals from measurements on the X-ray debye-scherrer spectrum. *Philos. Mag.* **1955**, *1*, 34–46. [CrossRef]

39. Neslušan, M.; Jurkovič, M.; Kalina, T.; Pitoňák, M.; Zgútová, K. Monitoring of S235 steel over-stressing by the use of Barkhausen noise technique. *Eng. Fail. Anal.* **2020**, *117*, 104843. [CrossRef]
40. Smallman, R.E.; Ngan, A.H.W. *Modern Physical Metalugry*, 8th ed.; Butterworth-Heinemann: Amsterdam, The Netherlands, 2014.
41. Roskosz, M.; Fryczowski, K.; Schabowicz, K. Evaluation of ferromagnetic steel hardness based on an analysis of the Barkhausen noise number of events. *Materials* **2020**, *13*, 2059. [CrossRef]

Disclaimer/Publisher's Note: The statements, opinions and data contained in all publications are solely those of the individual author(s) and contributor(s) and not of MDPI and/or the editor(s). MDPI and/or the editor(s) disclaim responsibility for any injury to people or property resulting from any ideas, methods, instructions or products referred to in the content.

Article

Ab Initio Studies of Mechanical, Dynamical, and Thermodynamic Properties of Fe-Pt Alloys

Ndanduleni Lesley Lethole * and Patrick Mukumba

Department of Physics, University of Fort Hare, Private Bag X1314, Alice 5700, South Africa; pmukumba@ufh.ac.za
* Correspondence: nlethole@ufh.ac.za

Abstract: The density functional theory (DFT) framework in the generalized gradient approximation (GGA) was employed to study the mechanical, dynamical, and thermodynamic properties of the ordered bimetallic Fe-Pt alloys with stoichiometric structures Fe_3Pt, FePt, and $FePt_3$. These alloys exhibit remarkable magnetic properties, high coercivity, excellent chemical stability, high magnetization, and corrosion resistance, making them potential candidates for application in high-density magnetic storage devices, magnetic recording media, and spintronic devices. The calculations of elastic constants showed that all the considered Fe-Pt alloys satisfy the Born necessary conditions for mechanical stability. Calculations on macroscopic elastic moduli showed that Fe-Pt alloys are ductile and characterized by greater resistance to deformation and volume change under external shearing forces. Furthermore, Fe-Pt alloys exhibit significant anisotropy due to variations in elastic constants and deviation of the universal anisotropy index value from zero. The equiatomic FePt showed dynamical stability, while the others showed softening of soft modes along high symmetry lines in the Brillouin zone. Moreover, from the phonon densities of states, we observed that Fe atomic vibrations are dominant at higher frequencies in Fe-rich compositions, while Pt vibrations are prevalent in Pt-rich.

Keywords: Fe-Pt alloys; elastic constants; Debye temperature; phonon dispersion curves; stability; mechanical stability; thermodynamic properties

Citation: Lethole, N.L.; Mukumba, P. Ab Initio Studies of Mechanical, Dynamical, and Thermodynamic Properties of Fe-Pt Alloys. *Materials* **2024**, *17*, 3879. https://doi.org/10.3390/ma17153879

Academic Editor: Grega Klančnik

Received: 2 July 2024
Revised: 29 July 2024
Accepted: 1 August 2024
Published: 5 August 2024

Copyright: © 2024 by the authors. Licensee MDPI, Basel, Switzerland. This article is an open access article distributed under the terms and conditions of the Creative Commons Attribution (CC BY) license (https://creativecommons.org/licenses/by/4.0/).

1. Introduction

Ferromagnetic Fe-Pt intermetallic alloys are a significant group of solid-state materials, exhibiting intriguing physical and chemical properties that are inherently dependent on their specific chemical compositions. These materials find applications across various industries, including catalysis for chemical reactions, ultrahigh-density magnetic data storage such as hard disk drives and magnetic sensors, spintronic devices, and high-temperature mechanics due to their large uniaxial magnetocrystalline anisotropies (MCA) and high coercivity, amongst other desirable magnetic properties [1–4]. The large MCA presents the potential of decreasing the magnetic particle dimensions to a few nanometers of grain sizes without degrading the thermal stability of the magnetization axis direction. Crystallographically, they stabilize in two Strukturbericht symbols and three space groups: $L1_0$ (body-centered tetragonal $P4/mmm$-FePt) and $L1_2$ (face-centered cubic $Pm\bar{3}m$-Fe_3Pt, $Pm\bar{3}m$-$FePt_3$ and body-centered tetragonal $I4/mmm$-Fe_3Pt). The $L1_0$ phase features an alternating arrangement of Fe and Pt atoms which can undergo a disorder–order phase transition from face-centered cubic (fcc) to tetragonal at temperatures of approximately above 650 °C [5]. Müller et al. employed both analytic bond-order potential and lattice-based Monte Carlo simulations and reported that the observed disorder–order transition temperature decreases with particle grain size [6,7]. Furthermore, Yu et al. reported that the ferromagnetic orientation and dynamical stability break down above the threshold pressure of 96.7 GPa as a result of undergoing spontaneous magnetization [8]. The transformation has significant implications for understanding the dynamical properties of Fe-Pt alloys, as it influences the material's magnetic behavior band mechanical response. Furthermore, in $L1_0$ bimetallic compounds, the difference in atomic size between the two constituent atoms is less than 15%. Thus, the orientation of magnetization and magnitude of the local magnetic moment and stability are significantly influenced by the concentrations and atomic sizes of Fe and Pt [9]. These dissimilarities in crystal lattice arrangement and chemical compositions significantly impact the lattice dynamics of these alloys. The face-centered Fe-rich $Pm\bar{3}m$-Fe$_3$Pt has a similar intermetallic arrangement as the Cu$_3$Ag-ordered crystal structure. Temperature-based X-ray diffraction and transmission electron microscopy studies demonstrated that the $Pm\bar{3}m$-Fe$_3$Pt alloy fully crystallizes at annealing temperatures above 800 °C [10,11]. The grain size and lattice parameter increase until this temperature and remain constant thereafter, signifying that crystallization is fully converged and lattice has reached stability. Moreover, the Mössbauer spectrometry and magnetic characterization results showed that the structure conforms to a ferromagnetic state at temperatures below 120 K and paramagnetic state at 300 K and above. The $L1_2$ cubic Pt-rich $Pm\bar{3}m$-FePt$_3$ is an intermediate alloy that occurs below 300 °C during the formation of the $L1_0$ FePt [4]. Further structural arrangements were discussed in detail in our previous communication [3]. Moreover, recently revised phase diagram studies have revealed the existence of other novel ordered FePt$_2$ and Fe$_2$Pt alloys [12]. Yu et al. have since performed computational simulations on these alloys to investigate their various thermodynamic, mechanical, and dynamical stabilities [13,14]. For the FePt$_2$ composition, they reported on five different space groups, namely: $I4/mmm$, $P4/nmm$, $Immm$, $Cmcm$, and $P6_3/mmc$, which were predicted to be thermodynamically and mechanically stable, with the $I4/mmm$ being the most thermodynamically stable [13]. Furthermore, four space groups of Fe$_2$Pt, namely: $Cmcm$, $Immm$, $I4/mmm$, and $P\bar{3}m1$ were also reported to be low-energy ferromagnetic stable structures [14].

Phonon dispersion spectra measurements of Fe-Pt alloys to determine their dynamical properties have previously been reported using both experimental and density functional theory (DFT)-based first-principles studies. Pierron-Bohnes et al. measured the vibrational modes of the conventional unit cell of the ordered $L1_0$ $P4/mmm$-FePt structure at 300 K using the inelastic neutron scattering method [15]. They observed that this alloy is thermodynamically stable since the dispersion relations curves displayed only positive modes of vibrations. Earlier, Noda et al. also investigated and compared the dispersion spectra between the ferromagnetic $Pm\bar{3}m$-Fe$_3$Pt and paramagnetic $Pm\bar{3}m$-FePt$_3$ cubic $L1_2$ alloys using the inelastic neutron scattering technique at room temperature [16]. They observed that the acoustic modes for both alloys are similar, particularly in the lower-energy region, mainly because their lattice constants differ by only 3%. It has been also determined that the dispersion curves in Fe$_3$Pt and FePt$_3$ are influenced by the force constants associated with their respective nearest neighbors, namely Fe-Fe and Pt-Pt; thus, the nearest neighbor pairs have dominant roles in lattice dynamics. Sternik et al. corroborated the experimental measurements by conducting first-principles computations on the phonon dispersion curves of the 2 × 2 × 2 primitive supercells of $P4/mmm$-FePt, $Pm\bar{3}m$-Fe$_3$Pt, and $Pm\bar{3}m$-FePt$_3$ alloys using the direct method as embedded in the PHONON code [17]. They reported that the $P4/mmm$-FePt and $Pm\bar{3}m$-FePt$_3$ alloys show no imaginary modes along high symmetry directions of the Brillouin zone, while there are imaginary frequencies for the transversal acoustic mode on the $Pm\bar{3}m$-Fe$_3$Pt alloy. Furthermore, their investigation into the impact of magnetic ordering on the dispersion relations of $Pm\bar{3}m$-FePt$_3$ revealed that the frequencies of the antiferromagnetic orientation slightly surpass those of the ferromagnetic orientation. Nevertheless, the dispersion curves exhibit a remarkable similarity.

Investigation of mechanical properties such as elasticity, hardness, shear resistance, ductility, and anisotropy is a crucial prerequisite in solid-state materials before their design and development. DFT calculations of elastic constants have been performed for various binary and ternary alloys and were reported to reveal valuable insights at both ambient and elevated conditions. Phasha et al. investigated the structural and mechanical properties of the ordered Mg-Li binary alloys to predict their phase stability [18]. Their findings revealed that there is a correlation between the heats of formation and elastic constants in terms of predicting stability. Phases with negative heats of formation tend to satisfy the Born necessary stability conditions and have a positive tetragonal shear modulus C'. Recent DFT studies on various alloys and other compounds have also corroborated these pioneering conclusions. Botha et al. investigated the relationship between the mechanical, dynamical, and thermodynamic properties of Pt-Pd alloys by calculating their elastic constants, phonon dispersion spectra, and heats of formation, respectively [19]. Their results indicated that energetically favorable, i.e., with negative heats of formation, compositions satisfy Born stability conditions (mechanical stability) and depict only positive modes of vibrations in the phonon dispersion spectra (dynamical stability). Similar studies were performed on Co-alloyed MnPt alloys, revealing that energetically favorable compositions are mechanically stable [20,21]. It was further revealed that the Debye temperature of $CsGaSb_2$ calculated from the single-crystal elastic constants yields approximately a similar value to the one calculated from the phonon density of states, confirming a robust link between mechanical and dynamical properties. Moreover, the polycrystalline bulk modulus calculated from the elastic constants exhibited exceptional alignment with values derived from alternative equations of state [21]. DFT calculations on the elastic constants of Fe-Pt alloys ($P4/mmm$-FePt, $Pm\bar{3}m$-Fe$_3$Pt, and $Pm\bar{3}m$-FePt$_3$) have also been previously performed [11]. It was reported that $Pm\bar{3}m$-Fe$_3$Pt is mechanically unstable due to negative C' and C_{44} values, while others satisfied the Born stability conditions. On the contrary, earlier results revealed mechanical stability [7,22].

Despite the significant experimental and computational advances made in these alloys, there remains a notable gap regarding the origins of lattice dynamics and elastic behavior for optimizing their functionality and ensuring their long-term stability. Moreover, little work has been completed on the $I4/mmm$-Fe$_3$Pt alloy, prompting the need to understand and compare the mechanical, dynamical, and thermodynamical properties across the entire spectrum of existing Fe-Pt alloys. Thus, the current communication will put special emphasis on this alloy. These properties are critical in designing and developing new composition-dependent magnetic materials. Additionally, understanding these properties provides valuable insights in predicting the stability of these materials when subjected to industrial conditions. Building from our previous communication [3] where the structural, magnetic, and electronic properties of $P4/mmm$-FePt, $I4/mmm$-Fe$_3$Pt, $Pm\bar{3}m$-Fe$_3$Pt, and $Pm\bar{3}m$-FePt$_3$ were reported, the current study presents density functional theory (DFT)-based first-principles computations to establish a corroboration between the mechanical, dynamical, and thermodynamic properties of the same alloys at ambient conditions. Elemental contributions of Fe and Pt atoms on the phonon dispersion spectra are also determined by the partial phonon density of states. Moreover, this work is an extrapolation of the existing theoretical and inelastic neutron scattering measurements of magnetic, mechanical and lattice dynamical properties of Fe-Pt alloys which have reported TA-mode softening on $Pm\bar{3}m$-Fe$_3$Pt [11,17,23,24]. We found that the equiatomic $P4/mmm$-FePt alloy exhibits dynamical stability, as evidenced by only positive vibrations in the phonon dispersion spectrum. Moreover, the less-studied $I4/mmm$-Fe$_3$Pt alloy shows comparable mechanical stability, ductility, anisotropy, and Debye temperature with the other alloys. Our findings contribute to the growing body of knowledge in the field of alloy research and provide a solid foundation for future investigations. By unravelling the complexities of these ordered alloys, we can unlock their full potential and harness their properties for technological advancements.

2. Computational Procedure

All calculations were performed using the Cambridge Serial Total Energy Package (CASTEP) [25] simulation code which is incorporated in with the Materials Studio 2020 version suite. CASTEP is a widely used code in the field of computational chemistry and materials science due to its ability to accurately predict the properties and behaviors of materials at the atomic level. Its key strength lies in its ability to perform first-principles calculations based on density functional theory, which accurately describes the electronic structure of materials without the need for empirical parameters, thus allowing the study of properties such as elastic constants, phonon spectra, and thermodynamic properties. However, some calculations are time-consuming and computationally expensive, limiting the size of systems that can be studied and the timescales that can be simulated. We carried out collinear spin polarization density functional theory (DFT) computations within the generalized gradient approximation (GGA) utilizing the Perdew, Burke, and Ernzerhof (PBE) functional [26]. All structures were initially relaxed by performing a full geometry optimization while maintaining a fixed basis quality and allowing cell volume to change. The Coulomb interactions between the valence electrons of iron (Fe: $3d^6\ 4s^2$) and platinum (Pt: $4f^{14}\ 5d^9\ 6s^1$) atoms, and their respective pseudo-ionic cores, were accurately modelled using on-the-fly generated (OTFG) ultrasoft pseudopotentials. This approach enables calculations to be conducted with reduced energy cut-offs and ensures that the generated potentials are steady across solid-state and pseudo-atom calculations. Furthermore, it enhances the accurateness and dependability of the results by utilizing the same exchange-correlation functional throughout. A customized energy cut-off of 350 eV was adequate to minimize the total energy of the bulk structure until the difference between two successive low-memory Broyden–Fletcher–Goldfarb–Shanno (BFGS) iterations was within 0.001 eV. The low-memory Broyden–Fletcher–Goldfarb–Shanno (LBFGS) [27] algorithm was preferred due to its ability to accelerate geometry optimization and accuracy for large systems [28]. To sample the wave functions, we employed the Monkhorst-Pack grid parameters of $9 \times 9 \times 7$, $4 \times 4 \times 4$, $6 \times 6 \times 6$, and $8 \times 8 \times 8$ for $P4/mmm$-FePt, $I4/mmm$-Fe$_3$Pt, $Pm\overline{3}m$-FePt$_3$, and $Pm\overline{3}m$-Fe$_3$Pt, respectively. The calculations for all structures were conducted assuming the ferromagnetic (FM) arrangement of the local magnetic moments. Moreover, for the tetragonal $P4/mmm$-FePt and $I4/mmm$-Fe$_3$Pt, we computed the primitive tetragonal cell as opposed to the conventional cell due to minimal computational resources. As a result, $I4/mmm$-Fe$_3$Pt with lattice parameters $a \neq c$ was computed as $I4/mmm$-Fe$_3$Pt with $a = c$. The phonon dispersion spectra along high symmetry lines and the accompanying phonon density of states were calculated via the finite displacement method which was proven to be highly effective for metallic systems [29,30]. The finite displacement technique relies on numerically differentiating forces acting on atoms, which are calculated for multiple unit cells with atomic displacements. Phonon density of states also serves as a requirement for the computation of thermodynamic properties, which permits assignment of temperature factors during analysis. Lastly, the monocrystalline elastic constants were calculated using the stress–strain approach with a maximum strain amplitude of 0.003. The additional calculation criteria for elastic constants involved sustaining total energy of convergence below 2.0×10^{-6} eV/atom, making sure that the Hellman–Feynman force is maintained under 0.006 eV/Å, and keeping the ionic displacement within 2.0×10^{-4}.

3. Results and Discussion

3.1. Mechanical Properties

Before determining the mechanical and dynamical properties of Fe-Pt alloys, structural relaxation was performed to obtain ground-state lattice parameters. To validate the approach utilized, the calculated structural lattice parameters were compared with the existing experimental data. The purpose of this analysis is to ensure the accuracy and reliability of our results. The current GGA-calculated lattice parameters exhibit an impressive agreement of over 98% with the previously reported data, thereby demonstrating the robustness of the approach employed.

Table 1 presents the calculated lattice constants, elastic constants, moduli, Pugh ratio, anisotropy factor, Poisson ratio, Vickers hardness, and Debye temperature for the Fe-Pt alloys. Calculation of elastic constants is crucial in characterizing the mechanical behavior of materials and response to externally induced stress. The Taylor expansion of the total energy of a strained system [31,32] (see Equation (1)) was used to calculate the elastic constants (C_{ij}).

$$U(V, \varepsilon) = U(V_0, 0) + V_0 \left[\sum_i \tau_i \varepsilon_i \xi_i + \frac{1}{2} \sum_{ij} C_{ij} \varepsilon_i \xi_i \varepsilon_j \xi_j \right], \quad (1)$$

where $U(V_0, 0)$ is the energy of the unstrained system with equilibrium volume V_0; τ_i and ξ_i are elements in the stress tensor and a factor taking care of the Voigt index, respectively. Cubic and tetragonal crystal systems contain three (c_{11}, c_{12}, c_{44}) and six ($c_{11}, c_{33}, c_{44}, c_{66}, c_{12}, c_{13}$) independent elastic constants, respectively. Except for C_{12} in $P4/mmm$-FePt, the obtained monocrystalline elastic constants are in agreement with the previous semi-empirical Monte Carlo results obtained using the modified embedded atom method (MEAM) and angular dependent analytic bond-order potential (ABOP) formalism [6,22], further affirming the accuracy of the approach employed. Moreover, our C_{12} value is in better agreement with the DFT value of 94 GPa reported by Zotov and Ludwig [11]. For cubic crystals ($Pm\bar{3}m$-Fe$_3$Pt and $Pm\bar{3}m$-FePt$_3$) to be deemed mechanically stable, the mandatory Born stability criterion in Equation (2) must be fulfilled [33,34].

$$C_{11} + 2C_{12} > 0, \quad C_{11} > |C_{12}| \text{ and } C_{44} > 0 \quad (2)$$

Table 1. Calculated lattice constants, elastic constants, and moduli for the four ordered bimetallic Fe-Pt alloys. Available experimental data are provided in parentheses.

Parameter	$Pm\bar{3}m$-Fe$_3$Pt	$I4/mmm$-Fe$_3$Pt	$P4/mmm$-FePt	$Pm\bar{3}m$-FePt$_3$
Lattice Parameters (Å)				
a (Å)	3.735 (3.72) [10]	5.276	2.725 (2.728)	3.914 (3.87) [22]
c (Å)	-	-	3.784 (3.85) [22]	-
Bond Length (Å) [4]				
Fe-Pt	2.641	2.638	2.700 (2.667)	2.768
Fe-Fe	2.641	2.638	-	-
Pt-Pt	-	-	-	2.768
Elastic Constants (GPa) [22]				
C_{11}	206.344 (238.8)	259.280	346.778 (304.2)	301.055 (325.8)
C_{33}		209.513	292.106 (242.0)	
C_{44}	85.957 (90.46)	68.757	113.855 (106.5)	105.108 (90.4)
C_{66}		35.522	48.38190 (40.8)	
C_{12}	170.625 (184.1)	90.790	73.457 (222.6)	183.094 (230.1)
C_{13}		152.871	161.090 (197.3)	
C'	17.859	84.245	136.66	58.981
B	182.531	168.908	197.102	222.415
G	46.379	51.670	87.423	83.360
λ	151.612	134.461	138.820	166.842
E	128.274	140.667	228.488	222.306
σ	0.383	0.361	0.307	0.333
A^U	1.624	1.082	0.891	0.412
θ_T	276.697	298.023	330.790	291.904
H	2.931	3.907	8.650	6.906
G/B	0.254	0.306	0.444	0.375

The mechanical stability conditions for tetragonal systems ($P4/mmm$-FePt and $I4/mmm$-Fe$_3$Pt) are delineated based on Equation (3) [35]. We note that the elastic constants for both the cubic and tetragonal systems are satisfied, demonstrating mechanical stability, which corroborates the thermodynamic stability reported previously [3]. Additionally, we have noted a positive C' value, which further confirms the mechanical stability of the systems.

$$C_{11} - C_{12} > 0; \quad C_{11} + C_{33} - 2C_{13} > 0;$$
$$C_{ii} > 0 \ (i = 1, 3, 4); \quad 2C_{11} + C_{33} + 2C_{12} + 4C_{13} > 0 \tag{3}$$

The elastic constants of the tetragonal systems can further be analyzed by examining their similarities and discrepancies. Significant differences are observed between the calculated C_{ij} values, suggesting that the tetragonal alloys are highly anisotropic. Particularly, C_{11} and C_{33} are significantly higher than C_{44} and C_{66}, indicating greater resistance to unidirectional compression compared to shear. Furthermore, there exists a substantial difference between the following pairs of C_{ij}s, C_{11} and C_{33}; C_{44} and C_{66}; C_{12} and C_{13}, which has an impact on the behavior of the material. C_{11} represents the stiffness of the material in the direction perpendicular to the crystallographic planes, while C_{33} represents the stiffness parallel to the crystallographic planes, namely [100]; [010]; and [001]. C_{11} values are significantly higher than those of C_{33}, indicating greater resistance to compression in the a-axis in comparison to the c-axis or greater linear compressibility along the c-axis compared to the a-axis when the material is subjected to external forces. Thus, the interatomic chemical bonds within the (001) plane are more robust compared to the bonding along the crystal direction [001]. The relationship $C_{44} > C_{66}$, indicates that shear deformation is more prominent when a net stress is imposed along the [100] crystallographic direction of the (010) crystallographic plane, as opposed to when the same stresses are exerted in the [100] direction within the plane (001). This indicates that the alloys exhibit anisotropic behavior with greater resistance to shear deformation in one direction compared to another. Moreover, the significant difference between C_{12} and C_{13} demonstrates that when stress is imposed along the a-axis, the subsequent strain will be more pronounced along the a-axis than the c-axis. This further highlights the anisotropic response of these alloys under stress, as evidenced by the high A^U values. When assessing the level of anisotropy, we have chosen to utilize the universal anisotropy index A^U in Equation (4) as postulated by Ranganathan and Ostoja-Starzewski [36]. This index is particularly suitable as it considers both shear and bulk contributions, acknowledging the inherent anisotropy present in all single crystals. The magnitude of the deviation of A^U from zero is the measure of the degree of anisotropy exhibited by a single crystal, with a value of zero indicating a locally isotropic crystal.

$$A^U = 5\frac{G^V}{G^R} + \frac{B^V}{B^R} - 6 \tag{4}$$

where G^V, G^R, B^V, and B^R are the shear and bulk moduli Voigt and Reuss estimates, respectively. We observe that A^U decreases with Fe content, thus $Pm\overline{3}m$-Fe$_3$Pt and $I4/mmm$-Fe$_3$Pt are highly anisotropic over $P4/mmm$-FePt and $Pm\overline{3}m$-FeP$_3$, respectively.

The calculation of independent elastic constants allowed us to estimate the macroscopic elastic bulk (B), shear (G), and Young's (E) moduli utilizing the Voigt [37] and Reuss [38] estimates. These estimates establish simple and linear relationships between the isotropic bulk and shear moduli of the polycrystalline material. Additionally, Hill demonstrated that the Voigt and Reuss expressions serve as upper and lower bounds, respectively, and proposed an arithmetic average modulus value [39]. The bulk modulus signifies the material's ability to withstand compression under applied hydrostatic pressure, while the shear modulus indicates its resistance to deformation from external forces at a constant volume. Young's modulus, on the other hand, measures the material's stiffness by determining the ratio of vertical linear stress to linear strain.

$$G_V = \left(\frac{C_{11} - C_{12} + 3C_{44}}{5}\right); \quad G_R = \left(\frac{C_{44}(C_{11} - C_{12})}{4C_{44} + 3(C_{11} - C_{12})}\right); \quad G_H = \left(\frac{G_V + G_R}{2}\right) \tag{5}$$

$$B_V = \tfrac{1}{9}[2(C_{11} + C_{12}) + C_{33} + 4C_{13}]; B_R = \tfrac{C^2}{M}; C^2 = (C_{11} + C_{12})C_{33} - 2C_{13}^1;$$
$$B_H = \left(\tfrac{B_V + B_R}{2}\right) \tag{6}$$

$$C' = \left(\frac{C_{11} - C_{12}}{2}\right) \tag{7}$$

The current DFT calculations revealed that Fe-Pt alloys have intrinsic mechanical hardness, stiffness, and shear resistance, i.e., are highly resistant to changes from external mechanical stress due to relatively large magnitudes of B, G, and E. The shear modulus is less than the bulk in all the alloys, suggesting they exhibit greater resistance to volumetric change than shear deformation, and that the shear modulus is the parameter limiting stability. $Pm\bar{3}m$-FePt$_3$ possesses the highest bulk modulus value, suggesting the greatest mechanical hardness and compression resistance over $P4/mmm$-FePt, $Pm\bar{3}m$-Fe$_3$Pt, and $I4/mmm$-Fe$_3$Pt, respectively. This is consistent with a previous DFT study on Pt/Pd alloys, which showed that the bulk modulus is higher in Pt-rich compositions [19]. The $Pm\bar{3}m$-Fe$_3$Pt system possesses the lowest shear modulus value of 46.379 GPa, indicating the greatest susceptibility to shear deformation compared to $I4/mmm$-Fe$_3$Pt, $Pm\bar{3}m$-FePt$_3$, and $P4/mmm$-FePt, respectively. The Young's modulus suggests that $P4/mmm$-FePt adopts highest stiffness over, $Pm\bar{3}m$-FePt$_3$, $Pm\bar{3}m$-Fe$_3$Pt, and $I4/mmm$-Fe$_3$Pt, respectively.

To assess ductility and brittles, we computed the Pugh shear to the bulk modulus (K) [40] and Poisson (v) [41] ration of solid-state materials. Pugh postulated that a material is considered ductile if K is less than the critical value 0.5 and brittle when greater than 0.5. We noticed that K values are less than 0.5 for all Fe-Pt alloys, demonstrating ductility. Ductile materials exhibit greater resistance to thermal shock and are easier to machine. Moreover, we noted that the Fe-rich $Pm\bar{3}m$-Fe$_3$Pt and $I4/mmm$-Fe$_3$Pt systems possess lower K values, due to their low shear modulus. The Poisson's ratio associated with volume changes in directions perpendicular to the applied tension during deformation typically ranges from −1 to 0.5. An indication of ductility is present when the value exceeds 0.26, otherwise the material is considered brittle. The Poisson ratio for these alloys is greater than 0.26, further confirming ductility. Interestingly, the less-studied $I4/mmm$-Fe$_3$Pt shows excellent ductility over $Pm\bar{3}m$-FePt$_3$ and $P4/mmm$-FePt, respectively. Furthermore, the Poisson ratio was used to analyze bonding type in Fe-Pt alloys. Crystals are considered covalent if v is approximately 0.1, ionic if $v = 0.25$, and metallic if $v > 0.33$ [42,43]. We note that v values are greater than 0.33 for $Pm\bar{3}m$-Fe$_3$Pt, $I4/mmm$-Fe$_3$Pt, and $Pm\bar{3}m$-FePt$_3$, implying metallic bonding. On the other hand, $P4/mmm$-FePt shows a ratio less than 0.33 but greater than 0.25, indicating half-metallic behavior. This is in good agreement with the density of states predictions reported in our previous communication [3]. To gain insight into the thermal properties of Fe-Pt alloys, we determined the Debye temperature (θ_D). This property represents the point at which the atoms within a solid material cease to vibrate independently, but instead begin to move in a synchronized manner. It signifies the highest normal mode of vibration and serves to establish a correlation between elastic constants and phonon dispersions, specific heat, thermal expansion, and thermal conductivity. Equation (8) was utilized in the computation of the Debye temperature [44].

$$\theta_D = \frac{\hbar v}{k_B}\left(6\pi^2 n\right)^{1/3}, \tag{8}$$

where n, v, and k_B are the atom concentration, phase velocity, and Boltzmann constant, respectively. Our DFT calculations show relatively huge magnitudes of Debye temperature (~300 K), indicating high vibrational modes in the phonon dispersion spectra (see Figure 1) and consequently high thermal conductivity. Laureti et al. employed the Correlated Debye Model (CDMT) to measure the Debye temperature of $P4/mmm$-FePt and reported $\theta_D = 340$ K [4], which is consistent with our results (330.79 K). Materials with higher Debye temperatures also have stiffer and more rigid lattices, while those with lower Debye temperatures exhibit softer and more flexible structures [45,46]. To estimate the Vickers hardness of Fe-Pt alloys, we employed the microscopic bond resistance model proposed by Tian et al. [47] as shown in Equation (9).

$$H_V = 0.92 k^{1.137} G^{0.708}, \qquad (9)$$

where k is the G/B ratio and G is the shear modulus. Our findings show that the metallic Fe-Pt alloys relatively show low hardness. In principle, this is expected for ductile materials with low (<0.5) G/B values, as proposed by Chen et al. [48].

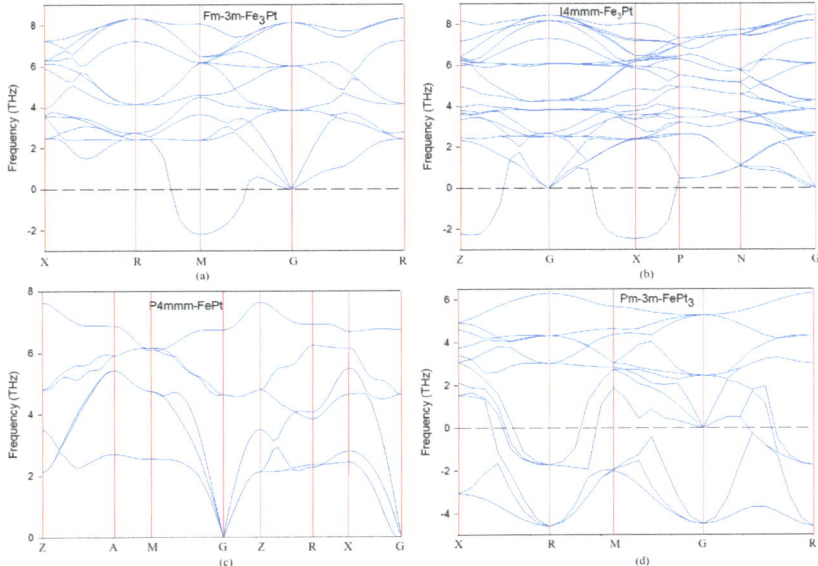

Figure 1. Phonon dispersion curves for (**a**) $Fm\bar{3}m$-Fe$_3$Pt, (**b**) $4mmm$-Fe$_3$Pt, (**c**) $P4/mmm$-FePt, and (**d**) $Pm\bar{3}m$-FePt$_3$ alloys.

3.2. Phonon Dispersion Curves

The phonon dispersion relations along high symmetry lines in the Brillouin zone were calculated to determine the lattice dynamic stability of Fe-Pt alloys and are presented in Figure 1. Phonon dispersion frequencies are calculated as the analysis of forces associated with a systematic set of atomic displacements from equilibrium positions in a crystal system. Materials are considered stable if there are no negative frequencies (soft modes) along high symmetry lines. We observe that the ordered L1$_0$ $P4/mmm$-FePt displays only positive phonon vibrational modes along the high symmetry directions, indicating that all the eigenvalues are real and the alloy is dynamically stable. This finding aligns well with previous theoretical and experimental research [15,17,49]. The other compositions, in particular $Pm\bar{3}m$-FePt$_3$, show imaginary frequencies of vibration along high symmetry lines in the spectrum, reflecting deviation dynamical instability and possible structural deformation which results in lowering of the symmetry [50]. The anomalous behavior shows that the frequencies of vibration decrease with the increasing wave vector as opposed to increasing. This may be attributed to the presence of complex interactions between the Fe and Pt atoms in the lattice, leading to the formation of localized vibrations that propagate through the material in a non-traditional manner. Dynamical instability is not pronounced for $Pm\bar{3}m$-Fe$_3$Pt and $I4/mmm$-Fe$_3$Pt since the negative vibrations are not along the center of the Brillouin zone G.

The modes of vibration for $Pm\bar{3}m$-Fe$_3$Pt and $I4/mmm$-Fe$_3$Pt as depicted in Figure 1a,b share a related frequency range and similarities along G, which can be attributed to their identical chemical composition and stoichiometry. Moreover, we note that the separation between the acoustic and optical modes is not enough to create a gap in the dispersion relations of all the alloys, suggesting seamless continuous energy transfer between the phonon modes. The disparities in mass among the alloys are evident through the observation that the frequencies of vibrations (optical modes) are higher in Fe-rich (with smaller mass) alloys and lower in Pt-rich (with larger mass) alloys.

3.3. Phonon Density of States

To analyze the distinctive elemental vibrational contributions, we conducted calculations of the phonon density of states (DOS) as illustrated in Figure 2. The spectra can be categorized into two frequency regions: one where Fe vibrations predominate and another where Pt vibrations are more prominent. In Fe-rich compositions, Fe vibrations are prevalent at higher frequencies, whereas Pt-rich compositions exhibit dominance of Pt vibrations. Additionally, a significant level of anisotropy in lattice dynamics is apparent in all structures, particularly in $P4/mmm$-FePt (Figure 2c). The lowest frequency band (1.9–3.6 THz) comprises vibrations of the heavier platinum atoms, while frequencies above 4 THz are attributed to Fe vibrations. These findings align well with similar DFT calculations reported in previous studies [17,50]. In the Fe-rich $Pm\bar{3}m$-Fe$_3$Pt and $I4/mm$-Fe$_3$Pt compositions, Pt movement prevails up to 4 THz with minimal contribution from Fe, indicating that heavier Pt movement is responsible for the presence of imaginary soft modes leading to dynamical instability. Conversely, in the $Pm\bar{3}m$-FePt$_3$ composition, Fe vibrations dominate in the −5 to 2 THz range, while Pt is more prominent in the 4 to 6 THz region. Therefore, the negative frequencies observed in the phonon dispersion curves of this alloy stem from the movement of multiple Fe atoms. Moreover, it should be noted that the elemental partial DOSs of Fe and Pt were sampled for only one atom each. Hence, there are significant differences between the sum (red) and atomic plots (blue and green) in the non-equiatomic compositions since other atoms are not included.

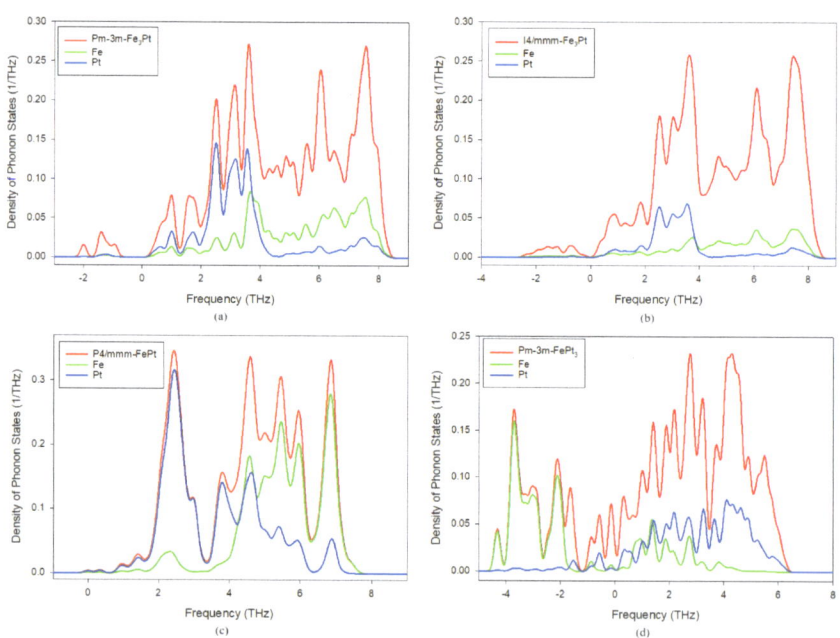

Figure 2. Calculated phonon density of states (**a**) $Pm\bar{3}m$-Fe$_3$Pt, (**b**) $I4/mmm$-Fe$_3$Pt, (**c**) $P4/mmm$-FePt, and (**d**) $Pm\bar{3}m$-FePt$_3$ alloys.

3.4. Thermodynamic Properties

Thermodynamic (temperature-dependent) quantities have been evaluated from the phonon density of states (DOS) in the temperature range of 0 to 1000 K based on the relations developed by Baroni et al. as listed in Table 2 [51]. All quantities were calculated at ground state, that is, geometry optimization was fully converged and all the phonon eigenfrequencies were real and non-negative.

Table 2. Formulae for thermodynamic quantities.

Formula	Quantity
$E(T) = E_{tot} + E_{zp} + \int \frac{\hbar\omega}{e^{(\frac{\hbar\omega}{kT})}-1} F(\omega) d\omega$	Enthalpy
$F(T) = E_{tot} + E_{zp} + kT \int F(\omega) \left[1 - e^{(-\frac{\hbar\omega}{kT})}\right] d\omega$	Free energy
$S(T) = k \left\{ \int \frac{\frac{\hbar\omega}{kT}}{e^{(\frac{\hbar\omega}{kT})}-1} F(\omega) d\omega - \int F(\omega) \ln\left[1 - e^{(\frac{\hbar\omega}{kT})}b\right] d\omega \right\}$	Entropy
$C_v(T) = k \int \frac{(\frac{\hbar\omega}{kT})^2 e^{(\frac{\hbar\omega}{kT})}}{\left[e^{(\frac{\hbar\omega}{kT})}-1\right]^2} F(\omega) d\omega$	Heat capacity
$C_V^D = 9Nk \left(\frac{T}{\theta_D}\right)^3 \int_0^{\theta_D/T} \frac{x^4 e^x}{(e^x-1)^2} dx$	Debye temperature

where E_{zp}, k, $\hbar$, $F(\omega)$, θ_D, and N are zero-point vibrational energy, Boltzmann constant, Planck constant, phonon density of states, Debye temperature, and number of atoms per cell, respectively.

Enthalpy, free energy, and entropy are fundamental thermodynamic properties that govern the behavior of solid-state materials. In the context of solid-state materials, enthalpy plays a crucial role in understanding phase transitions, such as melting and crystallization. For example, the enthalpy change associated with the melting of a solid reflects the energy required to overcome intermolecular forces and disrupt the crystalline structure, leading to a transition from a solid to a liquid state. Free energy is a key parameter in predicting the stability and equilibrium of different crystal structures. By comparing the free energy of different crystal phases, we can determine the phase that is thermodynamically favored under specific conditions. Entropy is closely related to the organization of atoms and molecules in a crystal lattice. Higher entropy is associated with increased disorder and greater freedom of movement for particles, whereas lower entropy corresponds to a more ordered and structured arrangement which is associated with favorable thermal conductivity and resistance to thermal expansion and deformation.

These three thermodynamic potentials as a function of temperature are plotted in Figure 3. From our graphical representations, we observed that entropy and enthalpy rise with temperature, while the free energy decreases. The $I4/mmm$-Fe$_3$Pt alloy shows the largest entropy values, alluding to the high mobility of atoms within the lattice structure, while the $P4/mmm$-FePt has the lowest, suggesting a more compact structural arrangement and thermal stability and conductivity. This is consistent with the phonon dispersion spectra predictions where $P4/mmm$-FePt showed dynamical stability. The free energies are negative and continue to decrease with temperature, implying thermodynamic stability [49], which corroborates the heats of formation in the previous communication [3]. Figure 4a shows the variation in isobaric heat capacity (C_v) as a function of temperature (0–1000 K). Heat capacities are critical in determining the thermal behavior of solid-state materials and are strongly dependent on intermolecular bonds. Materials with strong intermolecular bonds tend to have higher heat capacities as more energy is required to disrupt these bonds and increase the temperature. As the temperature increases, the heat capacities of Fe-Pt alloys show uneven characteristics. At temperatures below 200 K, C_v increases exponentially and gradually scales off toward the Dulong–Petit boundary at higher temperatures. This behavior is common in most solid-state materials [19,21,52]. In the temperature range, $I4/mmm$-Fe$_3$Pt (~46 cal/cell·K) shows the highest C_v value, followed by $Fm\overline{3}m$-Fe$_3$Pt (~23 cal/cell·K), $Fm\overline{3}m$-FePt$_3$ (~17 cal/cell·K), and $P4/mmm$-FePt (~12 cal/cell·K), respectively. Interestingly, in temperatures below 80 K, the heat capacities are consistent with the bond length (see Table 1). An alloy with the shortest bond length ($I4/mmm$-Fe$_3$Pt) exhibits the highest C_v value, while an alloy with the longest bond length ($P4/mmm$-FePt) has the lowest value, further confirming a compact lattice structure on the latter.

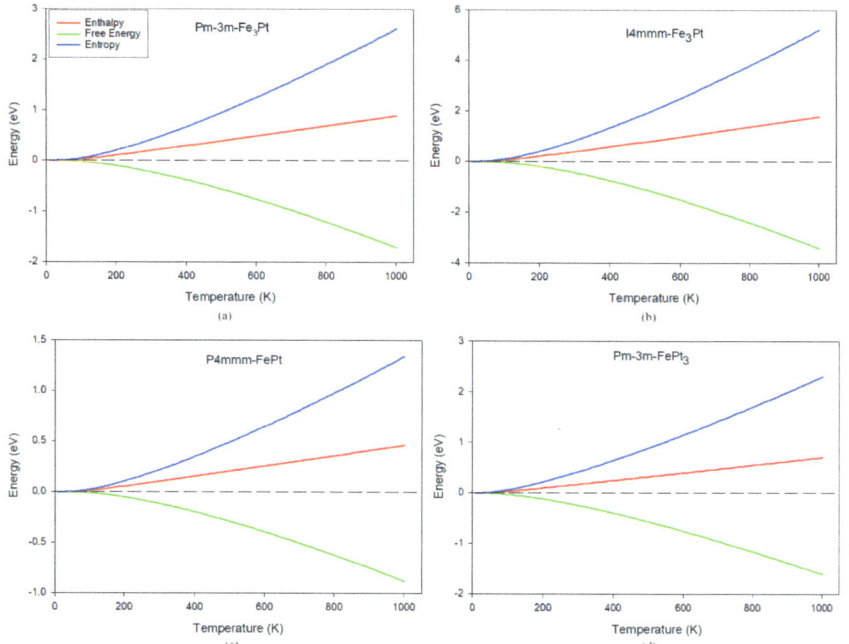

Figure 3. Thermodynamic properties of Fe-Pt alloys. (**a**) $Pm\overline{3}m$-Fe$_3$Pt (**b**) $I4/mmm$-Fe$_3$Pt (**c**) $P4/mmm$-FePt (**d**) $Pm\overline{3}m$-FePt$_3$.

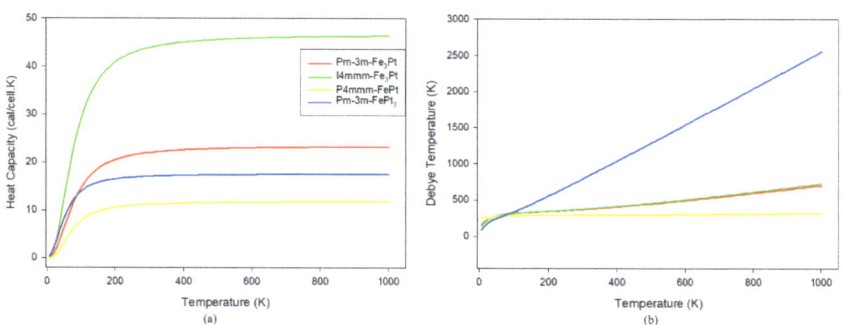

Figure 4. (**a**) Isochoric heat capacities and (**b**) Debye temperature vs. temperature for Fe-Pt alloys.

The variation in the Debye temperature (θ_D) as a function of temperature is shown in Figure 4b. Our results show a very strong linear increment relation between θ_D and T for $Pm\overline{3}m$-FePt$_3$ and a moderate linear increase for $I4/mmm$-Fe$_3$Pt and $Pm\overline{3}m$-Fe$_3$Pt. Interestingly, $P4/mmm$-FePt shows a nearly constant θ_D vs. T relationship with a small standard deviation within 14, suggesting that the modes of vibration are compact and thermal stability is achieved for $P4/mmm$-FePt, which is in good agreement with the phonon dispersion spectra and entropy predictions. The Debye temperature decreases with Pt content, which is in good agreement with related DFT studies [19,53]. Moreover, θ_D of $Pm\overline{3}m$-FePt$_3$ is highly affected by temperature, indicating the increase in atomic mobility and vibrations, which is consistent with the negative phonon modes of vibration.

4. Conclusions

The current study has successfully conducted ab initio simulations on the bimetallic Fe-Pt alloys to gain insights into the intrinsic factors underlying their mechanical, dynamical, and thermodynamic behavior, as well as to derive their stability trends. Moreover, we have established a link between the lattice dynamic and thermodynamic properties. All Fe-Pt alloys were predicted to be mechanically stable since they satisfied the Born stability conditions for cubic and tetragonal crystal lattices. Moreover, Fe-Pt alloys are characterized by the presence of sizeable anisotropy rising from the discrepancies in elastic constants and deviation of the universal anisotropy index from zero, which may give rise to a type of anisotropy called magnetocrystalline anisotropy. The prediction of mechanical stability and anisotropy is in good agreement with the thermodynamic stability and sizeable magnetocrystalline anisotropy energies reported in our previous communication. Computation of phonon dispersion spectra showed that the equiatomic $L1_0$ $P4/mmm$-FePt is dynamically stable at ambient conditions, while the other compositions show soft modes along high-symmetry directions of the Brillouin zone. This study revealed that the negative vibrational modes predominantly emanate from Fe atoms in the highly dynamically unstable $Pm\overline{3}m$-FePt$_3$ alloy, while dynamical instability is not pronounced in $I4/mmm$-Fe$_3$Pt since the negative vibrations are not along the center of the Brillouin zone. Moreover, $I4/mmm$-Fe$_3$Pt shows excellent mechanical stability, ductility, and thermal properties. Furthermore, the temperature versus entropy plots indicated that dynamically unstable alloys exhibit high atomic mobility as entropy exponentially increases with temperature, leading to the disordering of the lattice structure. This was corroborated by the heat capacity and Debye temperature plots. The free energies were predicted to be negative and to continue to decrease with temperature, implying thermodynamic stability in all the alloys. In addition to the current conclusions, it is of interest to perform ternary alloying of FePt with other transition metals such as Mn, Co, and Ni to determine any potential enhancement of the desired properties.

Author Contributions: Conceptualization, N.L.L. and P.M.; methodology, N.L.L.; software, N.L.L. and P.M.; validation, N.L.L. and P.M.; formal analysis, N.L.L.; investigation, N.L.L.; resources, N.L.L. and P.M.; data curation, N.L.L.; writing—original draft preparation, N.L.L.; writing—review and editing, N.L.L. and P.M.; visualization, N.L.L.; supervision, N.L.L. and P.M.; project administration, N.L.L.; funding acquisition, N.L.L. and P.M. All authors have read and agreed to the published version of the manuscript.

Funding: This research received no external funding.

Institutional Review Board Statement: Not applicable.

Informed Consent Statement: Not applicable.

Data Availability Statement: The generated data can be obtained from nlethole@ufh.ac.za or ndandulethole@gmail.com.

Acknowledgments: This work was performed at the University of Fort using the computing resources provided by the National Integrated Cyberinfrastructure System (NICIS). The financial support was provided by the Renewable Energy—Wind Research Niche Area of the Govern Mbeki Research and Development Centre (GMRDC), University of Fort Hare.

Conflicts of Interest: The authors declare no conflicts of interest.

References

1. Wang, J.P. FePt Magnetic Nanoparticles and Their Assembly for Future Magnetic Media. *Proc. IEEE* **2008**, *96*, 1847. [CrossRef]
2. Alsaad, A.; Ahmad, A.A.; Obeidat, T.S. Structural, electronic and magnetic properties of the ordered binary FePt, MnPt, and CrPt3 alloys. *Heliyon* **2020**, *6*, e03545. [CrossRef] [PubMed]
3. Lethole, N.; Ngoepe, P.; Chauke, H. Compositional Dependence of Magnetocrystalline Anisotropy, Magnetic Moments, and Energetic and Electronic Properties on Fe-Pt Alloys. *Materials* **2022**, *15*, 5679. [CrossRef] [PubMed]

4. Laureti, S.; D'Acapito, F.; Imperatori, P.; Patrizi, E.; Varvaro, G.; Puri, A.; Cannas, C.; Capobianchi, A. Synthesis of highly ordered L10 MPt alloys (M = Fe, Co, Ni) from crystalline salts: An in situ study of the pre-ordered precursor reduction strategy. *J. Mater. Chem. C* **2023**, *11*, 16661. [CrossRef]
5. Crisan, A.D.; Bednarcik, J.; Michalik, Š.; Crisan, O. In situ monitoring of disorder–order A1–L10 FePt phase transformation in nanocomposite FePt-based alloys. *J. Alloys Compd.* **2014**, *615*, S188. [CrossRef]
6. Müller, M.; Erhart, P.; Albe, K. Thermodynamics of L10 ordering in FePt nanoparticles studied by Monte Carlo simulations based on an analytic bond-order potential. *Phys. Rev. B* **2007**, *76*, 155412. [CrossRef]
7. Müller, M.; Albe, K. Lattice Monte Carlo simulations of FePt nanoparticles: Influence of size, composition, and surface segregation on order-disorder phenomena. *Phys. Rev. B* **2005**, *72*, 094203. [CrossRef]
8. Yu, G.L.; Cheng, T.; Zhang, X. Effect of pressure on the magnetic, mechanical, and dynamical properties of L10-FePt alloy. *J. Appl. Phys.* **2023**, *134*, 085902. [CrossRef]
9. Whang, S.H.; Feng, Q.; Gao, Y.Q. Ordering, deformation and microstructure in L10 type FePt. *Acta Mater.* **1998**, *46*, 6485. [CrossRef]
10. Crisan, O.; Crisan, A.D.; Randrianantoandro, N. Temperature-Dependent Phase Evolution in FePt-Based Nanocomposite Multiple-Phased Magnetic Alloys. *Nanomaterials* **2022**, *12*, 4122. [CrossRef]
11. Zotov, N.; Ludwig, A. First-principles calculations of the elastic constants of FeePt alloys. *Intermetallics* **2008**, *16*, 113. [CrossRef]
12. Cabri, L.J.; Oberthür, T.; Schumann, D. The Mineralogy of Pt-Fe alloys and phase relations in the Pt–Fe binary system. *Can. Miner.* **2022**, *60*, 331. [CrossRef]
13. Yu, G.; Cheng, T.; Zhang, X. Exploring the structural stability and related physical properties of FePt$_2$ alloys. *Phys. B Condens. Matter.* **2024**, *680*, 415833. [CrossRef]
14. Yu, G.; Cheng, T.; Zhang, X. Exploration of novel structures and related physical properties of Fe$_2$Pt ordered alloys. *Solid. State Sci.* **2023**, *146*, 107380. [CrossRef]
15. Pierron-Bohnes, V.; Montsouka, R.V.P.; Goyhenex, C.; Mehaddene, T.; Messad, L.; Bouzar, H.; Numakura, H.; Tanaka, K.; Hennion, B. Atomic migration in bulk and thin film L1 0 alloys: Experiments and molecular dynamics simulations. *Defect Diffus. Forum* **2007**, *1*, 263.
16. Noda, Y.; Endoh, Y.; Katano, S.; Iizumi, M. Lattice dynamics of FePt alloys of AB3 type ordered structure. *Phys. B* **1983**, *120*, 317. [CrossRef]
17. Sternik, M.; Couet, S.; Łazewski, J.; Jochym, P.T.; Parlinski, K.; Vantomme, A.; Temst, K.; Piekarz, P. Dynamical properties of ordered FeePt alloys. *J. Alloys Compd.* **2015**, *651*, 528. [CrossRef]
18. Phasha, M.J.; Ngoepe, P.E.; Chauke, H.R.; Pettifor, D.G.; Nguyen-Mann, D. Link between structural and mechanical stability of fcc- and bcc-based ordered MgeLi alloys. *Intermetallics* **2010**, *18*, 2083. [CrossRef]
19. Botha, L.M.; Ouma, C.N.M.; Obodo, K.O.; Bessarabov, D.G.; Sharypin, D.L.; Varyushin, P.S.; Plastinina, E.I. Ab Initio Study of Structural, Electronic, and Thermal Properties of Pt/Pd-Based Alloys. *Condens. Matter.* **2023**, *8*, 76. [CrossRef]
20. Diale, R.G.; Ngoepe, P.E.; Moema, J.S.; Phasha, M.J.; Moller, H.; Chauke, H.R. A computational study of the thermodynamic and magnetic properties of Co alloyed MnPt. *MRS Adv.* **2023**, *8*, 651. [CrossRef]
21. Al-Essa, S.; Essaoud, S.S.; Bouhemadou, A.; Ketfi, M.E.; Omran, S.B.; Chik, A.; Radjai, M.; Allali, D.; Khenata, R.; Al-Douri, Y. A Comprehensive Ab Initio Study of the Recently Synthesized Zintl Phase CsGaSb$_2$ Structural, Dynamical Stability, Elastic and Thermodynamic Properties. *J. Inorg. Organomet. Polym. Mater.* **2024**. [CrossRef]
22. Kim, J.; Koo, Y.; Lee, B.J. Modified embedded-atom method interatomic potential for the Fe–Pt alloy system. *J. Mater. Res.* **2006**, *21*, 199. [CrossRef]
23. Tajima, K.; Endoh, Y.; Ishikawa, Y.; Stirling, W.G. Acoustic-Phonon Softening in the Invar Alloy Fe$_3$Pt. *Phys. Rev. Lett.* **1976**, *37*, 519. [CrossRef]
24. Kastner, J.; Neuhaus, J.; Wassermann, E.; Petry, W.; Hennion, B.; Bach, H. TA1 [110] phonon dispersion and martensitic phase transition in ordered alloys Fe$_3$Pt. *Eur. Phys. J. B* **1999**, *11*, 75. [CrossRef]
25. Clark, S.J.; Segall, M.D.; Pickard, C.J.; Hasnip, P.J.; Probert, M.J.; Refson, K.; Payne, M.C. First principles methods using CASTEP. *Z. Krist. Cryst. Mater.* **2005**, *220*, 567. [CrossRef]
26. Perdew, J.P.; Burke, K.; Ernzerhof, M. Generalized gradient approximation made simple. *Phys. Rev. Lett.* **1996**, *77*, 3865. [CrossRef] [PubMed]
27. Hao, D.; He, X.; Roitberg, A.E.; Zhang, S.; Wang, J. Development and Evaluation of Geometry Optimization Algorithms in Conjunction with ANI Potentials. *J. Chem. Theory Comput.* **2022**, *18*, 978. [CrossRef] [PubMed]
28. Packwood, D.; Kermode, J.; Mones, L.; Bernstein, N.; Wooley, J.; Gould, N.; Ortner, C.; Csanyi, G. A universal preconditioner for simulating condensed phase materials. *J. Chem. Phys.* **2016**, *144*, 164109. [CrossRef] [PubMed]
29. Togo, A. First-principles Phonon Calculations with Phonopy and Phono3py. *J. Phys. Soc. Jpn.* **2023**, *92*, 012001. [CrossRef]
30. Togo, A.; Tanaka, I. First principles phonon calculations in materials science. *Scr. Mater.* **2015**, *108*, 1. [CrossRef]
31. Chen, H.S. *Elastic Anisotropy of Metal*; Metallurgy Industry Press: Beijing, China, 1996.
32. Fast, L.; Wills, J.M.; Johansson, B.; Eriksson, O. Elastic constants of hexagonal transition metals: Theory. *Phys. Rev. B* **1995**, *51*, 17431. [CrossRef] [PubMed]
33. Karki, B.B.; Ackland, G.J.; Crain, J. Elastic instabilities in crystals from ab initio stress—Strain relations. *J. Phys. Condens. Matter.* **1997**, *9*, 8579. [CrossRef]

34. Kittel, C. *Introduction to Solid State Physics*, 8th ed.; John Wiley & Sons Inc.: Hoboken, NJ, USA, 2005.
35. Born, M.; Huang, K. *Dynamical Theory of Crystal Lattices*; Clarendon: Oxford, UK, 1956.
36. Ranganathan, S.I.; Ostoja-Starzewski, M. Universal Elastic Anisotropy Index. *Phys. Rev. Lett.* **2008**, *101*, 055504. [CrossRef] [PubMed]
37. Voigt, W. *Lehrbuch der Kristallphysik*; Taubner: Leipzig, Germany, 1928.
38. Reuss, A. Calculation of the flow limits of mixed crystals on the basis of the plasticity of monocrystals. *Z. Angew. Math. Mech.* **1929**, *9*, 55.
39. Hill, R. The Elastic Behaviour of a Crystalline Aggregate. *Proc. Phys. Soc. A* **1952**, *65*, 349. [CrossRef]
40. Pugh, S.F. XCII. Relations between the elastic moduli and the plastic properties of polycrystalline pure metals. The London, Edinburgh, and Dublin Philosophical Mag. *J. Sci.* **1954**, *45*, 823.
41. Frantsevich, I.N.; Voronov, F.F.; Bokuta, S.A. *Elastic Constants and Elastic Moduli of Metals and Insulators*; Frantsevich, I.N., Ed.; Naukova Dumka: Kiev, Ukraine, 1983; p. 60.
42. Murtaza, G.; Gupta, S.K.; Seddik, T.; Khenata, R.; Alahmed, Z.A.; Ahmed, R.; Khachai, H.; Jha, P.K.; Omran, S.B. Structural, electronic, optical and thermodynamic properties of cubic REGa3 (RE = Sc or Lu) compounds: Ab initio study. *J. Alloys Compd.* **2014**, *597*, 36. [CrossRef]
43. Haines, J.; Léger, J.M.; Bocquillon, G. Synthesis and Design of Superhard Materials. *Annu. Rev. Mater.* **2001**, *31*, 1. [CrossRef]
44. Kiejna, A.; Wojciechowski, K.F. The surface of real metals. In *Metal Surface Electron Physic*; Alden Press: Oxford, UK, 1996; p. 19.
45. Jiang, D.; Xiao, W.; Liu, D.; Liu, S. Structural stability, electronic structures, mechanical properties and debye temperature of W-Re alloys: A first-principles study. *Fusion Eng. Des.* **2021**, *162*, 112081. [CrossRef]
46. Tohei, T.; Kuwabara, A.; Oba, F.; Tanaka, I. Debye temperature and stiffness of carbon and boron nitride polymorphs from first principles calculations. *Phys. Rev. B* **2006**, *73*, 064304. [CrossRef]
47. Tian, Y.; Xu, B.; Zhao, Z. Microscopic theory of hardness and design of novel superhard crystals. *Int. J. Refract. Hard. Met.* **2012**, *33*, 93. [CrossRef]
48. Chen, X.Q.; Niu, H.; Li, D.; Li, Y. Modeling hardness of polycrystalline materials and bulk metallic glasses. *Intermetallics* **2011**, *19*, 1275. [CrossRef]
49. Zerrougui, Z.; Bouferrache, K.; Ghebouli, M.A.; Slimani, Y.; Chihi, T.; Ghebouli, B.; Fatmi, M.; Benlakhdar, F.; Mouhammad, S.A.; Algethami, N.; et al. Study of structural, elastic, mechanical, electronic and magnetic properties of FeX (X=Pt, Pd) austenitic and martensitic phases. *Solid State Sci.* **2023**, *141*, 107211. [CrossRef]
50. Piekarz, P.; Lazewski, J.; Jochym, P.T.; Sternik, M.L.; Parlinski, K. Vibrational properties and stability of FePt nanoalloys. *Phys. Rev. B* **2017**, *95*, 134303. [CrossRef]
51. Baroni, S.; de Gironcoli, S.; Corso, A.D.; Giannozzi, P. Phonons and related crystal properties from density-functional perturbation theory. *Rev. Mod. Phys.* **2001**, *73*, 515. [CrossRef]
52. Dima, R.S.; Maleka, P.M.; Maluta, N.E.; Maphanga, R.R. Structural, Electronic, Mechanical, and Thermodynamic Properties of Na Deintercalation from Olivine NaMnPO4: First-Principles Study. *Materials* **2022**, *15*, 5280. [CrossRef] [PubMed]
53. Tang, K.; Wang, T.; Qi, W.; Li, Y. Debye temperature for binary alloys and its relationship with cohesive energy. *Phys. B Condens. Matter.* **2018**, *531*, 95. [CrossRef]

Disclaimer/Publisher's Note: The statements, opinions and data contained in all publications are solely those of the individual author(s) and contributor(s) and not of MDPI and/or the editor(s). MDPI and/or the editor(s) disclaim responsibility for any injury to people or property resulting from any ideas, methods, instructions or products referred to in the content.

Article

The Influence of Slide Burnishing on the Technological Quality of X2CrNiMo17-12-2 Steel

Tomasz Dyl [1], Dariusz Rydz [2], Arkadiusz Szarek [3], Grzegorz Stradomski [2,*], Joanna Fik [4] and Michał Opydo [2]

1. Department of Marine Maintenance, Faculty of Marine Engineering, Gdynia Maritime University, Morska Street 81-87, 81-225 Gdynia, Poland; t.dyl@wm.umg.edu.pl
2. Faculty of Production Engineering and Materials Technology, Czestochowa University of Technology, 19 Armii Krajowej Av., 42-201 Czestochowa, Poland; dariusz.rydz@pcz.pl (D.R.); michal.opydo@pcz.pl (M.O.)
3. Faculty of Mechanical Engineering and Computer Science, Department of Technology and Automation, Czestochowa University of Technology, 21 Armii Krajowej Av., 42-201 Czestochowa, Poland; arkadiusz.szarek@pcz.pl
4. Faculty of Science and Technology, Jan Dlugosz University in Czestochowa, Armii Krajowej Street 13/15, 42-200 Czestochowa, Poland; j.fik@ujd.edu.pl
* Correspondence: grzegorz.stradomski@pcz.pl; Tel.: +48-343-250-782

Citation: Dyl, T.; Rydz, D.; Szarek, A.; Stradomski, G.; Fik, J.; Opydo, M. The Influence of Slide Burnishing on the Technological Quality of X2CrNiMo17-12-2 Steel. *Materials* 2024, 17, 3403. https://doi.org/10.3390/ma17143403

Academic Editors: Wojciech Borek and Chih-Chun Hsieh

Received: 21 May 2024
Revised: 26 June 2024
Accepted: 8 July 2024
Published: 10 July 2024

Copyright: © 2024 by the authors. Licensee MDPI, Basel, Switzerland. This article is an open access article distributed under the terms and conditions of the Creative Commons Attribution (CC BY) license (https://creativecommons.org/licenses/by/4.0/).

Abstract: Metal products for the metallurgical and machinery industries must meet high requirements in terms of their performance, including reliability, accuracy, durability and fatigue strength. It is also important that materials commonly used to manufacture such products must meet specific requirements. Therefore, various techniques and technologies for modifying the surface layer are becoming more and more widely used. These include burnishing, which may be dynamic or static. This article studies the process of slide burnishing of surfaces of cylindrical objects. The burnishing was performed using a slide burnisher with a rigid diamond-tipped clamp on a general-purpose lathe. The tests were performed for corrosion-resistant steel X2CrNiMo17-12-2. The aim of the research was to determine the impact of changes in burnishing conditions and parameters—feed rate, burnisher depth and burnishing force at a constant burnishing speed—on the surface roughness and hardness. Additionally, the microstructure was assessed in the critical areas: the surface and the core. Another phenomenon observed was surface cracking, which would be destructive due to the occurrence of indentation. In the paper, it was stated that the microstructure, or rather the grains, in the area of the surface layer was oriented in the direction of deformation. It was also observed that in the area of the surface layer, no cracks or other flaws were revealed. Therefore, slide burnishing not only reduces the surface roughness but hardens the surface layer of the burnished material.

Keywords: slide burnishing; degree of relative hardening; roughness; microstructure

1. Introduction

Contemporary manufacturers strive to offer more and more reliable products. They are constantly improving their methods of producing materials and finishing end products. Because they are trying to meet increasing requirements while keeping production costs low, we can see rapid development in various fields of technical science, including material engineering, which improves both manufacturing processes and the functional properties of their outcomes. New corrosion-resistant steel grades are an example of such achievements. The origin of these materials dates back to the beginning of the 20th century. Metallurgists in Sheffield (Great Britain) discovered that steel with an approximately 13% admixture of chromium was more resistant to electrochemical corrosion. The New York Times mentioned the production of the first "non-rusting steel" in 1915 [1]. However, a patent for chrome–nickel stainless steel was filed and accepted by German Kaiserliches Patentamt earlier in 1912 [2]. Chrome–nickel steels of the "18-8" type were developed in the 1920s. Stainless steels are today a group of materials that are very important from an economic point of view. Due to the well-mastered technology of stainless steel production, an important aspect is reducing costs while staying in control of the functional properties of the material. The steel industry is committed to continuous improvement and modernization of its production technologies to meet the increasingly higher requirements of the market. However, quality improvement should not be followed by an excessive increase in product prices. The need to balance these aspects has been a strong driver for development in many fields of science including mechanical and material engineering. This ongoing process is aimed mainly at developing newer technologies to improve production processes and their outcomes [3–8]. One of them is burnishing, which has slightly sunk into oblivion.

Innovative products of the metallurgical and machinery industries must meet high requirements in terms of their performance, especially reliability, precision and durability. It is also important that the materials commonly used for the manufacture of machine components—such as uniform and stepped shafts, drive rotors, etc.—should have specific features [9,10]. Generally, burnishing can be dynamic or static [10–13]. The former method involves hitting the workpiece with a tool, which exerts variable pressure on the metal surface. In the case of static burnishing, the tool stays in contact with the surface, so the pressure is constant.

The main criteria used to distinguish between burnishing methods include the following [10,13–15]:

- The number and shape of the burnishing elements of the tool;
- The timing of contact between the tool and the workpiece: steady vs. pulsed;
- The mode of applying pressure to the workpiece: rigid or elastic;
- The type of friction between the tool and the workpiece.

The slide burnishing process described in this article is a specific type of static burnishing. There is sliding friction between the burnishing tip of the tool and the workpiece. The distinguishing feature of slide burnishing is the slippage between the burnishing tip and the machined surface [7,8,11,16]. An example configuration of the slide burnishing process is shown in Figure 1. The burnishing force should be optimal for each type of burnishing treatment. An excessive force may cause surface flaking—a sharp increase in roughness. However, if the force is too small, it will not be able to smoothen the surface [7].

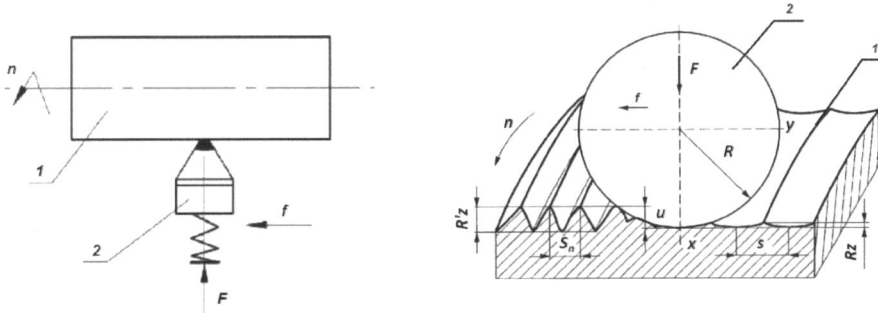

Figure 1. Slide burnishing arrangement [7]—(1) workpiece, (2) tool.

Figure 1 shows the slide burnishing for smoothness with a spherical-tipped tool, where F—burnishing force [N], f—feed rate [mm/rev], n—rotational speed [rpm], $R'z$—surface roughness ridge height before burnishing, Rz—surface roughness ridge height after burnishing, R—tool circular cross-section radius, u—burnisher depth, S_n—surface roughness ridge pitch before burnishing and s—surface roughness ridge pitch after burnishing.

Figure 2 shows the influence of the feed rate and burnisher depth on the surface roughness after slide burnishing. The differences between the roughness profiles depend on the feed rate and burnisher depth. In the first case (a), the feed rate is larger than the length of contact of the tool face with the machined surface (in the tool travel direction). Areas that are not fully plastically deformed remain on the machined surface. However, the surface ridge height in these areas is the same as before the burnishing. Such machining conditions are unacceptable. In the second case (b), where the feed rate is lower than the contact length and the burnisher depth is small (less than half the roughness ridge height), the surface after burnishing is smoother than before (but not the best achievable). Therefore, the metal extruded by the tool from the ridges does not fully fill the valleys. In the third case (c), where the feed rate is lower than the contact length and the burnisher depth is large (sufficient to form a "headwave" and a "tailwave" of plasticized metal), the surface ridge height and shape are different and can be predicted using geometric and kinematic relationships. Also, this approach to machining is undesirable because, again, the result is less than optimal. Case (d) represents the conditions optimal for surface smoothing: the feed rate is smaller than the contact length and the burnisher depth is such that the ridges are flattened and their material is pressed into the valleys without any tailwave. This scenario may be considered optimal for slide burnishing. The ultimate surface quality strongly depends on the careful selection of the machining parameters. Based on the literature, it was determined that a feed rate (f) in the range of 0.03–0.08 mm/rev and a burnishing force (F) in the range of 20–200 N are recommended [7,9–12,14].

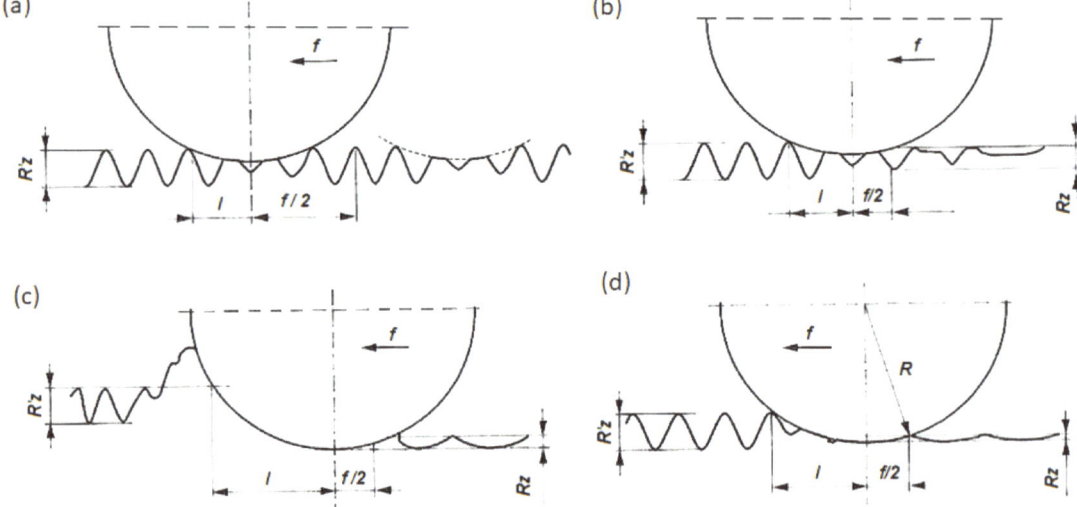

Figure 2. Slide burnishing performance as a function of feed rate and burnisher depth [10]: (**a**) feed rate larger than the contact length and small burnisher depth, (**b**) feed rate smaller than the contact length and small burnisher depth, (**c**) feed rate smaller than the contact length and large burnisher depth, (**d**) feed rate smaller than the contact length and burnisher depth. f—feed rate [mm/rev], l—contact length [mm], R′z—surface roughness ridge height before burnishing, Rz—surface roughness ridge height after burnishing, R—tool circular cross-section radius.

When designing a process for the production or reworking of machine elements, we should choose a burnishing method, machining conditions, geometries and number of burnishers depending on the method of applying pressure to the surface, which can be elastic or rigid. The reliability of machines and devices is a very important issue and is discussed in many studies. As an example, they can be used to repair individual components of ship machinery. Such work is often undertaken during long cruises. Burnishing processes are also used when restoring the outer surfaces of rolls intended for plastic forming. The selection of burnishing parameters is made based on preliminary calculations of burnishing unit forces and pressures, the results of experimental tests of materials with similar properties, general-purpose nomograms and specialized standards [15–17]. In the absence of sufficiently credible calculation formulas and nomograms, and for burnishing combined with cutting, the selection of burnishing conditions is made based on preparatory tests [16,17].

The burnishing process intended to minimize surface roughness should use quite a strong burnishing force. The burnishing rate and the feed rate should be relatively small. The burnishing process itself, which is intended to strengthen the surface layer of machine components, among others, by increasing hardness, should be characterized by the use of high burnishing forces at low feed rate and burnishing speed values. The burnishing force should be optimal for each burnishing process type. Applying a force stronger than optimal may result in surface flaking, accompanied by a sharp increase in roughness. On the other hand, a force weaker than optimal cannot provide satisfactory smoothness [15]. The corrosion resistance of burnished elements depends on the degree of deformation and smoothness of the surface. It is therefore important to correctly determine the parameters of burnishing depending on whether it is to be a smoothening or strengthening treatment.

The burnishing process, even though it was developed many years ago, is still used in many applications. The evidence of this is the number of publications devoted to this topic. This is indicated by the example of a review of publication databases such as scopus.com, which registered 403 publications on this topic in the years 2020–2024. The scope of use of burnishing technology is wide and allows it to be used for special alloys such as Inconel 718 [18,19] or aluminum alloys [20], not to mention duplex cast steels [5]. Currently, the problem of reducing the weight of structures makes it necessary to also look for new solutions for materials such as magnesium alloys [21,22]. For example, in paper [23], the authors employed mechanical burnishing, shot peening and ball burnishing to treat the surface of the high-strength magnesium alloy AZ80. They stated that the surface after ball burnishing not only obtained good fatigue performance but also showed superior corrosion fatigue properties in a 3.5% NaCl solution. Also, P. Zhang et al. [24] conducted tests on the AZ80 alloy and showed that, under a static pressure of 200 N, the fatigue strength of the alloy after the ball burnishing treatment increased by 110%.

Products are goods, objects and products that must be initially designed, then manufactured and finally sold, used and disposed of, or possibly reprocessed. Currently, efforts are being made to continuously improve and modernize manufacturing technologies in such a way as to meet the increasingly higher requirements of high-quality products from customers. Modern design of the product and its surface layer must be focused on construction integrated with production and then on operation. For example, burnishing treatment of the surface layer may be used, among other reasons, for the production and regeneration of pump drive shafts in place of seals.

2. Methodology of the Experimental Research

The aim of the research was to obtain satisfactory surface quality of the outer surfaces of cylinders (48 mm in diameter) made of corrosion-resistant X2CrNiMo17-12-2 steel. The assumption of the work was to carry out the burnishing process in such a way as to not only obtain a smoothing effect of the working surface after processing but also to potentially increase the service life of the component. The journals were prepared before burnishing by turning on a general-purpose TU250 × 1000 lathe. Turning parameters: cutting depth ap 0.5 mm, feed rate f 0.2 mm/rev, rotational speed n 355 rpm and cutting speed vc = 55 m/min. To carry out slide burnishing, it is important to select optimal parameters and processing conditions. The feed rate should not be greater than the contact length of the burnishing tool with the workpiece because such a setting of the feed rate prevents the skipping of surface irregularities. The surface roughness in these omitted places would be the same as before machining. When setting the feed rate, special attention should be paid to the tool's circular cross-section radius because this dimension is decisive for the tool contact length [8,10]. The second very important, perhaps the most important, parameter is the burnishing force, which is one of the factors responsible for the burnisher depth. To obtain the best smoothness, the tool should reach at least half of the height of the roughness ridges. Otherwise, the squashed ridges will not completely fill the valleys. In order to obtain better surface smoothness, the burnisher depth should be greater than half the height of the ridges. On the other hand, the penetration should not be deep, as this would create the headwave and the tailwave seen in plasticized metal. A less-than-optimal depth setting can produce new surface irregularities and the overall roughness may not be minimized. Another unwanted result of using too much force (i.e., going too deep into the material) can be an excessive crumpling of the surface layer, leading to such unfavorable effects as a decrease in the plasticity of the material and an increase in its internal stress [5,7]. If the burnishing parameter settings are suboptimal, the desired improvement in smoothness or hardness will not be achieved. This also has an economic dimension—repeating the process will increase the cost of machining. It is a good idea to perform this entire process in one pass of the tool over the surface of the workpiece because each subsequent step is less effective and, in addition, takes time. The feed rate and the burnishing force should be set in such a way that the squashed roughness ridges completely fill the valleys. The process of slide burnishing of the cylindrical surfaces was performed on a general-purpose TU250 × 1000 lathe, using a YAMATO YDB900652-003 slide burnisher (Sasso Marconi, Italy) with a rigid diamond-tipped clamp with a radius of 2.5 mm [11,17]. Based on the authors' own research and recommendations provided in the YAMATO catalog [16], the following parameter values were used: burnisher depth an 0.01–0.03 mm, burnishing force F = 30–90 N, feed rate f 0.04–0.08 mm/rev, rotational speed n 355 rpm and burnishing speed vn 53 m/min. Additionally, corrosion-resistant steels, including X2CrNiMo17-12-2, are difficult to machine, so it was necessary to use machine oil to cool and lubricate the tools. The chemical composition of the steel samples was determined using a SOLARIS-CCD PLUS optical spark emission spectrometer (New Taipei City, Taiwan). It is presented in Table 1.

Table 1. Chemical composition of corrosion-resistant steel X2CrNiMo17-12-2.

C [%]	Cr [%]	Ni [%]	Mo [%]	Cu [%]	Mn [%]	Si [%]	S [%]	P [%]	Nb [%]	Co [%]	V [%]	W [%]
0.024	16.449	9.286	2.058	0.544	0.946	0.381	0.024	0.019	0.03	0.218	0.091	0.022

This composition meets the requirements of the standard [25]. The tests were carried out on a cylindrical sample made of X2CrNiMo17-12-2 steel with a diameter of 48 mm and other dimensions, as shown in Figure 3. Nine journals were made for testing in one pass, which prevented the occurrence of errors likely to be caused by machine reconfiguration. For each new sample, a completely new burnishing tool was used, so the tool was used only once. This approach was intended to eliminate potential defects caused by wear.

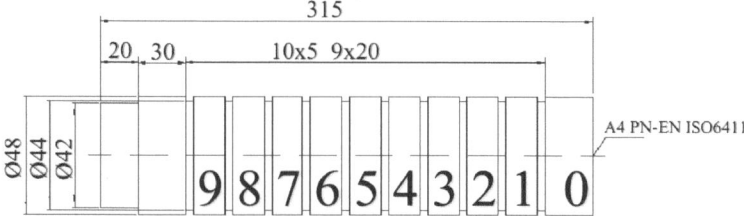

Figure 3. The arrangement of test samples with a diameter of 48 mm made of X2CrNiMo17-12-2 steel.

The study measured the impact of changes in individual processing parameters on the properties of the machined material surface. The variable parameters in the study were the feed rate and the burnishing force. This force was measured using a static FT-5304M/A/16 strain gauge load meter. It was determined that 1 mm of burnisher depth corresponds to 3 kN of burnishing force. The slide burnishing was carried out using a YAMATO YDB900652-003 tool on a general-purpose TU250 × 1000 universal lathe (Figure 4).

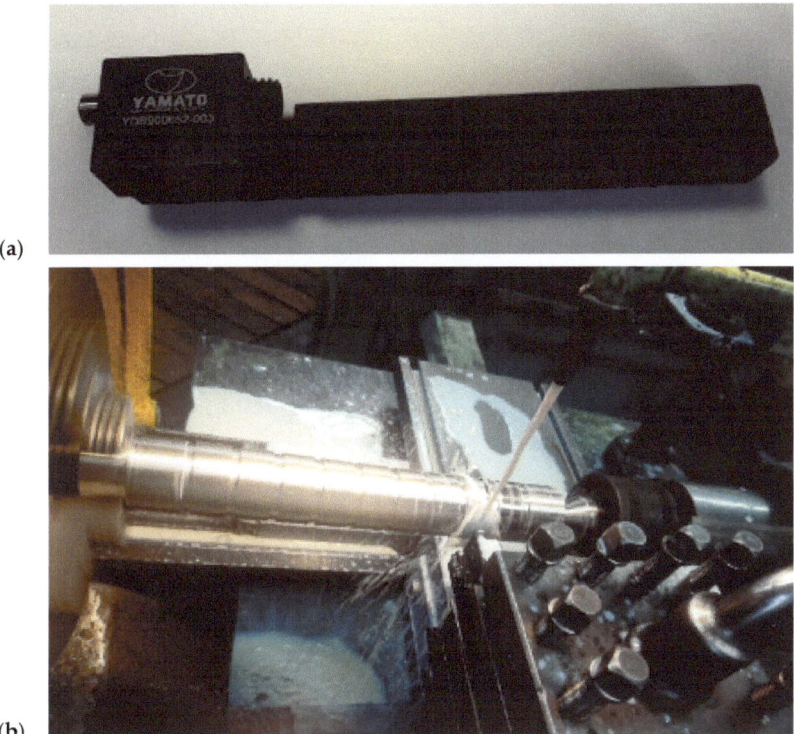

Figure 4. (**a**) YAMATO YDB900652-003 burnisher with a radius of 2.5 mm, and (**b**) the burnisher and shaft on the universal lathe.

A Qness Q250M hardness tester was used to measure hardness (Figure 5). It allows for very accurate and fully automatic measurement using the Vickers, Brinell, Rockwell and Knoop methods. The hardness measurement was performed using the Vickers method according to the standard [26] for a burnishing force of 50 N at three measurement points on each journal.

The roughness measurement was performed using a HOMMEL-ETAMIC W20 profilometer (Villingen-Schwenningen, Germany) with a TKL 300 L head (Figure 6) in accordance with the standard [27] for a measuring section of 4.8 mm and for an elementary section of 0.8 mm. The arithmetic mean of the profile deviation from the mean line (Ra) and the roughness ridge height according to ten profile points (Rz) were measured. The measurement was performed at four points on each journal.

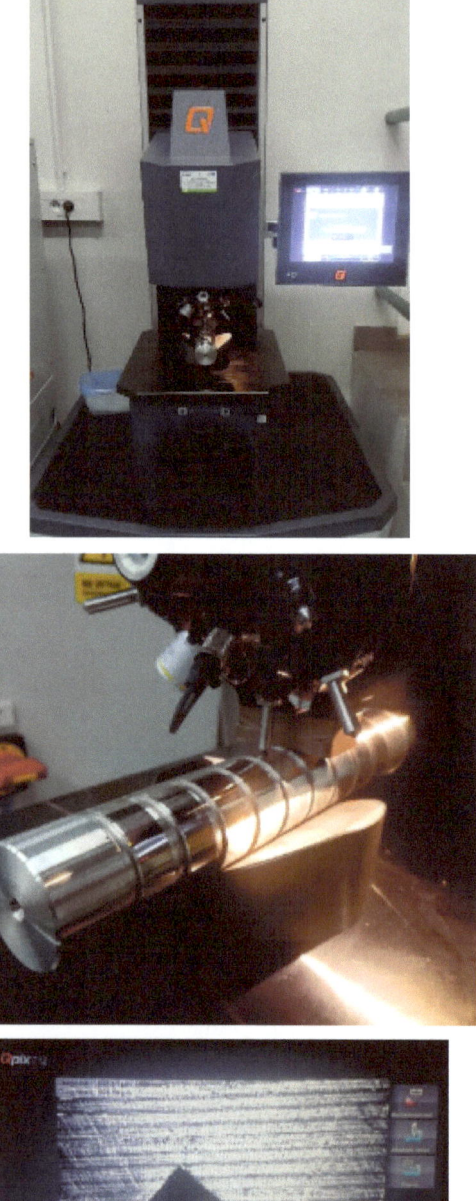

Figure 5. Qness Q250M hardness tester: (**a**) general view, (**b**) view during measurement, (**c**) surface with visible indent.

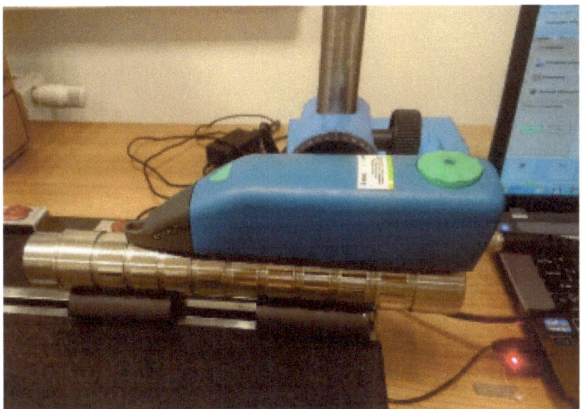

Figure 6. HOMMEL-ETAMIC W20 roughness profilometer.

3. Results of Experimental Research

The aim of the experimental research was to determine the influence of burnishing parameters and conditions on the surface roughness and hardness of the surface layer of corrosion-resistant steel. Table 2 presents the process parameters and results of the roughness measurements after burnishing (pin 0 after turning, pins 1–9 after burnishing).

Table 2. Results of roughness measurements and indicators of roughness reduction after turning and burnishing with the parameters of slide burnishing.

Nr	F [N]	f [mm/rev]	Rz [μm]	Ra [μm]	K_{Ra} [–]	K_{Rz} [–]
0	-	0.2	4.92	0.84	-	-
1	30	0.04	1.43	0.16	5.25	3.44
2	60	0.04	1.48	0.15	5.60	3.32
3	90	0.04	1.51	0.14	6.00	3.26
4	30	0.06	1.05	0.13	6.46	4.66
5	60	0.06	1.02	0.12	6.72	4.80
6	90	0.06	1.01	0.11	7.64	4.85
7	30	0.08	1.85	0.21	3.91	2.66
8	60	0.08	1.60	0.22	3.81	3.07
9	90	0.08	1.52	0.20	4.25	3.23

Figure 7 shows examples of surface roughness profiles plotted by the HOMMEL-ETAMIC W20 profilometer after turning and after slide burnishing.

Based on the analysis of the test results presented in Figure 7, it can be seen that burnishing significantly reduced the surface roughness. The results show that slide burnishing was carried out correctly, as the intention was to reduce the roughness. Also, the results of the calculations performed using the equations [5,11] were analyzed:

$$K_{Ra} \frac{R'a}{Ra} \qquad (1)$$

$$K_{Rz} \frac{R'z}{Rz} \qquad (2)$$

where
$R'a$, $R'z$—roughness after turning;

Ra, Rz—roughness after slide burnishing.

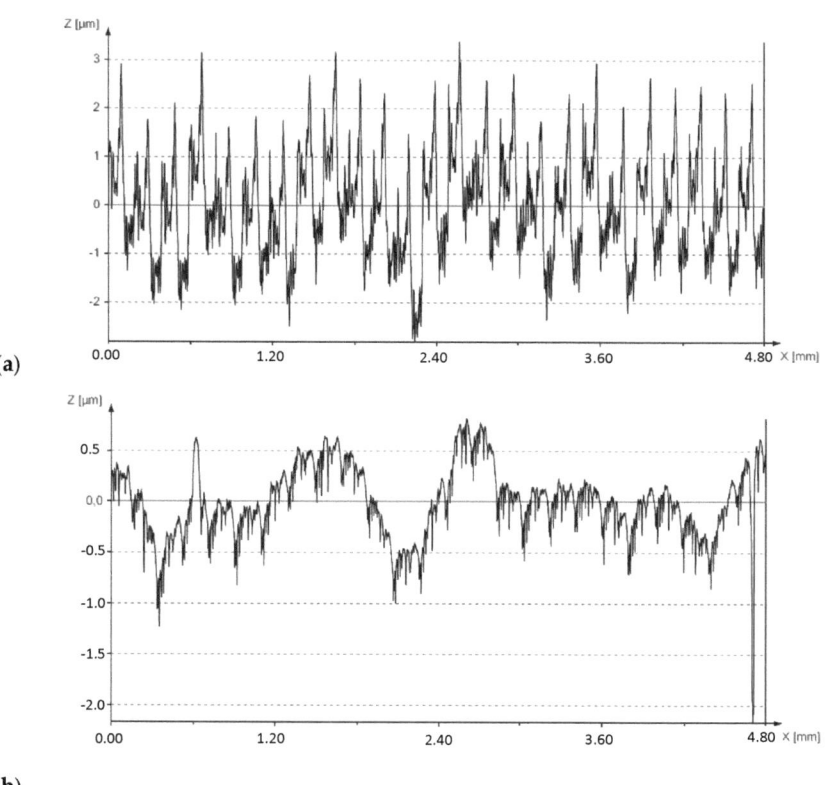

(a)

(b)

Figure 7. Examples of surface roughness profiles plotted by the HOMMEL-ETAMIC W20 device: (a) after turning, (b) after slide burnishing.

Figure 8 shows the dependence of the surface roughness Ra and Rz on the change in the feed rate and the burnishing force. Within the given force range, it can be seen that as the burnishing force increases, the surface roughness value decreases. The lowest values of Ra and Rz were obtained when the burnishing force F 90 N and the feed rate f = 0.06 mm/rev. When the feed rate f 0.08 mm/rev, the values of Ra and Rz after slide burnishing were the highest.

The surface roughness reduction rates determined using Formulas (1) and (2) are presented in Table 2 and Figure 9. It can be seen (Figure 9) that with an increase in force in the range of F 30–90 N and the feed rate f 0.04–0.06 mm/rev, the roughness reduction rate for Ra and Rz increases. The lowest values of Ra 0.11 μm and Rz 1.01 μm, and the greatest reduction in surface roughness KRa 7.64 and KRz 4.85, were obtained for the feed rate f 0.06 mm/rev and the burnishing force F 90 N.

Table 3 shows the hardness measurement results after slide burnishing. The degree of relative strengthening of the surface layer was determined by the following equation [5,11]:

$$S_u \quad \frac{HV - HV'}{HV'} \quad 100\% \qquad (3)$$

where

HV', HV—surface layer hardness before and after burnishing

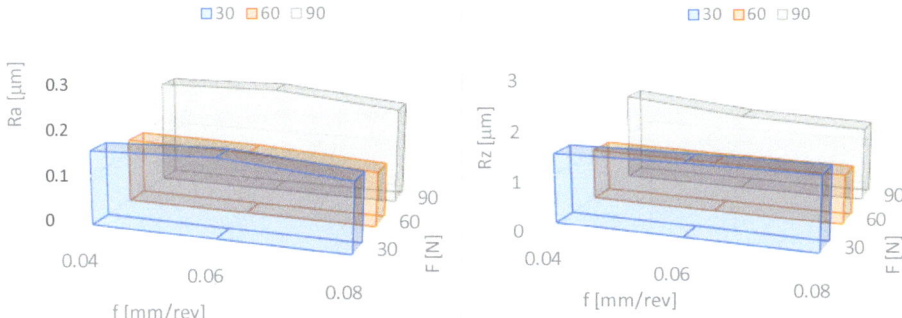

Figure 8. Dependence of surface roughness on the burnishing force (F) and the feed rate (f). Ra—the arithmetic mean roughness value from the amounts of all profile values, Rz—maximum height of profile average value of the five measurements.

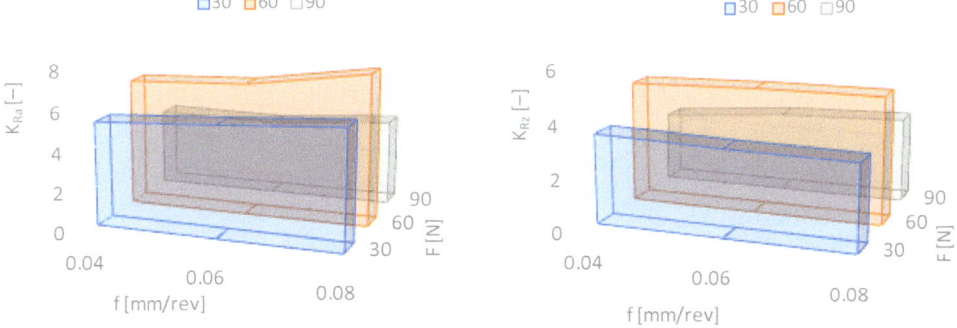

Figure 9. Dependences of the surface roughness reduction rates (K_{Ra}—arithmetic mean roughness value from the amounts of all profile values, K_{Rz}—maximum height of profile average value of the five measurements) on the feed rate (f) and the burnishing force (F).

Table 3. The degree of relative strengthening and hardness after slide burnishing.

Nr	F [N]	F [mm/obr]	S_u [%]	HV	Stand. Dev. for HV
0	-	0.20	-	263	1.49
1	30	0.04	1.91	268	0.96
2	60	0.04	5.71	278	1.60
3	90	0.04	7.22	282	1.71
4	30	0.06	5.32	277	1.29
5	60	0.06	6.46	280	1.80
6	90	0.06	7.61	283	2.50
7	30	0.08	3.42	272	2.69
8	60	0.08	3.81	273	2.99
9	90	0.08	4.18	274	3.14

As can be seen from Table 3, an increase in hardness was observed. This increase is up to approximately 20 Vickers units. Although this increase is not significant, it guarantees less friction in combination with the reduction in roughness. It should therefore be concluded that slide burnishing extends the service life of the working surface by not only reducing the surface roughness but also by increasing its hardness. The highest degree of relative hardening, Su 7.61%, occurs when the feed rate f 0.06 mm/rev and the burnishing force F 90 N. The smallest increase in the hardness of the surface layer occurs for the feed rate f 0.08 mm/rev. The degree of relative strengthening is several percent.

From Table 3 and Figure 10, it can be seen that the increase in hardness was not significant but occurred in every journal (in every case). The research methodology and selection of technological parameters for burnishing treatment were carried out on the basis of the literature [28–30] and our own previous research. Increasing hardness along with reducing surface roughness also improved the resistance parameters to tribological wear, especially in the case of sliding friction in the presence of lubricant. Sliding burnishing is used as an anti-fatigue and anti-friction treatment, as well as for smoothness, in place of polishing and grinding, mainly due to lower burnishing costs and also due to a less harmful impact on the environment.

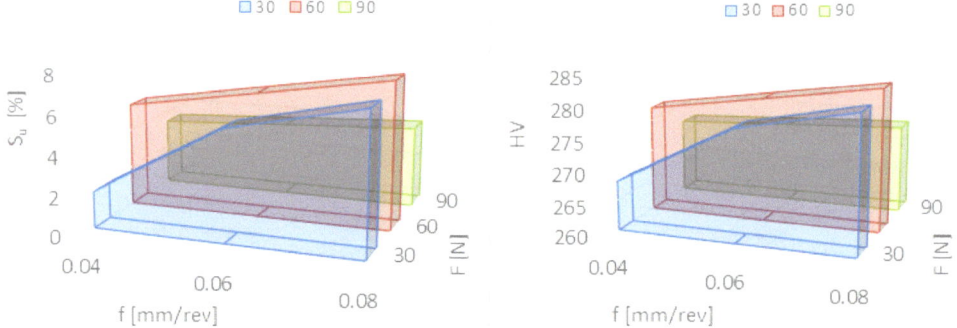

Figure 10. Dependence of the degree of relative strengthening (Su) and hardness (HV) on the burnishing force (F) and the feed rate (f).

This indicates that slide burnishing not only reduces the surface roughness but also hardens the surface layer of the material. As the burnishing force increases in the range of F = 30–90 N, the hardness of the surface layer of steel increases. The highest degree of relative hardening, Su = 7.61%, occurs when the feed rate f = 0.06 mm/rev and the burnishing force F 90 N. The smallest increase in the hardness of the surface layer occurs for the feed rate f 0.08 mm/rev. The degree of relative strengthening is several percent. This is not a very large increase in hardness but it is visible and has a significant impact on the operational wear of machine elements. However, slide burnishing is used in particular for smoothing purposes.

After experimental tests of the slide burnishing of cylindrical elements made of corrosion-resistant steel X2CrNiMo17-12-2, it can be concluded that a significant reduction in surface roughness and an increase in the degree of relative strengthening of the material occurs for the feed rate in the range of f 0.04–0.06 mm/rev and the burnishing force in the range of F 30–90 N. Figure 11 shows the results of macroscopic and microscopic observations after burnishing. For convenience, the observation areas are marked in red (Figures 11 and 12).

Deformations in the surface layer are clearly visible and their measured depth is 400 μm. A clear grain orientation can be seen in the surface area, i.e., the area subjected to the plastic forming process. The deformed elongated grains have the same orientation with homogenous mechanical properties in the surface layer, which is particularly important in terms of the potentially negative impact of the burnishing process, as described by the authors of publications [5,11,13,17], among others. The core area is characterized by a clearly equiaxed microstructure and high grain size uniformity. One of the advantages of the burnishing process, apart from the relatively low cost [31–33], is the fact that there is no change in the chemical composition of the material [34–37]. Therefore, the authors performed scanning electron microscopy with energy-dispersive X-ray spectroscopy (SEM/EDX) to confirm this fact. Figure 12 shows the results of the analysis of the deformed surface and core areas.

As can be seen, the chemical composition is similar in both areas, which indicates that the burnishing process ran correctly, without contamination of the material surface with solids from the burnisher. Grains clearly deformed in the direction of the application of force are visible in the area of the surface layer. No cracks or tears were observed in this area, which also indicates that the proposed parameters are correct. The core is characterized by an equiaxed microstructure.

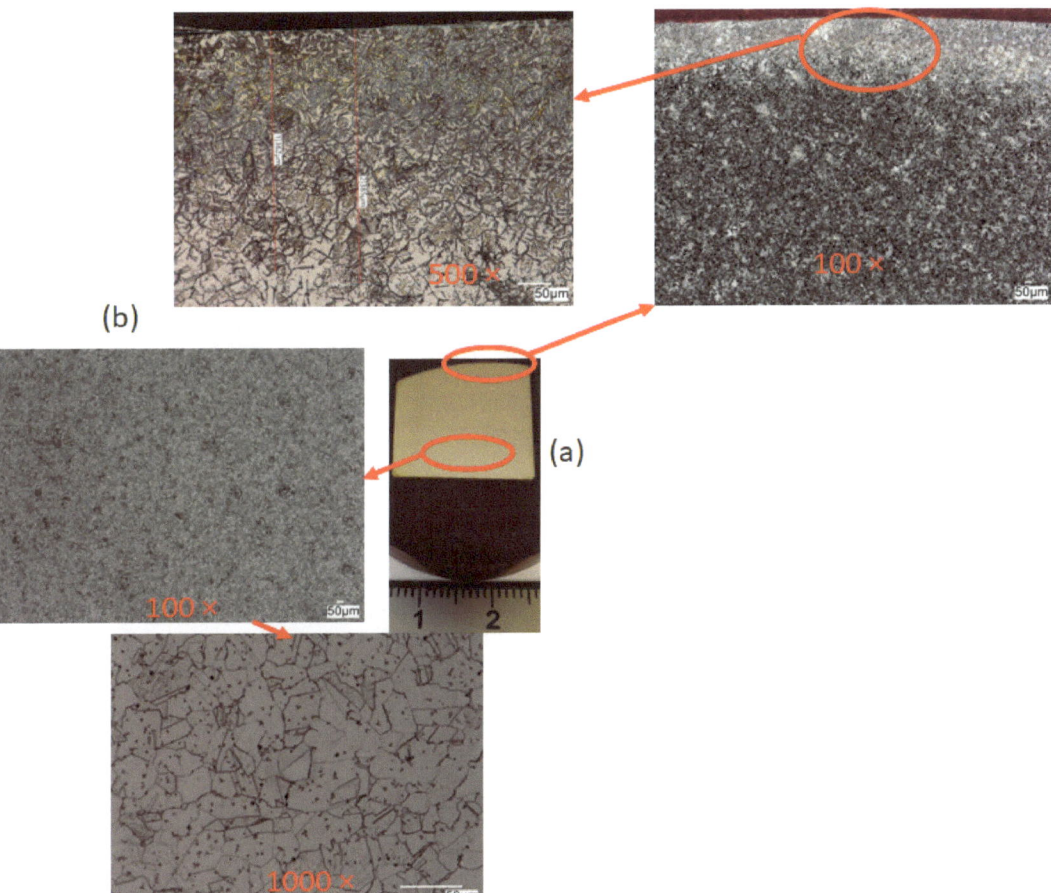

Figure 11. Cross-sectional view after burnishing: (**a**) macroscopic sample view, (**b**) microscopic observation.

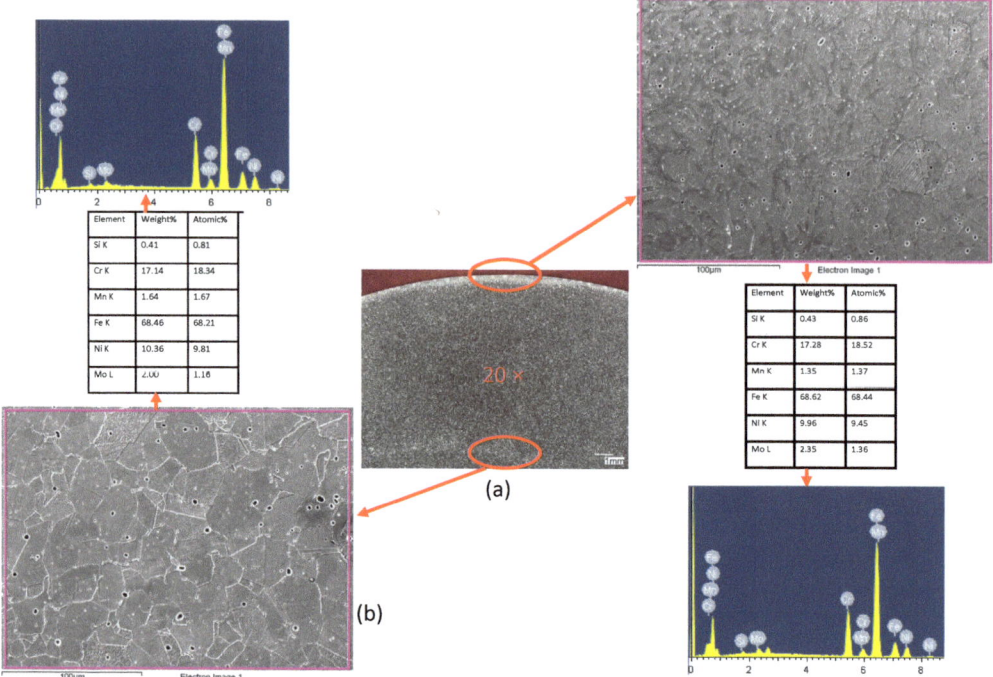

Figure 12. Cross-sectional view after burnishing: (**a**) macroscopic observation, (**b**) microscopic observation.

4. Discussion

The article describes the influence of changes in burnishing conditions and parameters—such as the feed rate and the burnisher depth (pressure force) at a constant burnishing speed—on the surface roughness and hardness of the surface layer of the material [38–40]. Modern design of the product and its surface layer must be focused on construction integrated with production and then on operation. Burnishing of the surface layer may be used, among other techniques, for the production and regeneration of pump drive shafts in place of seals, crankshaft journals, pins and intake and outlet valves. The main mechanism influencing hardness in the burnishing process is strengthening. Due to the fact that the process is carried out cold, the temperature increase in the contact zone is negligible, not more than about 25 degrees (based on the authors' previous research), and there are no microstructure reconstruction phenomena. This is clearly visible in the form of elongated grains in the direction of deformation. Of course, during plastic deformation during burnishing, the same phenomena occur in the surface layer as during cold rolling.

Burnishing offers many possibilities like reducing the surface roughness, but also, the machined object becomes more resistant to abrasion and has better resistance to fatigue [41–43]. The surface layer is squashed during burnishing, while the deeper layers of the material are not affected. As a result, the hardness on the surface of the workpiece increases, but the material still retains its properties. The main aim of this work was to investigate the influence of changes in burnishing parameters on the final quality of the machined surface. The focus was on parameters such as the feed rate and the burnishing force, and the surface was checked for changes in its roughness and hardness. It can be concluded from the results of the study that the changed parameters have a significant impact on the quality of the machined surface [44–46]. Changes in the feed rate have a very strong impact on the roughness of the surface layer, but its impact on hardness is much smaller. It can be noticed that increasing the feed rate causes minor changes up to a certain point, but after exceeding it, the surface roughness increases dramatically. This is caused by the burnishing tool leaving part of the surface unprocessed. In turn, the infeed force is the opposite—when changing the roughness of the workpiece, it has a small effect, but changes the hardness of the surface layer to a much greater extent. With the increase in the feed force, the final hardness increases, but it cannot be increased indefinitely. One of the limitations is the tailwave formed behind the tool, which negatively affects the final smoothness. Another cause is that too much pressure on the diamond tip can accelerate its wear, which will significantly increase the cost of processing. It follows that the selection of burnishing parameters depends on the purpose for which we use this treatment. The main parameter for smoothness is the feed rate. In order to obtain a specific hardness, an appropriate burnishing force must be selected. Because small feed rates and burnishing forces are recommended for slide burnishing with a rigid diamond-tipped clamp, their values were set at f 0.04–0.08 mm/rev and F 30–90 N, respectively. After analyzing the test results for slide burnishing of outer cylindrical surfaces made of corrosion-resistant steel X2CrNiMo17-12-2, it was found that the greatest reduction in surface roughness (KRa 7.64 and KRz 4.85) and the greatest relative degree of material strengthening (Su 7.61%) occurred for the feed rate f 0.06 mm/rev and the burnishing force F = 90 N. Therefore, in order to obtain the appropriate quality of cylindrical elements made of corrosion-resistant steel X2CrNiMo17-12-2, it seems reasonable to use a feed rate in the range of f 0.04–0.06 mm/rev and a burnishing force in the range of F 30–90 N. The selection of the technological process parameters used should depend on the purpose of the burnishing treatment being used. The technological process aimed at obtaining low surface roughness should be carried out with the greatest possible force pressing the working element to the processed surface, while the burnishing speed and feed should be as low as possible. The lowest values of the Ra and Rz parameters occur when a burnishing force of 90 N and a feed rate of 0.06 mm/rev are applied. When using a feed of 0.08 mm/rev, the values of these roughness parameters after slide burnishing are the highest. This is due to the fact that at low forces (30 N), the feed is greater than the contact length and a small recess occurs, and thus, the roughness is greater than when the lower feed is used (0.04–0.06 mm/rev). However, when using high force values (90 N) and high feed values (0.08 mm/rev), the feed is smaller than the contact length with a large tool cavity. A so-called wave is created in front of the burnishing element, and therefore, the roughness is greater compared to the same force (90 N) but at a lower feed (0.04–0.06 mm/rev). Experimental tests have confirmed the possibility of improving some functional properties, including reducing roughness and increasing hardness, by strengthening the surface layer. A surface with a low roughness shows greater resistance to corrosion than a surface with a higher roughness, with the same chemical composition. However, the surface layer with higher hardness and at the same time low roughness, according to the literature data [10,22,43], should be characterized by higher resistance to tribological wear while maintaining high fatigue strength.

5. Conclusions

Based on the research and obtained results, the following conclusions could be formulated:

- The burnishing process can provide a final roughness similar to that produced by smooth finishing.
- The microstructure, or rather the grains, in the area of the surface layer is oriented in the direction of deformation.
- The analysis of the microstructure in the area of the surface layer did not reveal any cracks or other flaws, which proves that the recommended process is correct.
- The slide burnishing not only reduces the surface roughness but hardens the surface layer of the burnished material.
- The greatest reduction in the surface roughness (KRa 7.64 and KRz 4.85) and the greatest relative degree of material strengthening (Su 7.61%) occur for the feed rate f 0.06 mm/rev and the burnishing force F 90 N.

- It is best to use a burnishing feed rate in the range of f 0.04–0.06 mm/rev and a burnishing force in the range of F 30–90 N.
- The use of a feed rate of 0.08 mm/rev increased the surface roughness; the feed is smaller than the contact length with a large tool cavity, and a so-called wave is created in front of the burnishing element.

Author Contributions: Methodology, T.D. and G.S.; Validation, J.F. and M.O.; Formal analysis, Dariusz Rydz, A.S., G.S. and J.F.; Investigation, T.D. and M.O.; Resources, T.D. and D.R.; Data curation, A.S. and M.O.; Writing—original draft, T.D. and G.S.; Writing—review & editing, D.R. and A.S.; Visualization, T.D., G.S. and J.F.; Supervision, G.S.; Project administration, D.R.; Funding acquisition, A.S. All authors have read and agreed to the published version of the manuscript.

Funding: The APC was funded by Czestochowa University of Technology.

Institutional Review Board Statement: Not applicable.

Informed Consent Statement: Not applicable.

Data Availability Statement: Data are contained within the article.

Conflicts of Interest: The authors declare no conflict of interest.

References

1. A Non-Rusting Steel. Sheffield Invention Especially Good for Table Cutlery. *New York Times*. Available online: https://www.nytimes.com/1915/01/31/archives/a-nonrusting-steel-sheffield-invention-especially-good-for-table.html (accessed on 16 April 2024).
2. Kaiserliches Patentamt-Patentschrift Nr 304126, 18 October 1912. Available online: https://depatisnet.dpma.de/DepatisNet/depatisnet?window=1&space=menu&content=treffer&action=pdf&docid=DE000000304126A (accessed on 16 April 2024).
3. Burakowski, T.; Wierzchoń, T. Surface Engineering of Metals: Principles, Equipment, Technologies. In *Materials Science and Technology*; CRC Press LLC.: London, UK; New York, NY, USA; Washington, DC, USA, 1999.
4. Chlebus, E. *Techniki Komputerowe CAx w Inżynierii Produkcji*; Wydawnictwo Naukowo-Techniczne: Warszawa, Poland, 2000.
5. Dyl, T.; Rydz, D.; Stradomski, G. Nagniatanie staliwa typu dupleks w aspekcie zwiększenia twardości i zmniejszenia chropowatości powierzchni. *Zesz. Nauk. Akad. Morskiej W Gdyni* **2017**, 76–86.
6. Knosala, R. *Inżynieria Produkcji. Kompendium Wiedzy*; Polskie Wydawnictwo Ekonomiczne: Warszawa, Poland, 2017.
7. Konefal, K.; Korzyński, M.; Byczkowska, Z.; Korzyńska, K. Improved corrosion resistance of stainless steel X6CrNiMoTi17-12-2 by slide diamond burnishing. *J. Mater. Process. Technol.* **2013**, *213*, 1997–2004. [CrossRef]
8. Przybylski, W. *Technologia Obróbki Nagniataniem*; Wydawnictwo Naukowo-Techniczne: Warszawa, Poland, 1987.
9. Dyl, T. The influence of burnishing parameters on the hardness and roughness on the surface layer stainless steel. *J. Achiev. Mater. Manuf. Eng.* **2017**, *82*, 63–69. [CrossRef]
10. Korzyński, M. *Nagniatanie Ślizgowe*; Wydawnictwo Naukowo-Techniczne: Warszawa, Poland, 2007.
11. Czechowski, K.; Polowski, W.; Wrońska, I.; Wszołek, J. Nagniatanie ślizgowe jako metoda obróbki wykończeniowej powierzchni. *Proj. I Konstr. Inżynierskie* **2009**, 21–27.
12. Polowski, W.; Stóś, J.; Bednarski, P.; Czechowski, K.; Wszołek, J. Nagniatanie ślizgowe narzędziami diamentowymi. *Mechanik* **2010**, *83*, 965–967.
13. Zaleski, K.; Skoczylas, A.; Bławucki, S. *Obróbka Gładkościowa i Umacniająca*; Politechnika Lubelska: Lublin, Poland, 2017.
14. Dzierwa, A.; Markopoulos, A.P. Influence of Ball-Burnishing Process on Surface Topography Parameters and Tribological Properties of Hardened Steel. *Machines* **2019**, *7*, 11. [CrossRef]
15. Jerez-Mesa, R.; Fargas, G.; Roa, J.J.; Llumà, J.; Travieso-Rodriguez, J.A. Superficial Effects of Ball Burnishing on TRIP Steel AISI 301LN Sheets. *Metals* **2021**, *11*, 82. [CrossRef]
16. YAMATO Product Catalog: "Diamond Burnishers". 2022. Available online: https://euroband.pl/img/cms/Catalogs/diamond/Diamond-General_catalog.pdf (accessed on 20 May 2024).
17. Dyl, T.; Koczurkiewicz, B.; Biś, J. Analysis of the impact of burnishing on the steel sleeves technological quality in a corrosive environment. *Ochr. Przed Korozją* **2020**, *8*, 242–245. [CrossRef]
18. Naif, A. Interaction of electric current with burnishing parameters in surface integrity assessment of additively manufactured Inconel 718. *Meas. J. Int. Meas. Confed.* **2024**, *230*, 114474. [CrossRef]
19. Amanov, A.; Karimbaev, R.; Li, C.; Wahab, M.A. Effect of surface modification technology on mechanical properties and dry fretting wear behavior of Inconel 718 alloy fabricated by laser powder-based direct energy deposition. *Surf. Coat. Technol.* **2023**, *454*, 129175. [CrossRef]
20. Harish; Shivalingappa, D.; Raghavendra, N.; Ganesh, V. Impact of Ball Burnishing Process on Residual Stress Distribution in Aluminium 2024 Alloy Using Experimental and Numerical Simulation. *IOP Conf. Ser. Mater. Sci. Eng.* **2021**, *1189*, 012002. [CrossRef]

21. Buldum, B.B.; Cagan, S.C. Study of Ball Burnishing Process on the Surface Roughness and Microhardness of AZ91D Alloy. *Exp. Tech.* **2018**, *42*, 233–241. [CrossRef]
22. Zhang, P.; Lindemann, J.; Ding, W.J.; Leyens, C. Effect of Roller Burnishing on Fatigue Properties of the Hot-Rolled Mg–12Gd–3Y Magnesium Alloy. *Mater. Chem. Phys.* **2010**, *124*, 835–840. [CrossRef]
23. Hilpert, M.; Wagner, L. Corrosion Fatigue Behavior of the High-Strength Magnesium Alloy AZ 80. *J. Mater. Eng. Perform.* **2000**, *9*, 402–407. [CrossRef]
24. Zhang, P.; Lindemann, J. Effect of Roller Burnishing on the High Cycle Fatigue Performance of the High-Strength Wrought Magnesium Alloy AZ80. *Scr. Mater.* **2005**, *52*, 1011–1015. [CrossRef]
25. *EN 10088-1*; Stainless Steels—Part 1: List of Stainless Steels. CEN: Brussels, Belgium, 2014.
26. *PN-EN ISO 6507-3:2018-05*; Metale—Pomiar Twardości Sposobem Vickersa. Polish Committee for Standardization: Warsaw, Poland, 2018.
27. *EN ISO 4287*; Geometrical Product Specifications (GPS)—Surface Texture: Profile Method—Terms, Definitions and Surface Texture Parameters. ISO: Geneva, Switzerland, 1997.
28. Korzynski, M.; Dudek, K.; Korzynska, K. Effect of Slide Diamond Burnishing on the Surface Layer of Valve Stems and the Durability of the Stem-Graphite Seal Friction Pair. *Appl. Sci.* **2023**, *13*, 6392. [CrossRef]
29. Skoczylas, A.; Zaleski, K.; Matuszak, J.; Ciecielag, K.; Zaleski, R.; Gorgol, M. Influence of Slide Burnishing Parameters on the Surface Layer Properties of Stainless Steel and Mean Positron Lifetime. *Materials* **2022**, *15*, 8131. [CrossRef] [PubMed]
30. Dyl, T.; Charchalis, A.; Stradomski, G.; Rydz, D. Impact of processing parameters on surface roughness and strain hardening of two-phase stainless steel. *J. KONES* **2019**, *26*, 37–44. [CrossRef]
31. Chen, X.D.; Wang, L.W.; Yang, L.Y.; Tang, R.; Yu, Y.Q.; Cai, Z.B. Investigation on the impact wear behavior of 2.25Cr–1Mo steel at elevated temperature. *Wear* **2021**, *476*, 203740. [CrossRef]
32. Ovali, İ.; Akkurt, A. Comparison of Burnishing Process with Other Methods of Hole Surface Finishing Processes Applied on Brass Materials. *Mater. Manuf. Process.* **2011**, *26*, 1064–1072. [CrossRef]
33. Chi, H.; Liu, J.; Zhou, J.; Ma, D.; Gu, J. Influence of Microstructure on the Mechanical Properties and Polishing Performance of Large Prehardened Plastic Mold Steel Blocks. *Metals* **2024**, *14*, 477. [CrossRef]
34. Łastowska, O.; Starosta, R.; Jabłońska, M.; Kubit, A. Exploring the Potential Application of an Innovative Post-Weld Finishing Method in Butt-Welded Joints of Stainless Steels and Aluminum Alloys. *Materials* **2024**, *17*, 1780. [CrossRef] [PubMed]
35. Hadad, M.; Attarsharghi, S.; Dehghanpour Abyaneh, M.; Narimani, P.; Makarian, J.; Saberi, A.; Alinaghizadeh, A. Exploring New Parameters to Advance Surface Roughness Prediction in Grinding Processes for the Enhancement of Automated Machining. *J. Manuf. Mater. Process.* **2024**, *8*, 41. [CrossRef]
36. Piotrowski, A.; Zaborski, A.; Tyliszczak, A. Computer-Aided Analysis of the Formation of the Deformation Zone in the Burnishing Process. *Appl. Sci.* **2024**, *14*, 1062. [CrossRef]
37. Maximov, J.; Duncheva, G.; Anchev, A.; Dunchev, V.; Anastasov, K.; Daskalova, P. Effect of Roller Burnishing and Slide Roller Burnishing on Surface Integrity of AISI 316 Steel: Theoretical and Experimental Comparative Analysis. *Machines* **2024**, *12*, 51. [CrossRef]
38. Kułakowska, A.; Bohdal, Ł. Surface Characterization of Carbon Steel after Rolling Burnishing Treatment. *Metals* **2024**, *14*, 31. [CrossRef]
39. Labuda, W.; Wieczorska, A.; Charchalis, A. The Influence of the Burnishing Process on the Change in Surface Hardness, Selected Surface Roughness Parameters and the Material Ratio of the Welded Joint of Aluminum Tubes. *Materials* **2024**, *17*, 43. [CrossRef] [PubMed]
40. Yin, D.; Zhao, H.; Chen, Y.; Chang, J.; Wang, Y.; Wang, X. Modification of Johnson–Cook Constitutive Parameters in Ball Burnish Simulation of 7075-T651 Aluminum Alloy. *Metals* **2023**, *13*, 1992. [CrossRef]
41. Skoczylas, A.; Kłonica, M. Selected Properties of the Surface Layer of C45 Steel Samples after Slide Burnishing. *Materials* **2023**, *16*, 6513. [CrossRef]
42. Swirad, S. Changes in Areal Surface Textures Due to Ball Burnishing. *Materials* **2023**, *16*, 5904. [CrossRef]
43. Stradomski, G.; Fik, J.; Lis, Z.; Rydz, D.; Szarek, A. Wear Behaviors of the Surface of Duplex Cast Steel after the Burnishing Process. *Materials* **2024**, *17*, 1914. [CrossRef] [PubMed]
44. Dzyura, V.; Maruschak, P.; Slavov, S.; Dimitrov, D.; Semehen, V.; Markov, O. Evaluating Some Functional Properties of Surfaces with Partially Regular Microreliefs Formed by Ball-Burnishing. *Machines* **2023**, *11*, 633. [CrossRef]
45. Slavov, S.; Dimitrov, D.; Konsulova-Bakalova, M.; Van, L.S.B. Research of the Ball Burnishing Impact over Cold-Rolled Sheets of AISI 304 Steel Fatigue Life Considering Their Anisotropy. *Materials* **2023**, *16*, 3684. [CrossRef] [PubMed]
46. Maximov, J.; Duncheva, G. The Correlation between Surface Integrity and Operating Behaviour of Slide Burnished Components—A Review and Prospects. *Appl. Sci.* **2023**, *13*, 3313. [CrossRef]

Disclaimer/Publisher's Note: The statements, opinions and data contained in all publications are solely those of the individual author(s) and contributor(s) and not of MDPI and/or the editor(s). MDPI and/or the editor(s) disclaim responsibility for any injury to people or property resulting from any ideas, methods, instructions or products referred to in the content.

Article

The Influence of the Second Phase on the Microstructure Evolution of the Welding Heat-Affected Zone of Q690 Steel with High Heat Input

Huan Qi [1], Qihang Pang [1,*], Weijuan Li [1,*] and Shouyuan Bian [2]

[1] School of Materials and Metallurgy, University of Science and Technology Liaoning, Anshan 114051, China; henryhuanhuan@163.com
[2] Anshan Iron and Steel Group, Anshan 114006, China; 322301@ustl.edu.cn
* Correspondence: qihang25@163.com (Q.P.); liweijuan826@163.com (W.L.)

Citation: Qi, H.; Pang, Q.; Li, W.; Bian, S. The Influence of the Second Phase on the Microstructure Evolution of the Welding Heat-Affected Zone of Q690 Steel with High Heat Input. *Materials* **2024**, *17*, 613. https://doi.org/10.3390/ma17030613

Academic Editor: Wojciech Borek

Received: 12 December 2023
Revised: 23 January 2024
Accepted: 24 January 2024
Published: 27 January 2024

Copyright: © 2024 by the authors. Licensee MDPI, Basel, Switzerland. This article is an open access article distributed under the terms and conditions of the Creative Commons Attribution (CC BY) license (https://creativecommons.org/licenses/by/4.0/).

Abstract: Q690 steel is widely used as building steel due to its excellent performance. In this paper, the microstructure evolution of the heat-affected zone of Q690 steel under simulated high heat input welding conditions was investigated. The results show that under the heat input of 150–300 kJ/cm, the microstructures of the heat-affected zone are lath bainite and granular bainite. The content of lath bainite gradually decreased with the increase in heat input, while the content of granular bainite steadily increased. The proportion of large-angle grain boundaries decreased from 51.1% to 40.3%. Overall, the average size of original austenite increased, and the precipitates changed from Ti (C, N) to Cr carbides. During the cooling process, the nucleation position of bainitic ferrite was from high to low according to the nucleation temperature, and in order of inclusions at grain boundaries, triple junctions, intragranular inclusions, bainitic ferrite/austenite phase boundaries, twin boundaries, grain boundaries, and intragranular inclusions at the bainitic ferrite/austenite phase interface. The growth rate of bainitic ferrite nucleated at the phase interface, grain boundary, and other plane defects was faster, while it was slow at the inclusions. Moreover, it was noted that the Mg-Al-Ti-O composite inclusions promote the nucleation of lath bainitic ferrite, while the Al-Ca-O inclusions do not facilitate the nucleation of bainitic ferrite.

Keywords: Q690 steel; second phase; high heat input; heat affected zone; microstructure

1. Introduction

With the large-scale of equipment, the traditional multi-pass welding method has seriously hindered the production efficiency, and the high heat input welding technology of using >50 kJ/cm heat input to achieve single pass welding of thick plates has been widely concerned, which greatly reduces the production hours. However, the higher peak temperature and longer cooling time brought about by the increase in heat input significantly reduced the properties of the heat affected zone (HAZ) [1,2]. Therefore, improving the microstructure and properties of the coarse grain heat affected zone (CGHAZ) with the weakest performance in the HAZ has become the focus of research [3,4]. Different techniques have been explored to improve low temperature toughness of CGHAZ, in particular oxide metallurgy, which uses non-metallic oxides in steel to promote acicular ferrite (AF) nuclei and refine grains by obtaining large amounts of AF. For example, Luo et al. increased the impact energy of pearlite ferrite steel CGHAZ from 13 J to 127 J at $-20\,°C$ by adding Ti and Ti oxides to induce the nucleation of AF [5]. However, for quenched and tempered high-strength steel, the nucleation of AF is challenging due to the increase in alloy element content.

The use of high-strength steel in production can effectively reduce the weight of the structure and improve the reliability of the product [6]. Compared with traditional carbon steel, high-strength Q690 steel has higher yield strength and tensile strength, which can usually reach 690 MPa or even higher. This high strength allows Q690 steel to withstand greater loads and stresses in structural design, which is particularly important for applications where high-strength materials are required, such as long-span bridges, high-rise buildings and heavy machinery. At the same time, Q690 steel also has good toughness and weldability, and can be connected by conventional multi-pass welding and laser welding methods, which is convenient for construction and maintenance [7–9]. For the wide application of Q690 steel, the evolution of its CGHAZ has been studied for many years. Zhang et al. examined the influence of welding peak temperature on the heat-affected zone of Q690 steel [10]. The results showed that from 800 °C to 1150 °C, the peak temperature increases and the microstructure changes, while at the low temperature, impact toughness decreases. Chiew et al. found that HAZ is generally accompanied by several detrimental characteristics, such as large prior austenite grain size, upper bainite, martensite-austenite (M/A) constituents, and microalloy precipitates, which may lead to the lowest toughness in the heat-affected zone. Among these microstructural features, the M/A constituent (crack susceptibility) plays an important role in decreasing joint toughness [11]. The results of Hung-Wei Yen et al. show that the embrittlement in conventional S690Q steel with higher carbon content is primarily explained by the formation of lenticular martensite along prior austenite grain boundaries or packet boundaries [12]. Similarly, Li et al.'s study believed that M/A(diameter greater than 2 μm) was the main source of crack initiation [13].

However, traditional studies are usually based on very small heat input (single digits), and the welding of thick steel plates usually requires hundreds of kilojoules per centimeter of heat input, which means the evolution mechanism of HAZ of Q690 steel under the condition of high heat input welding is still immature, especially the influence of inclusions in Q690 steel on the evolution of tissues. In this paper, the microstructure evolution of Q690 steel under high heat input was studied, especially the influence of the second relative bainite ferritic core and growth.

2. Materials and Methods

2.1. Experimental Materials

The steel grade used in the experiment is Q690, and it was obtained directly from the steel plant in China. The base metal was hot rolled, quenched, and tempered. The thickness of the base metal is 40 mm, and its yield strength at room temperature reaches 800 MPa. According to GB/T 1591-2008 [14], the specific chemical composition and mechanical properties of Q690C used in this research are given in Tables 1 and 2. Mechanical properties were measured three times and averaged. As shown in Figure 1, the microstructure at 1/4 of the thickness of the experimental steel is tempered martensite.

Table 1. Chemical composition of experimental steel (%).

C	Si	Mn	P	S	Al	Nb	Cr	Ni	Cu	Ca	Ti	Mg	N	O
0.07	0.22	1.57	0.01	0.0024	0.03	0.01	0.24	0.46	0.32	0.0026	0.01	0.002	0.0036	0.002

Table 2. Mechanical properties of experimental steel.

State	Yield Strength (MPa)	Tensile Strength (MPa)	Elongation (%)	Impact Toughness at −20 °C (J)
Quench + temper	800	836	20.5	295

 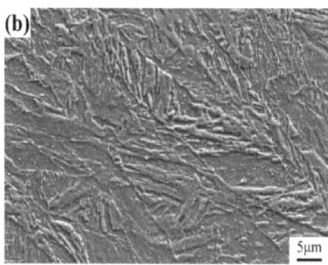

Figure 1. Original structure of experimental steel: (**a**) OM, (**b**) SEM.

2.2. Simulated Welding Thermal Cycle Experiment

To study the change rule of microstructure and mechanical properties at the welding heat affected zone, the simulated welding thermal cycle experiment was conducted using a thermecmaster- 100 kN thermal simulation experimental machine. The sample was cut along the perpendicular in the rolling direction of the steel plate with dimensions of 11 × 11 × 140 mm³. The $t_{8/5}$ value of different heat inputs was calculated by an empirical formula, and the welding thermal cycle curve was established. $t_{8/5}$ represents the time from 800 °C to 500 °C during sample cooling. The calculations show the cooling times as 53 s, 71 s, 89 s, and 106 s, with corresponding cooling speeds of 5.6 °C/s, 4.2 °C/s, 3.4 °C/s, and 2.8 °C/s, and the heat inputs of 150 kJ/cm, 200 kJ/cm, 250 kJ/cm, and 300 kJ/cm, respectively. The analysis was performed at 1350 °C at a heating rate of 100 °C/s and then cooled to room temperature with different heat inputs and holding for 2 s. The $t_{8/5}$ empirical formula is shown in Equation (1) [15]:

$$t_{8/5} = \frac{E}{2\pi\lambda}\left[\frac{1}{500-t_0} - \frac{1}{800-t_0}\right] \quad (1)$$

where E is the welding heat input. λ is the thermal conductivity, 0.36. t_0 is the initial welding temperature, 20 °C.

2.3. Microstructure Observation

To observe the heat-affected zone of the simulated welding thermal cycle, the sample was cut where the thermocouple was placed, polished, and then etched with 4% nitric acid alcohol solution. Subsequently, an Axio vert A1 optical microscope and an Evo MA 10 scanning electron microscope were employed to observe surface morphology, and a nano-measurer plug-in was used to measure the grain size of original austenite. An Energy Dispersive Spectrometer (EDS) was used to analyze the composition of inclusions in steel. TEM carbon extraction samples [16] were prepared according to the conventional experimental procedures, and the samples were extracted from self-made welding samples. A VL2000DX laser confocal microscope was used to observe the experimental steel in situ. In the experiment, the steel was heated to 1350 °C at a heating rate of 10 °C/s, then held for 2 s, and finally cooled to room temperature at a cooling rate of 2.8 °C/s. The laser confocal sample was cut to 1/4 of the thickness of the experimental steel. The sample diameter was 7.5 mm with a length of 2.5 mm. The samples used for EBSD analysis were intercepted on simulated welding samples, processed by mechanical vibration polishing and ion etching at room temperature, and analyzed by EBSD using a nordlysnano microscope.

3. Experimental Results and Analysis

3.1. Microstructure of Welding Heat Affected Zone

The microstructure of the simulated welding heat affected zone under the heat inputs of 150–300 kJ/cm is shown in Figure 2. From Figure 2, It can be observed that the microstructure under different heat input conditions is lath bainite and granular bainite. The bainitic ferrite in lath bainite is in the form of a plate, and the M/A islands are distributed between laths in the form of long strips. The bainitic ferrite in the granular bainite is irregularly massive, and the M/A islands are also irregularly distributed on the ferrite matrix. There are many bainite blocks with different orientations in the same original austenite grain, which can be composed of lath bainite or granular bainite. With the increase in heat input, lath bainite decreases, and the granular bainite increases because the formation temperature of granular bainite is marginally higher. Due to the increase in heat input, the residence time of austenite in the high-temperature stage is prolonged, and granular bainite is developed at higher temperatures [17], leading to the inconspicuous lath characteristics.

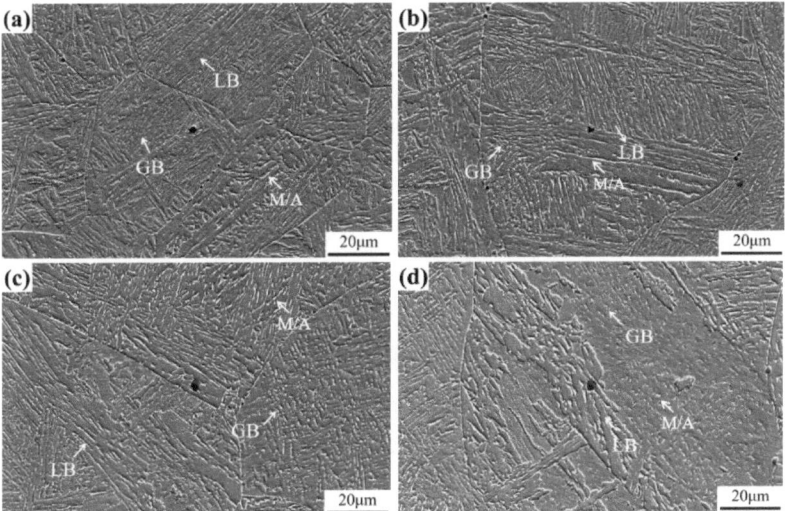

Figure 2. Microstructure of simulated welding heat affected zone: (**a**) Heat input—150 kJ/cm, (**b**) Heat input—200 kJ/cm, (**c**) Heat input—250 kJ/cm, (**d**) Heat input—300 kJ/cm.

Figure 3 shows the statistical results of the original austenite grain size under different heat input conditions. With the increase in heat input, the original austenite grain size increased from 89.6 μm to 104.5 μm. When the heat input is greater than 250 kJ/cm, the original austenite grain size grows appreciably. This is because when the heat input is high, the cooling rate at the high-temperature stage is low, and the second phase particles that prevent the growth of austenite grains may aggregate and grow or dissolve in austenite, thus losing the function of inhibiting the grain growth, hence the austenite grains grow rapidly.

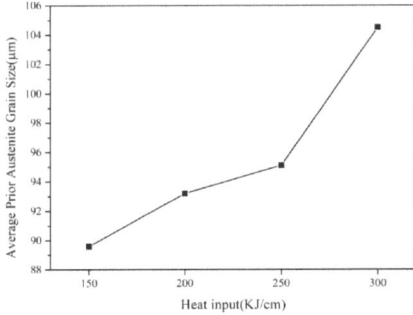

Figure 3. Original austenite grain size of simulated weld heat affected zone.

The grain boundary orientation distribution of samples with the heat input of 150 kJ/cm and 300 kJ/cm is shown in Figure 4. Orientation difference greater than 15° is defined as a large angle grain boundary, and 2°–15° is defined as the low angle grain boundary [17]. In the figure, the black line is a high-angle grain boundary, and the green line stands for a small-angle grain boundary. The proportion of large angle grain boundaries in the 150 kJ/cm sample is 51.1%, and that in the 300 kJ/cm sample is 40.3%. That means the heat input increases from 150 kJ/cm to 300 kJ/cm and the proportion of high-angle grain boundaries decreases accordingly. This is because the original austenite grain coarsens with the increase in heat input; the content of lath bainite decreases and the content of granular bainite increases. The original austenite grain boundary is a high-angle grain boundary, and the adjacent bainite blocks in austenite are oriented at high angles [18]. Therefore, with an increase in the heat input, the proportion of high-angle grain boundaries decreases. It must be noted that the toughness of materials is associated with the number of high-angle grain boundaries. In crack propagation, when the crack touches the high angle grain boundary, it changes its propagation path and consumes more energy, hence the toughness is improved.

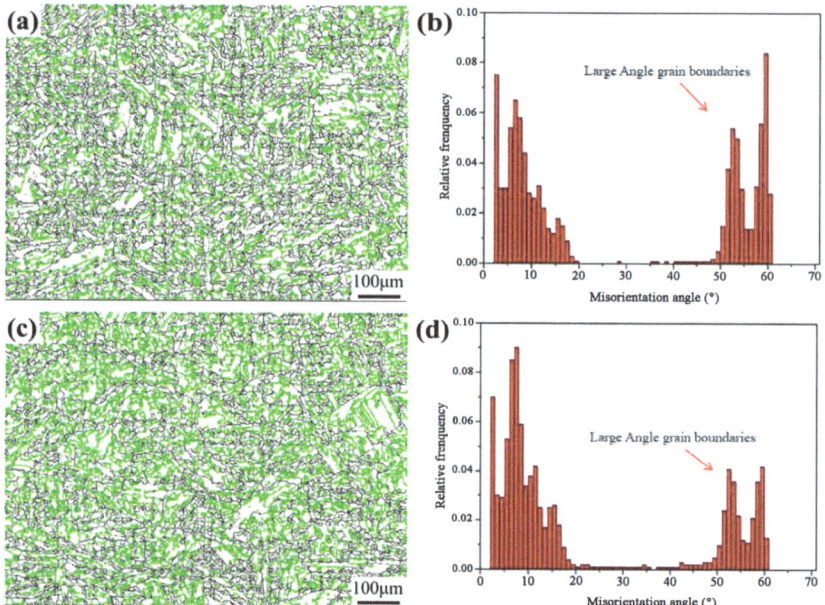

Figure 4. Grain boundary orientation distribution in simulated welding heat affected zone: (**a**,**b**) heat input—150 kJ/cm; (**c**,**d**) heat input—300 kJ/cm.

3.2. Second Phase in Welding Heat Affected Zone

Figure 5 shows a TEM image of precipitates at the simulated weld heat-affected zone under different heat input conditions. When the heat input is 150 kJ/cm, the Ti precipitates (C, N) exist with a diameter in the range of 15–65 nm and an average diameter of 29 nm (Figure 5a–c). When the heat input is 300 kJ/cm, the precipitates are Cr carbides with a diameter in the range of 16–140 nm and an average diameter of 62 nm (Figure 5d–f). The results show that the second phase coarsens and its average size increases with the increase in heat input; simultaneously, the phase composition also changes. When the heat input is 150 kJ/cm, the second phase particles do not grow substantially, and the average size is small due to the fast-cooling rate in the high-temperature stages, and its phase composition is Ti (C, N). Research shows that TiN dissolves at 1300–1350 °C, and a large lag effect is revealed during the dissolution of TiN. At a slower cooling rate, it continuously dissolves TiN when cooled below 1300 °C [16]. The carbide dissolution temperature of Cr_7C_3 is about 1890 °C, and the precipitation morphology depends on its nucleation and growth behavior at different cooling rates [19]. At 300 kJ/cm heat input, Ti (C, N) particles easily dissolve due to the slow cooling rate in the high-temperature stage, hence no Ti (C, N) particles are observed in TEM, and the second phase shows the Cr carbide. Simultaneously, the fine Cr carbides grow up at elevated temperatures, which increases their average size.

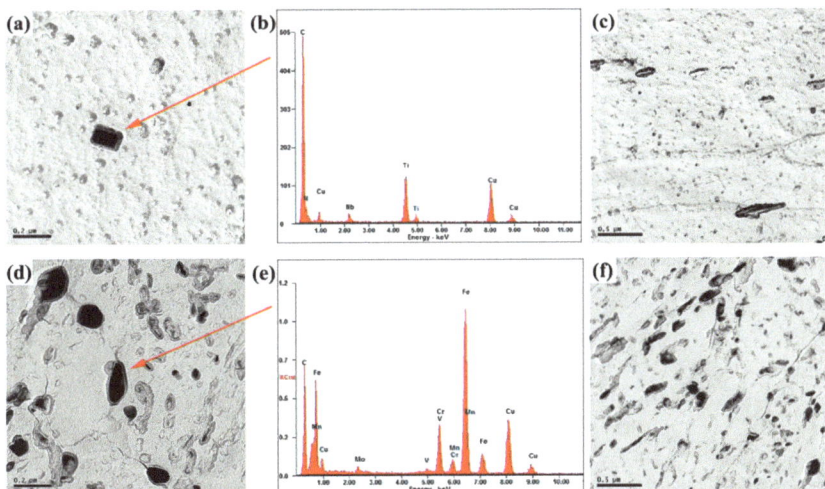

Figure 5. TEM observation of precipitates in welding heat affected zone: (**a**–**c**) heat input—150 kJ/cm; (**d**–**f**) heat input 300 kJ/cm.

With the increase in heat input, the average size of the second phase in the heat-affected zone keeps increasing, the inhibition effect on the growth of austenite grain decreases, and the continuous coarsening of the austenite grain is accelerated. This effect is consistent with the experimental results, as shown in Figures 2 and 3.

3.3. CLSM In Situ Observation

The dissolution and precipitation of inclusions during the heating and cooling process of the welding heat affected zone with 300 kJ/cm heat input were observed with a laser confocal microscope, as shown in Figure 6. When heated to 1002 °C, numerous fine inclusions are distributed on the austenite matrix (as shown in Figure 6a). When the temperature rises to 1098 °C, several fine inclusions dissolve and disappear, and some larger inclusions are refined (Figure 6b). By increasing the temperature to 1203 °C, the small-sized inclusions further dissolve, and only a few large-sized inclusions are observed on the substrate (Figure 6c). When the temperature continues rising to 1300 °C, the fine inclusions keep on dissolving, and the existing larger inclusions are coarsened (Figure 6d). This can be related to the Ostwald ripening process of inclusions during heating [20]. Small particles dissolve and large particles grow; the driving force of Ostwald ripening explains that the interface energy between the inclusion particles and the matrix keeps decreasing. During the cooling, when the temperature reduces to 1137 °C, the inclusions precipitated in the austenite grain and on the grain boundaries are observed. By cooling to 950 °C, the number of inclusions precipitated in austenite further increases. The final distribution is shown in Figure 6f.

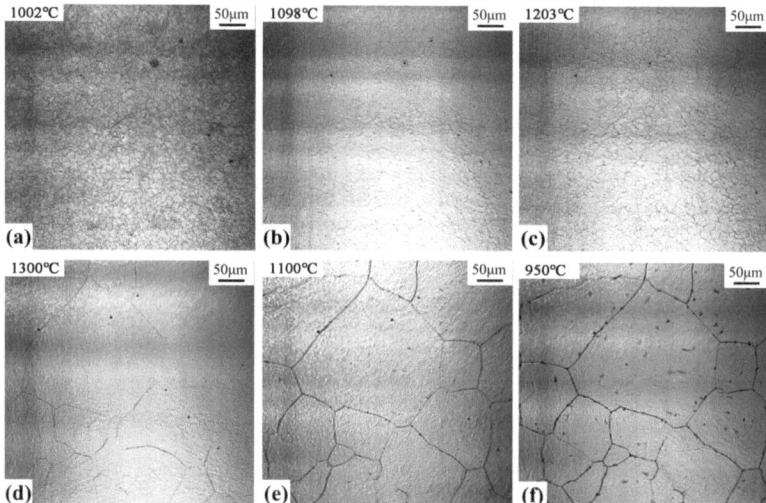

Figure 6. Dissolution and precipitation of inclusions during heating and cooling at 300 kJ/cm heat input: (**a**) Temperature rise—1002 °C, (**b**) Temperature rise—1098 °C, (**c**) Temperature rise—1203 °C, (**d**) Temperature rise—1300 °C, (**e**) Temperature drop—1100 °C; (**f**) Temperature drop: −950 °C.

Figure 7 shows the nucleation process of the bainitic ferrite in the welding heat-affected zone when 300 kJ/cm heat input is observed by a laser confocal microscope. In the image, the nucleation of bainitic ferrite is shown by the formation of surface bumps. Figure 7a is the image of the sample when the temperature decreases to 595 °C. By decreasing the temperature to 580 °C (Figure 7b), the bainite ferrite nucleation is observed at the inclusions on the austenite grain boundary. Similarly, when the temperature drops to 563 °C (Figure 7c), bainite ferrite nucleates are observed at the triple grain boundary of austenite; when the temperature drops to 554 °C (Figure 7d), strip-shaped nuclei of the bainite ferrite plate are observed on the intragranular inclusions; at the same time, bainitic ferrite rapidly forms at the phase interface between the bainitic ferrite lath formed earlier and austenite (Figure 7e). When the temperature drops further to 552 °C (Figure 7f), bainite ferrite nucleates at the twin boundary of austenite; and when the temperature drops to 549 °C (Figure 7g), bainite ferrite nucleates again at the austenite grain boundary. At 506 °C (Figure 7h), the bainitic ferrite lath formed in the austenite crystal collided with the inclusions; when the temperature marginally decreased to 503 °C (Figure 7i), the inclusion (as the nucleation core) promoted the nucleation of bainitic ferrite.

From the experimental results, it can be analyzed that bainite ferrite nucleation occurs at 7 positions in the austenite structure during the cooling process. According to the nucleation temperature (from high to low), inclusions exist at grain boundaries, trigeminal grain boundaries, intragranular inclusions, phase interfaces of bainitic ferrite/austenite, twin boundaries, and grain boundaries. The intragranular inclusions exist at bainitic ferrite/austenite phase interfaces. It can be concluded that (1) Inclusions promote the nucleation of bainite and ferrite. However, due to the various positions of inclusions at the grain boundary or in the crystal, the degree of nucleation is different because of the nucleation temperature. Normally, bainite nucleation occurs preferentially in the grain boundary because of many factors, including the high lattice distortion at the grain boundaries, the high energy, the low diffusion activation energy, the rapid migration of carbon atoms, and the formation of carbon-rich and carbon-poor regions. When the temperature of the carbon-poor region is lower than the transformation temperature of bainite ferrite, bainite ferrite begins to nucleate [21]. When the inclusions are at the grain boundary, they promote non-spontaneous nucleation [22] and the nucleation of bainitic ferrite. Therefore, the inclusions at the grain boundary become the most preferential position for the nucleation of bainitic ferrite. In addition, the inclusions in the crystal may promote the nucleation of bainitic ferrite to different degrees due to the different composition or the size of the inclusions (Figure 7d,i). The intragranular inclusions shown in Figure 7d promote the nucleation of bainitic ferrite, making its nucleation temperature higher. The intragranular inclusion shown in Figure 7i significantly promotes the nucleation of bainitic ferrite, but after collision with bainitic ferrite during its nucleation and growth, the inclusion performs as a nucleation core and promotes the nucleation of bainitic ferrite. This may be related to the further increase in strain energy around inclusions and the promotion of bainite ferrite nucleation by internal stress [23]; (2) The phase interface, grain boundary, and twin boundary promote the nucleation of bainite and ferrite. However, according to the experimental results shown in Figure 7, the priority of bainitic ferrite nucleation is in the order of trigeminal grain boundary, bainitic ferrite/austenite phase interface, twin boundary, and grain boundary. Compared with the ordinary grain boundaries, the trigeminal grain boundaries produce more distortion and higher interfacial energy. The phase boundary is a type of surface defect. Since there is a significant difference in composition and structure between the two sides of the interface, the interface energy is higher than the usual grain boundary energy. The interface energy of the twin boundary is usually low [24], and the interface energy of non-coherent Luan crystal is higher than that of coherent Luan crystal. Xu et al. explained that the priority order of the nucleation sites of bainite is the phase boundary, grain boundary, and intracrystal [25,26]. It can be inferred that the size of the phase interface, grain boundary, and twin boundary promoting the nucleation of bainite is related to the interface structure and energy. The larger interface energy produces a stronger effect on promoting nucleation and the bainite nucleates at higher temperatures.

The experimental results show that bainite nucleates and grows continuously during the cooling process, resulting in an increasing volume fraction of bainite. Bainitic ferrite formed successively at different nucleation positions shows different crystal orientations and high-angle grain boundaries, which assist in improving toughness [18]. If the nucleation position of bainite is increased by changing chemical composition and process control, more bainite blocks with different crystal orientations are expected in one austenite grain, and the mechanical properties of materials can be improved.

The kinetics of bainite formation depend on the nucleation and growth rate of bainite. Figure 8 shows the statistics of the growth rate of bainite nucleated at grain boundary inclusions, trigeminal grain boundaries, intragranular inclusions, bainite or austenite phase interface, twin boundary, grain boundary, and intragranular inclusions at bainite ferrite/austenite phase interface, which are 3.0 μm/s, 9.8 μm/s, 4.9 μm/s, 34 μm/s, 14.8 μm/s, 33.9 μm/s, and 50.4 μm/s. According to the statistical results, the growth rate of bainite from fast to slow is in the order of intragranular inclusions at the bainite ferrite/austenite phase interface, bainite/austenite phase interface, grain boundary, twin boundary, trigeminal grain boundary, intragranular inclusions, and grain boundary inclusions. This shows that the growth rate of bainite ferrite nucleated at the phase interface, grain boundary and other surface defects is fast, while the growth rate of bainite ferrite nucleated at the inclusions is slow.

According to the shear theory of bainite transformation [27], the growth of bainite is related to the diffusion of carbon atoms in ferrite and austenite. The larger diffusion coefficient of carbon atoms produces a faster growth of the bainite. Both crystal defects and temperature affect the diffusion of carbon atoms. The phase boundary, grain boundary, and other surface defects accelerate the diffusion of carbon atoms, and the bainitic ferrite nucleation is conducive to the diffusion of carbon atoms in ferrite and austenite, thus promoting the growth of bainitic ferrite. In addition, according to the experimental results, the nucleation temperature of bainitic ferrite at the surface defects (such as phase boundaries and grain boundaries) is lower than the inclusions. As the temperature decreases, the diffusion coefficient of carbon atoms decreases, and the growth rate of bainite drops. This shows that the growth rate of bainitic ferrite is affected by the phase boundary, grain boundary, other surface defects, and nucleation temperature, but also depends on the phase boundary, grain boundary, and other surface defects. The growth rate of bainite ferrite nucleated at the inclusion is slow, which promotes the refinement of bainite structure and enhances strength and toughness.

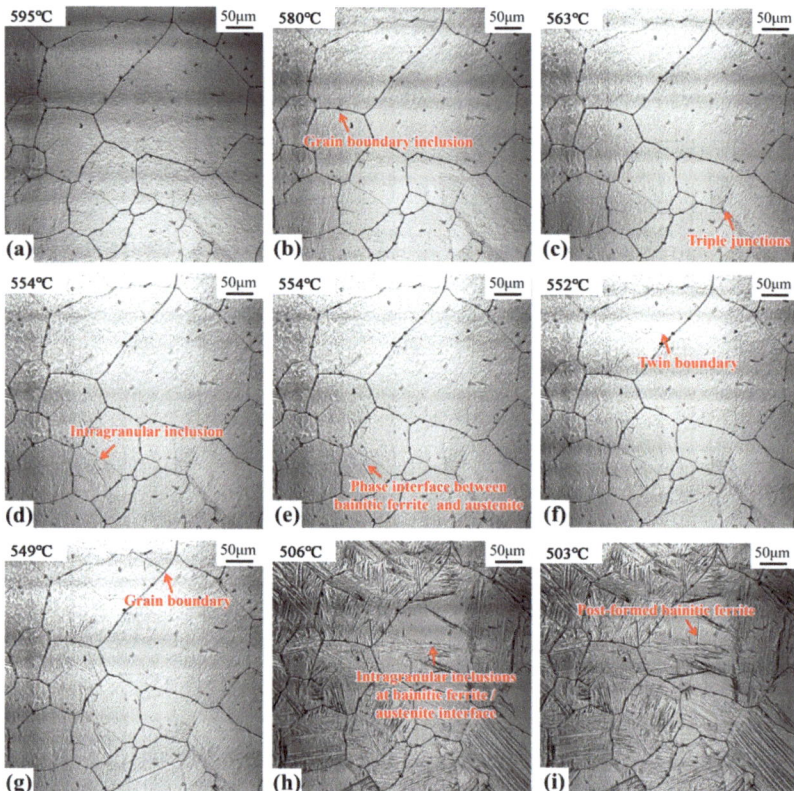

Figure 7. In situ observation of bainite ferrite formation in the weld heat affected zone at 300 kJ/cm heat input: (**a**) 595 °C; (**b**) 580 °C; (**c**) 563 °C; (**d**) 554 °C; (**e**) 554 °C; (**f**) 552 °C; (**g**) 549 °C; (**h**) 506 °C; (**i**) 503 °C.

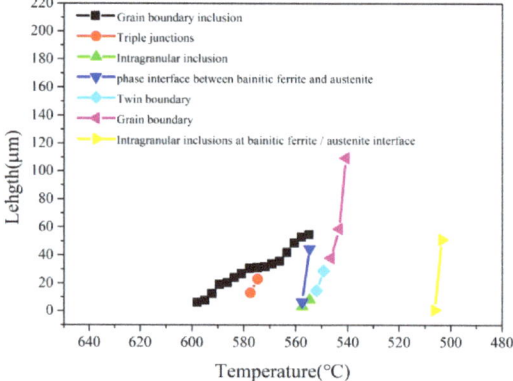

Figure 8. Relationship between bainite ferrite length and temperature change at each preferential nucleation position.

4. Influence of Inclusion Characteristics on Microstructure

Figure 9 shows the microstructure of the weld heat-affected zone under different heat inputs and the distribution of inclusions. The microstructure is composed of lath bainite and granular bainite. When the heat input is 150 kJ/cm (Figure 9a), lath bainite ferrite grows on the interface of composite inclusions. The main component of the black part in the composite inclusions is Al-Mg-O, and its maximum size is about 1.67 μm; the main component of the gray part is Al-Mg-Ti-O, and its maximum size is almost 1.45 μm; overall, the maximum size of inclusions is nearly 2.74 μm. Earlier published results showed that mg Al oxides are favorable for AF nucleation [28], and the acicular ferrite is considered a bainitic ferrite lath with a slightly higher formation temperature. The Mg-Al oxide also promotes the nucleation of lath bainite ferrite. When the heat input is 200 kJ/cm (Figure 9b), complex inclusions exist in the microstructure. The main component of the black part of the inclusions is Ti-Al-O, and its maximum size is around 1.35 μm; the main component of the gray part is Mg-Al-O, its maximum size is about 2.48 μm; and the overall maximum size of inclusions is about 2.18 μm. This composite inclusion does not directly promote the nucleation of bainitic ferrite but induces the formation of lath bainitic ferrite at the bainitic ferrite/austenite phase interface after collision with the bainitic ferrite. This observation is consistent with the results observed in Figure 7i. When the heat input is 250 kJ/cm (Figure 9c), a composite inclusion exists in the lath bainite structure. The main chemical composition of the black part of the inclusion is Fe-C, and its maximum size is abouthe 74 μm. The main chemical composition of the gray part is Al-Ca-O, and its maximum size is around 2.69 μm; the overall maximum size of inclusions is about 3.69 μm.

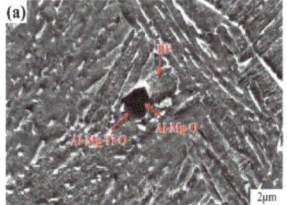 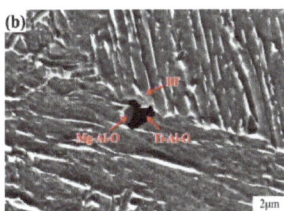 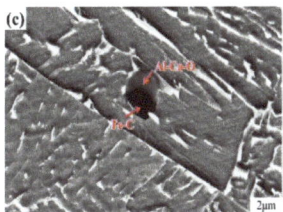

Figure 9. Microstructure and inclusions of welding heat affected zone: (**a**) heat input—150 kJ/cm; (**b**) heat input—200 kJ/cm; (**c**) heat input—250 kJ/cm.

The mechanisms by which the inclusions promote the nucleation of bainitic ferrite include the minimum mismatch mechanism, strain induction mechanism, cation vacancy mechanism, and Mn poor region mechanism [29]. In a certain temperature range, the promotion of bainite ferrite nucleation by inclusions is related to its chemical composition and size [30]. The mismatch between Mg-O and α-Fe is 3.9%, and the mismatch between Ti-O and α-Fe is 3.1%. Studies have shown that a mismatch of less than 6% effectively promotes the nucleation of bainite ferrite [31]. The composite inclusion shown in Figure 9b contains Mg-O and Ti-O, but it does not directly induce the nucleation of bainitic ferrite. Instead, the strain-inducing mechanism and the low mismatch mechanism work together and promote the nucleation of bainitic ferrite. The mismatch estimated between Al_2O_3, and α-Fe is 15.2%, and the mismatch between CaO and α-Fe is noted as 18.6%. The published studies [32] also show that Al-Ca-O inclusions do not promote the bainite ferritin core. This is also confirmed by the results shown in Figure 9c.

5. Conclusions

1. The microstructure of the HAZ of Q690 steel under the heat inputs of 150–300 kJ/cm is composed of lath bainite and granular bainite. With the increase in heat input, lath bainite decreases, granular bainite increases, and the proportion of large angle grain boundary decreases from 51.1% to 40.3%. Overall, the original austenite grain size increased from 89.6 μm to 104.5 μm.

2. When the heat input is 150 kJ/cm, the precipitates in the heat affected zone are Ti (C, N), with an average diameter of 29 nm; When the heat input is 300 kJ/cm, the precipitates are Cr carbides with an average diameter of 62 nm.
3. During the cooling process of austenite in the heat-affected zone, the nucleation positions of bainite ferrite from high to low exist according to the nucleation temperature, inclusions at the grain boundary, trigeminal grain boundary, intragranular inclusions, bainite ferrite/austenite phase interface, twin boundary, and grain boundary. The intragranular inclusions exist at the bainite ferrite/austenite phase interface.
4. The growth rate of the bainite ferrite nucleated at the phase interface, grain boundary, and other surface defects is faster, while the bainite ferrite nucleated at the inclusions is slower.
5. Mg-Al-Ti-O composite inclusions promote the nucleation of lath bainite ferrite, but Al-Ca-O inclusions do not facilitate this.

Author Contributions: Conceptualization, W.L.; Methodology, Q.P.; Investigation, S.B.; Writing—original draft, H.Q. All authors have read and agreed to the published version of the manuscript.

Funding: The authors appreciate the financial support by the National Natural Science Foundation of China (No. 52004122 and No. 52074152).

Institutional Review Board Statement: Not applicable.

Informed Consent Statement: Not applicable.

Data Availability Statement: All the raw data supporting the conclusion of this paper were provided by the authors.

Conflicts of Interest: The authors declare no conflict of interest.

References

1. Hu, J.; Du, L.; Wang, J.; Gao, C. Effect of welding heat input on microstructures and toughness in simulated CGHAZ of V-N high strength steel. *Mater. Sci. Eng. A* **2013**, *577*, 161–168. [CrossRef]
2. Wang, X.; Wang, C.; Kang, J.; Yuan, G.; Misra, R.D.K.; Wang, G. An in-situ microscopy study on nucleation and growth of acicular ferrite in Ti-Ca-Zr deoxidized low-carbon steel. *Mater. Charact.* **2020**, *165*, 110381. [CrossRef]
3. Chen, C.; Chiew, S.P.; Zhao, M.S.; Lee, C.K.; Fung, T.C. Welding effect on tensile strength of grade S690Q steel butt joint. *J. Constr. Steel Res.* **2019**, *153*, 153–168. [CrossRef]
4. Maraveas, C.; Fasoulakis, Z.C.; Tsavdaridis, K.D. Mechanical properties of high and very high steel at elevated temperatures and after cooling down. *Fire Sci. Rev.* **2017**, *6*, 3. [CrossRef]
5. Luo, X.; Niu, Y.; Chen, X.; Tang, H.; Wang, Z. High performance in base metal and CGHAZ for ferrite-pearlite steels. *J. Mater. Process. Technol.* **2017**, *242*, 101–109. [CrossRef]
6. Tuz, L. Determination of the causes of low service life of the air fan impeller made of high-strength steel. *Eng. Fail. Anal.* **2021**, *127*, 105502. [CrossRef]
7. Tuz, L. Evaluation of microstructure and selected mechanical properties of laser beam welded S690QL high-strength steel. *Adv. Mater. Sci.* **2018**, *18*, 34–42. [CrossRef]
8. Chiew, S.P.; Cheng, C.; Zhao, M.S.; Lee, C.K.; Fung, T.C. Experimental study of welding effect on S690Q high strength steel butt joints. *Ce/Papers* **2019**, *3*, 701–706. [CrossRef]
9. Li, B.; Xu, P.; Lu, F.; Gong, H.; Cui, H.; Liu, C. Microstructure characterization of fiber laser welds of S690QL high-strength steels. *Metall. Mater. Trans. B* **2018**, *49*, 225–237. [CrossRef]
10. Zhang, Y.; Xiao, J.; Liu, W.; Zhao, A. Effect of welding peak temperature on microstructure and impact toughness of heat-affected zone of Q690 high strength bridge steel. *Materials* **2021**, *14*, 2981. [CrossRef]
11. Chiew, S.P.; Chen, C.; Zhao, M.S.; Lee, C.K.; Fung, T.C. Post-welding behaviour of S690Q high strength steel butt joints. *J. Civ. Environ. Eng.* **2021**, *43*, 64–71.
12. Yen, H.W.; Chiang, M.H.; Lin, Y.C.; Chen, D.; Huang, C.Y.; Lin, H.C. High-temperature tempered martensite embrittlement in quenched-and-tempered offshore steels. *Metals* **2017**, *7*, 253. [CrossRef]
13. Li, Z.; Tian, L.; Jia, B.; Li, S. A new method to study the effect of M-A constituent on impact toughness of ICHAZ in Q690 steel. *Mater. Res.* **2015**, *30*, 1973–1978. [CrossRef]
14. Shi, G.; Zhu, X.; Ban, H. Material properties and partial factors for resistance of high-strength steels in China. *J. Constr. Steel Res.* **2016**, *121*, 65–79. [CrossRef]
15. Dong, L.; Qiu, X.; Liu, T.; Lu, Z.; Fang, F.; Hu, X. Estimation of cooling rate from 800 C to 500 C in the welding of intermediate thickness plates based on FEM simulation. *J. Mater. Sci. Eng. B* **2017**, *7*, 258–267.

16. Poorhaydari, K.; Ivey, D.G. Application of carbon extraction replicas in grain-size measurements of high-strength steels using TEM. *Mater. Charact.* **2007**, *58*, 544–554. [CrossRef]
17. Peng, K.; Yang, C.; Fan, C.; Lin, S. In situ observation and electron backscattered diffraction analysis of granular bainite in simulated heat-affected zone of high-strength low-alloy steel. *Sci. Technol. Weld. Join.* **2017**, *23*, 158–163. [CrossRef]
18. Gan, X.; Wan, X.; Zhang, Y.; Wang, H.; Li, G.; Xu, G.; Wu, K. Investigation of characteristic and evolution of fine-grained bainitic microstructure in the coarse-grained heat-affected zone of super-high strength steel for offshore structure. *Mater. Charact.* **2019**, *157*, 109893. [CrossRef]
19. Cai, Y.; Luo, Z.; Zeng, Y. Influence of deep cryogenic treatment on the microstructure and properties of AISI 304 austenitic stainless steel A-TIG weld. *Sci. Technol. Weld. Join.* **2016**, *22*, 236–243. [CrossRef]
20. Fu, W.; Li, C.; Duan, R.; Gao, H.; Di, X.; Wang, D. Formation mechanism of CuNiAl-rich multi-structured precipitation and its effect on mechanical properties for ultra-high strength low carbon steel obtained via direct quenching and tempering process. *Mater. Sci. Eng. A* **2022**, *833*, 142567. [CrossRef]
21. Ravi, A.M.; Sietsma, J.M.; Santofimia, J. Exploring bainite formation kinetics distinguishing grain-boundary and autocatalytic nucleation in high and low-Si steels. *Acta Mater.* **2016**, *105*, 155–164. [CrossRef]
22. Zou, X.; Sun, J.; Matsuura, H.; Wang, C. Documenting ferrite nucleation behavior differences in the heat-affected zones of EH36 shipbuilding steels with Mg and Zr additions. *Metall. Mater. Trans. A* **2019**, *50*, 4506–4512. [CrossRef]
23. Abson, D.J. Acicular ferrite and bainite in C-Mn and low-alloy steel arc weld metals. *Sci. Technol. Weld. Join.* **2018**, *23*, 635–648. [CrossRef]
24. Hong, J.; Kang, S.; Jung, J.; Lee, Y. The mechanism of mechanical twinning near grain boundaries in twinning-induced plasticity steel. *Scr. Mater.* **2020**, *174*, 62–67. [CrossRef]
25. Ricks, R.A.; Howell, P.R.; Barritte, G.S. The nature of acicular ferrite in HSLA steel weld metals. *J. Mater. Sci.* **1982**, *17*, 732–740. [CrossRef]
26. Sarma, D.S.; Karasev, A.V.; Jönsson, P.G. On the role of non-metallic inclusions in the nucleation of acicular ferrite in steels. *ISIJ Int.* **2009**, *49*, 1063–1074. [CrossRef]
27. Han, X.; Zhang, Z.; Rong, Y.; Thrush, S.J.; Barber, G.C.; Yang, H.; Qiu, F. Bainite kinetic transformation of austempered AISI 6150 steel. *J. Mater. Res. Technol.* **2020**, *9*, 1357–1364. [CrossRef]
28. Lin, C.; Pan, Y.; Su, Y.F.; Lin, G.; Hwang, W.; Kuo, J. Effects of Mg-Al-O-Mn-S inclusion on the nucleation of acicular ferrite in magnesium-containing low-carbon steel. *Mater. Charact.* **2018**, *141*, 318–327. [CrossRef]
29. Zhang, D.; Terasaki, H.; Komizo, Y. In situ observation of the formation of intragranular acicular ferrite at non-metallic inclusions in C-Mn steel. *Acta Mater.* **2010**, *28*, 1369–1378. [CrossRef]
30. Sun, L.; Li, H.; Zhu, L.; Liu, Y.; Hwang, J. Research on the evolution mechanism of pinned particles in welding HAZ of Mg treated shipbuilding steel. *Mater. Des.* **2020**, *192*, 108670. [CrossRef]
31. Lv, W.; Yan, L.; Pang, X.; Yang, H.; Qiao, L.; Su, Y.; Gao, K. Study of the stability of α-Fe/MnS interfaces from first principles and experiment. *Appl. Surf. Sci.* **2020**, *501*, 144017. [CrossRef]
32. Wang, X.; Wang, C.; Kang, J.; Yuan, G.; Misra, R.D.K.; Wang, G. Improved toughness of double-pass welding heat affected zone by fine Ti-Ca oxide inclusions for high-strength low-alloy steel. *Mater. Sci. Eng. A* **2020**, *780*, 139198. [CrossRef]

Disclaimer/Publisher's Note: The statements, opinions and data contained in all publications are solely those of the individual author(s) and contributor(s) and not of MDPI and/or the editor(s). MDPI and/or the editor(s) disclaim responsibility for any injury to people or property resulting from any ideas, methods, instructions or products referred to in the content.

Article

Research on Cold Roll Forming Process of Strips for Truss Rods for Space Construction

Xingwen Yang [1,2,*], Jingtao Han [1,3] and Ruilong Lu [1]

[1] School of Materials Science and Engineering, University of Science and Technology Beijing, Beijing 100083, China
[2] Industrial Training Center, Zhongyuan University of Technology, Zhengzhou 451191, China
[3] Guangzhou Sino Precision Steel Tube Industry Research Institute Co., Ltd., Guangzhou 511300, China
* Correspondence: yangxingwen1111@163.com; Tel.: +86-132-8507-0721; Fax: +86-010-6233-2572

Abstract: In this paper, a new technology for on-orbit cold forming of space truss rods is proposed. For the cold roll forming process of asymmetric cross sections of thin strips, the effects of roll gap and roll spacing on the forming of asymmetric cross sections of strips were investigated using ABAQUS simulation + experiments. The study shows the following. When forming a strip with a specific asymmetric cross section, the stresses are mainly concentrated in corners 2/4/6, with the largest strain value in corner 2. With increasing forming passes, when the roll gap is 0.3 mm, the maximum equivalent strain values are 0.09, 0.24, 0.64 sequentially. Roll gaps of 0.4 mm and 0.5 mm equivalent strain change amplitude are relatively similar, and their maximum equivalent strain values are approximately 0.07, 0.15, 0.44. From the analysis of the stress–strain history of the characteristic nodes in corners 2/4/6, it can be seen that the stress and strain changes in the deformation process mainly occur at the moment of interaction between the upper and lower rollers, where the stress type of node 55786 shows two tensile types and one compressive type, the stress type of nodes 48594 and 15928 shows two compressive and one tensile type, and the strain of the three nodes is in accordance with the characteristics of plane strain. When the roll gap is about 0.4 mm, the forming of the strip is relatively good. With increased roll spacing, the strip in the longitudinal stress peak through the rollers shows a small incremental trend, but the peak stresses are 380 Mpa or so. When the roll spacing is 120 mm, the longitudinal strain fluctuation of the strip is the most serious, followed by the roll spacing at 100 mm, and the minimum at 140 mm. Combined with the fluctuation in strip edges under different roll spacings, manufacturing cost and volume and other factors, a roll spacing of 100 mm is more reasonable. It is experimentally verified that when the roll gap is 0.4 mm and the roll spacing is 100 mm, the strip is successfully prepared in accordance with the cross-section requirements. When the rolling gap is 0.3 mm, due to stress–strain concentration, the strip is prone to edge waves in forming. The top of corner 2 of the flange triangular region is susceptible to intermittent tear defects, and the crack extension mechanism is mainly based on the cleavage fracture + ductile fracture.

Keywords: truss rods; cold roll forming process; roll gap; roll spacing; concentration of strain; tearing crack

Citation: Yang, X.; Han, J.; Lu, R. Research on Cold Roll Forming Process of Strips for Truss Rods for Space Construction. *Materials* **2023**, *16*, 7608. https://doi.org/10.3390/ma16247608

Academic Editor: Wojciech Borek

Received: 14 November 2023
Revised: 4 December 2023
Accepted: 8 December 2023
Published: 12 December 2023

Copyright: © 2023 by the authors. Licensee MDPI, Basel, Switzerland. This article is an open access article distributed under the terms and conditions of the Creative Commons Attribution (CC BY) license (https://creativecommons.org/licenses/by/4.0/).

1. Introduction

An important part of spacecraft structure, the space truss is mainly constructed by connecting a certain number of one-dimensional rods in three-dimensional space in a certain direction, and has been widely used in constructing deep space exploration bases and expanding the functions of space stations [1]. However, in current spacecraft, space truss structures are typically in the order of 10–1000 m in size for on-orbit operation [2,3]. Future space exploration will require large, lightweight, high-performing and cost-effective space trusses, such as in-orbit service platforms, which will reach geometric dimensions of 0.1–10 km when they are completed [4,5]. Obviously, the existing methods of deployment and assembly on orbit are very limited [6,7], and new ways of building are urgently needed. The use of various space manufacturing technologies, including on-orbit additive manufacturing [8,9] and on-orbit welding [10,11], to produce the rods is one of the effective ways to solve the bottleneck problem of large space trusses on orbit. In this paper, an on-orbit cold forming manufacturing technology is proposed for the production of space truss rods, which mainly uses small cold roll forming machines to coil the raw material into rods to realize on-orbit manufacturing. The method is simple in principle, easy to implement, and does not need to take into account a number of complex problems, such as the suspension of molten droplets in microgravity and solidification in space [2], and is expected to provide a new way of manufacturing truss rods on orbit.

The principle of using cold forming manufacturing technology to produce truss rods on orbit proposed in this paper is as follows. First, the strip in coils is passed through a small cold forming machine to form a strip with a specific cross section. Next, the strip, with a specific cross section, will be subjected to spiral molds and pressure wheels to achieve spiral bending, locking seam, and compaction, ultimately allowing for the manufacture of truss rods of any desired length. The principle is illustrated in Figure 1. In this technology, the initial cold roll forming of the strip to a specific cross section using multiple sets of rolls is an incredibly important step that has a direct impact on the subsequent success of the strip spiral bending and locking seam process. The partitioning and dimensions of the asymmetrical cross-section strip designed for this subject are shown in Figure 2. The investigation of the cold roll forming process mainly focuses on the production of cold-formed components with symmetrical cross sections [12]. These include items like hollow square tubes [13], U-shaped steels [14,15], C-beams [16,17], V-beams [18], W-section plates [19], hat-shaped plates [20,21], and corrugated plates [22,23]. Safdarian et al. [24] studied the formation process of square tubes by cold roll forming and found that the longitudinal strain of the strip edge was not substantially affected by the friction between the strip and the roll or by the rolling speed. Poursafar et al. [19] investigated the effect of each anisotropy of material plasticity and angular increment or pattern, strip width on springback, and longitudinal bending during forming in W-profile sheets. Najafabadi et al. [25] investigated the edge wrinkling mechanism during cold rolling of wide continuous U profiles. For asymmetric cross-section strips, due to the complex roll hole design and the need to consider force balance deformation and other issues specific to asymmetric cross sections, there has been limited research into the related aspects of cold roll forming. Wang et al. [26] investigated the effect of roll gap, friction coefficient, roll diameter increment and linear speed on the maximum longitudinal strain at the strip edge when forming asymmetric and deep complex cross sections, which mentioned that the effect of roll gap was very important, but did not conduct much in-depth research. The present investigation shows that strip thicknesses for cold roll forming are generally greater than 1 mm; however, this paper proposes a process where the strip thickness used is in the range of 0.3–0.5 mm for rod forming. When cold roll forming thin strips, the influence of the roll gap on the cold roll forming of the strip becomes more significant. Additionally, the design of thin strip cold roll forming equipment must also consider the roll spacing as an important parameter due to limitations on the equipment size imposed by the space station. Therefore, in order to form the target cross section of the strip, we developed an asymmetric cross-section strip roll gap self-adjusting equipment, focusing on the study of the thin strip in the process of the cold roll forming roll gap and roll spacing on the asymmetric cross section of the strip forming effect. ABAQUS numerical simulation was used to study the transverse stress and strain on the asymmetric cross section of a strip in the cold roll forming process, the change in stress and strain history in the corner regions of the cross section, the dimensional accuracy of the strip cross section, and the influence of the roll spacings on the longitudinal stress and strain of the strip to find the appropriate process parameters and to conduct experiments to verify the results. At the same time, the metallographic microstructure of corner 2 in the flange triangular part of the strip in the experiment was observed, and the cracking defects appearing in the strip during the forming process were analyzed using a scanning electron microscope, which revealed the reasons for the cracking of thin strips and their expansion mechanism. In this way, it provides a certain reference for the cold roll forming preparation of thin strips with an asymmetric cross section and contributes to the realization of space truss rods from strips to rods in on-orbit cold forming manufacturing technology.

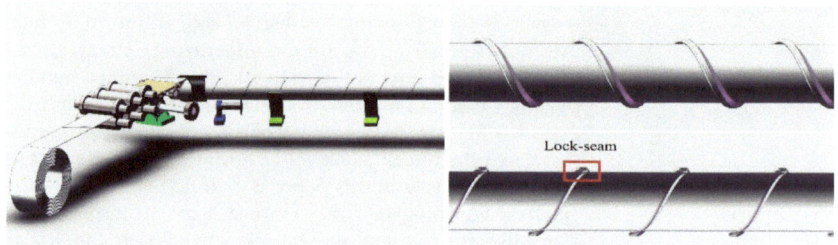

Figure 1. Schematic diagram of the production of truss rods by means of cold roll forming technology.

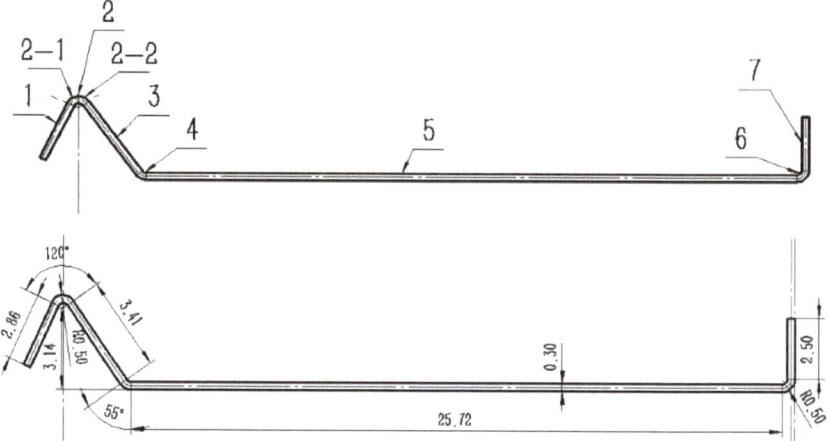

Figure 2. Partitioning and dimensions of the individual parts of the strip with a specific asymmetric cross section formed in the cold roll forming process. 1—partition 1, 2—partition 2 (Includes partitions 2-1 and 2-2), 3—partition 3, 4—partition 4, 5—partition 5, 6—partition 6, 7—partition 7.

2. Experimental Materials and Methods

2.1. Material Properties

As the cold roll forming process involves small diameter and large bending deformation of the strip, there are high demands on the plastic toughness of the material. A cold-rolled strip of Q195 galvanized steel was used as the test specimen, and its mechanical properties and chemical composition are shown in Table 1.

Table 1. Mechanical properties and chemical composition of Q195 galvanized steel strip.

Yield Strength/MPa	Tensile Strength/MPa	Elongation Rate/%	Quality Scores %							
			C	Si	Mn	S	P	N	Cr + Ni	Fe
403	441	27.5	0.12	0.3	0.5	0.035	0.035	0.012	0.1	Bal.

The strip is 0.3 mm thick and 36 mm wide. The strip is subjected to a longitudinal tensile test and the engineering stress–strain curve obtained is converted to a true stress–strain curve. After the strip enters necking, the Holloman hardening method is used to obtain $\sigma = 7227 \cdot \varepsilon^{0.183}$ based on the relevant data, the ABAQUS (Version 2018) model showed that the Q195 strip has a density of 7.85 g/cm^3, an elastic modulus E of 2.06×10^5 MPa and a Poisson ratio of 0.3.

2.2. Flower and Roll Design

As the equipment will be used in orbit on a space station, it needs to be miniaturized. On the basis of the dimensions of the strip in the target cross section, the number of forming passes is calculated using the method of calculation of the number of forming passes to be 4. The first pass is precompression, and deformation is mainly concentrated in passes 2–4. The distribution of the roll forming channel angle follows the principle of the cubic curve of the horizontal projection trajectory of the end of the vertical edge. The flower pattern of the designed target cross-section strip is shown in Figure 3. As the profile of the target cross section is asymmetrical, the angle of each set of rollers is designed by taking into account the variation index using Equation $cos\theta_i = 1 + (1 - cos\theta_o)\left[2\left(\frac{i}{N}\right)^{3+\kappa} - 3\left(\frac{i}{N}\right)^{2-\kappa}\right]$. The bending angle distribution of the rollers for each pass is shown in Table 2. The cubic curve equation obtained for the left side of the target cross-section strip (the flange triangular region) is:

$$f(x) = 8.464e^{-5}x^3 - 0.008047x^2 + 0.005208x + 7 \tag{1}$$

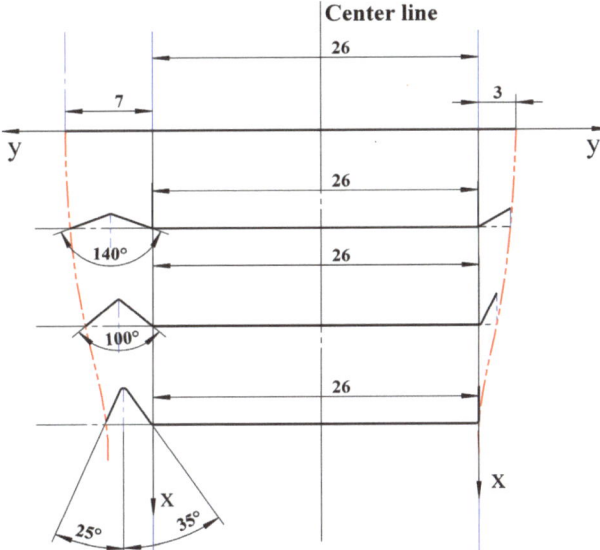

Figure 3. Flower diagram of a strip with the target cross section according to the principle of cubic curves of the horizontal projected trajectory of the end of the vertical edge.

Table 2. Roll bending angle distribution for each pass.

Number of Forming Passes	1	2	3	4
Corner 2-1 (partition 2-1)	0°	20°	40°	65°
The increment Δθ of corner 2-1	0°	20°	20°	25°
Corner 2-2 (partition 2-2)	0°	20°	40°	55°
The increment Δθ of corner 2-2	0°	20°	20°	15°
Corner 4 (partition 4)	0°	20°	40°	55°
The increment Δθ of corner 4	0°	20°	20°	15°
Corner 6 (partition 6)	0°	30°	60°	90°
The increment Δθ of corner 6	0°	30°	30°	30°

The cubic curve equation obtained for the right side of the target cross-section strip (right-angled region of the flange) is:

$$f(x) = 9.766e^{-5}x^3 - 0.007422x^2 - 0.003125x + 3 \qquad (2)$$

where $x = 8(n-1)$, n is the number of forming passes (since the first pass is a precompression-guided pass, the effective passes start from the 2nd pass). According to the roll bending angle distribution Table 2 for the roll set design, taking into account the volume factor, the roll design of its base circle diameter of 70 mm, the design of the strip cold roll forming three-dimensional model shown in Figure 4.

Figure 4. Schematic of the designed 3D model for cold roll forming of strips.

2.3. Experimental Scheme and Finite Element Modeling Process

In this paper, we mainly study the effect of different roll gaps and roll spacings on the cold roll forming process of thin strips with asymmetric cross sections and design a simulation experiment scheme, as shown in Table 3. The reasonableness of the simulation results and the accuracy of the rolled products are then verified experimentally.

Table 3. Experimental scheme of strip simulation process.

Simulation Parameter	1	2	3
The gap between the rolls	0.3 mm	0.4 mm	0.5 mm
The spacing of the rolls	100 mm	120 mm	140 mm

The strip cold roll forming process was modeled using ABAQUS as follows. In the cold roll forming process by means of four sets of forming rolls, the first set of rollers are prepress rollers with no significant deformation, so the model is built by biting into the strip directly from the second set of rollers adopting the ABAQUS/Explicit dynamic analysis model. A deformable body cell of cell type C3D8R (eight-node linear hexahedral cell, reduced integration, hourglass control) is used for the strip to accurately reflect the deformation process in each part of the strip. Hourglass control mode adopts stiffness hourglass control, because for plastic bending problems, better calculation results can be obtained by using stiffness-based hourglass control. Meanwhile, a reasonable mesh refinement is applied to the strip model. The artificial strain energy (ALLAE) of the strip after forming is found to be not more than 1.5% of the internal energy for the whole model (ALLIE) in numerical simulation, which indicates that the hourglass is controllable and the calculation results in this mode are accurate. The roller sets are set as discrete rigid bodies with unit-type C3D10M (ten-node modified quadratic tetrahedral unit). Construct the model using SolidWorks and import it to ABAQUS. The beveled section of the rollers is stitched to aid in the meshing of the roller sets. To investigate the impact of the roll gap, the roll gap was adjusted to 0.3 mm, 0.4 mm and 0.5 mm, respectively. The roller sets are equally spaced and modeled with roll spacing of 100 mm/120 mm/140 mm, respectively. Four sets of rollers (eight rollers) are rigidly fixed and the relative speed of the upper and lower rollers is 5 r/s. In the interaction module, the generic contact algorithm is selected. The contact properties of the interaction are set for normal behavior and tangential behavior.

The friction formula for tangential behavior is the penalty formula, which is applicable to most metal forming problems, the generic coulomb friction is selected, and the friction coefficient is set to 0.2. The pressure overclosure is set to hard contact for normal behavior, and the constraint enforcement is set to default mode. The length of raw strip is 800 mm. To precisely depict the distortion of the flange section in the strip's cross section, the mesh of the flange fragment requires refinement for mesh sizes between 0.15 mm and 0.375 mm. The deformation in the central section is minimal and the mesh is coarser, with a size of 3.9 mm. The meshed cold roll forming model of the strip is illustrated in Figure 5. The subdivided strip mesh is shown schematically in Figure 6.

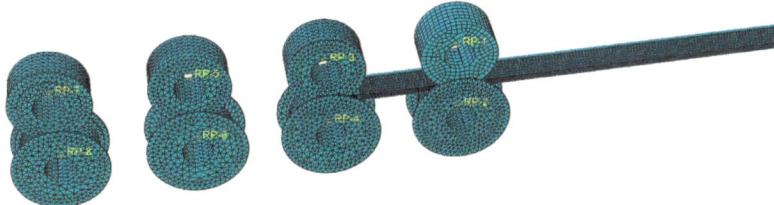

Figure 5. Cold roll forming model of strip with meshed view.

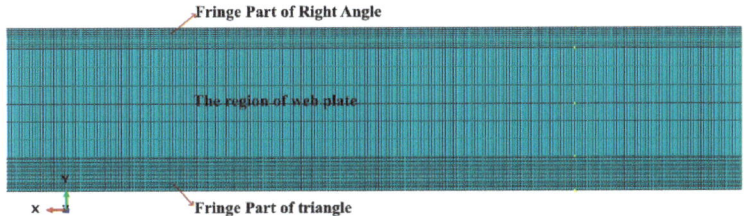

Figure 6. Schematic diagram of the subdivided strip mesh.

3. Results and Discussion

3.1. Influence of Roll Gap on Cold Roll Forming of Strip

3.1.1. Influence of Different Roll Gap on Strip Stress–Strain during Each Forming Pass

When the roll spacing is 120 mm, the stress and strain of each forming pass of the strip under different roll gaps (0.3 mm/0.4 mm/0.5 mm) are shown in Figures 7 and 8.

As can be seen in Figure 7, during the forming process of each pass, the peak equivalent stresses of the strip under different roll gaps are mainly concentrated in the regions of corners 2, 4 and 6. With the increase in the number of forming passes, there is a slight incremental trend in the stress of the above parts, and the stress value fluctuates around 403–515 MPa. In passes 2 and 3, the highest equivalent force is concentrated in corner 2 at the flange triangular region of the strip. However, when it comes to pass 4, the maximum equivalent force value begins to move from corner 2 (located at the top of flange triangle region of the strip) to corner 4 (situated at the bottom of the flange triangle region). Changing the roll gap value has no significant effect on the equivalent force value at the corners of the curve. However, for web plate 5 in the middle of the strip section, the equivalent stress value in this region is 60–280 MPa, and with the reduction in roll gap, the stress in this part is relatively low, showing a gentle fluctuation, and can be analyzed from the value of the force as mainly elastic stress. As can be seen in Figure 8, there are significant strain peaks in the region of corners 2, 4 and 6 where the equivalent stress values are greater. In corners 2 and 4 of the flange triangular region of the strip, the effect of the roll gap on the equivalent strain is apparent. When the roll gap is 0.3 mm, the maximum equivalent strain values in corner 2 increase with the number of passes, reaching 0.09, 0.24 and 0.64 in succession. The strain values at corner 4 near the bottom of the triangular region are significantly higher than the corresponding values, with roll gaps of 0.4 mm and 0.5 mm. In order, the equivalent strain values are 0.05, 0.12, and 0.39. The equivalent variation in roll gap of 0.4 mm/0.5 mm is relatively similar, and for corner 2, its equivalent variation is about 0.07/0.15/0.44, but corresponding to corner 4, when the roll gap is 0.4 mm, its equivalent variation is the lowest, only 0.2. For corner 6 of the flange right-angled region of the strip, in the second and third passes, the equivalent strain value is larger when the roll gap is 0.3 mm, but upon entering the fourth pass, the final equivalent strain values of each roll gap tend to be the same, which is approximately 0.2. The equivalent strain values in web plate 5 of the strip are all zero, which is also consistent with the equivalent stress values in Figure 7, indicating that the web plate region is mainly dominated by elastic strain and no plastic deformation occurs.

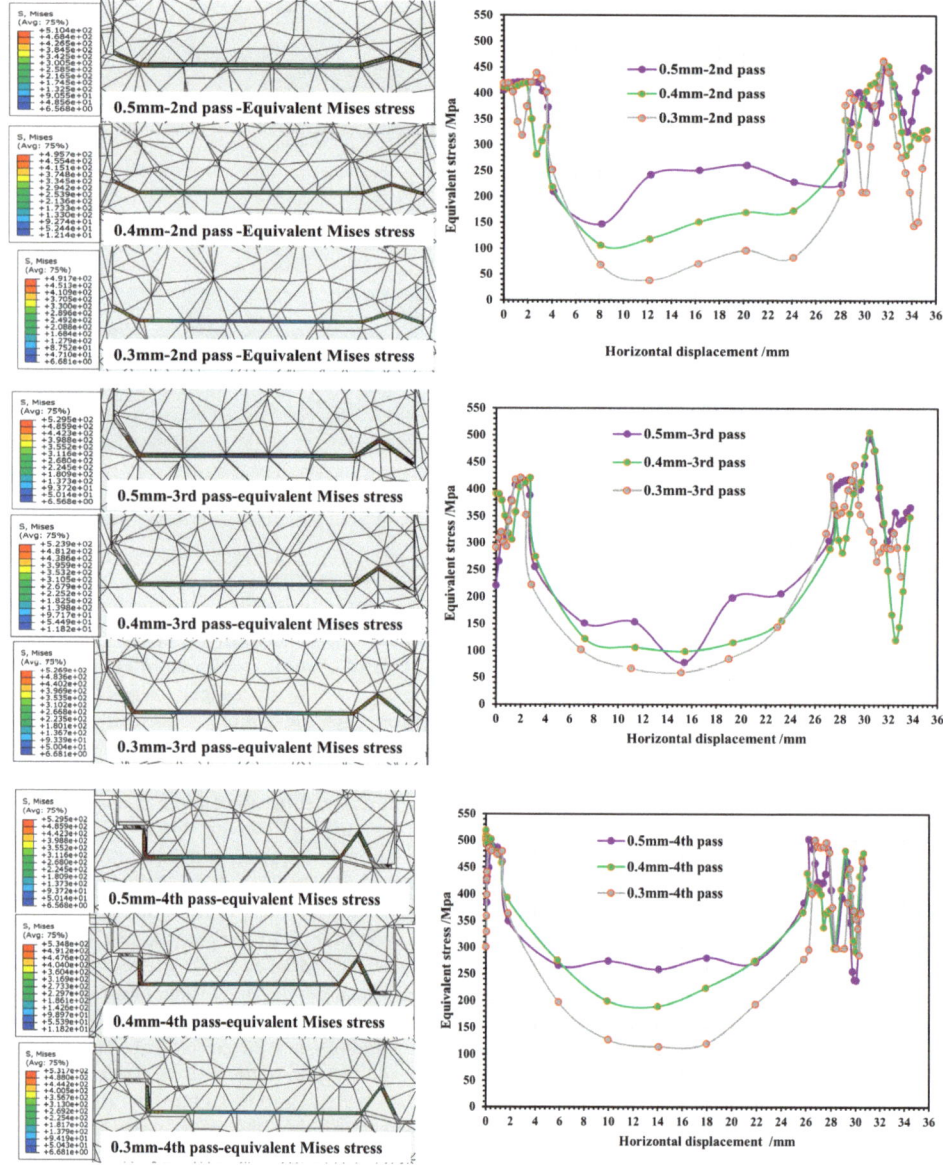

Figure 7. Stress variation in strip cross section at 2nd/3rd/4th pass with different roll gaps.

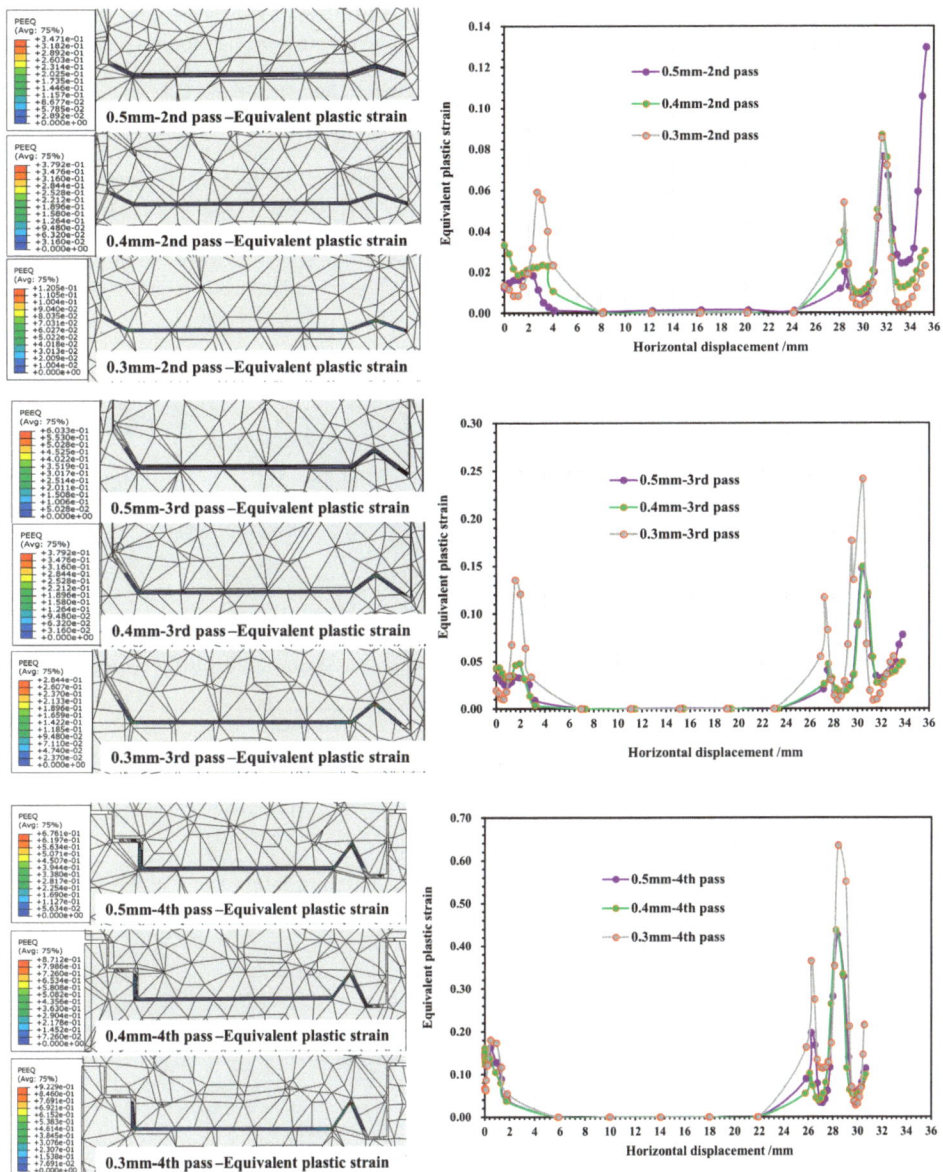

Figure 8. Strain variation in strip cross section at 2nd/3rd/4th pass with different roll gaps.

From the above analysis, it can be seen that the strip cross section in the deformation process, mainly for the strip cross section of the deformation of the corners, in the corners, since the bending angle of corner 2 at the flange triangular region of the strip is the smallest, the stress–strain value at this part is also the largest. As the roll gap decreases and the number of passes increases, all corner regions are subjected to varying degrees of plastic stress, resulting in an increase in strain as the bending angle of the corner region decreases. The strain surge occurs mainly in the fourth pass, and excessive strain values can lead to localized thinning of the strip in these regions, or even to wrinkling and cracking. When the gap between rolls is relatively large, the strip can obtain a larger deformation space during deformation, which can effectively alleviate the strain concentration phenomenon, but may have a greater effect on strip asymmetric cross-sectional forming accuracy.

3.1.2. Analysis of Changes in Stress–Strain History at Bend Corners

Taking the example of a roll gap of 0.4 mm, we selected the nodes with significant features, namely, 55786, 48594, and 15928, from the fourth forming roll located near corners 2, 4 and 6 to examine the changes in stress–strain history. Figure 9 shows a schematic diagram of the location of the selected characteristic nodes, Figure 10 shows the variation in stress–strain history of node 55786 (the top of corner 2), Figure 11 shows the variation in stress–strain history of node 48594 (inner side edge of corner 4 at the bottom of the flange triangular region), and Figure 12 shows the variation in stress–strain history of node 15928 (inner side edge of corner 6 at the flange right-angled region).

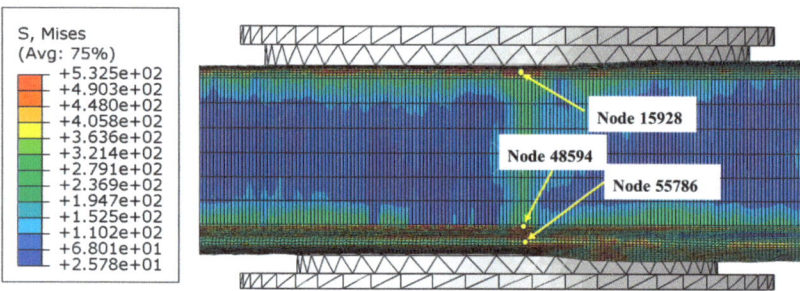

Figure 9. Schematic diagram of the location of the feature nodes in the selected corners.

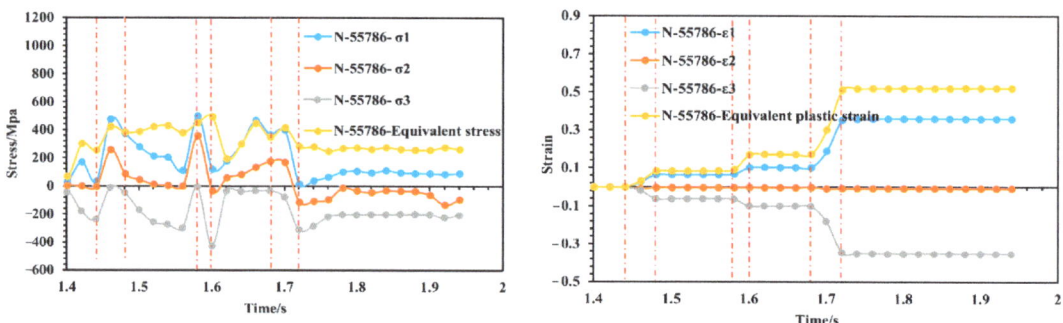

Figure 10. Stress–strain history variation of node 55786.

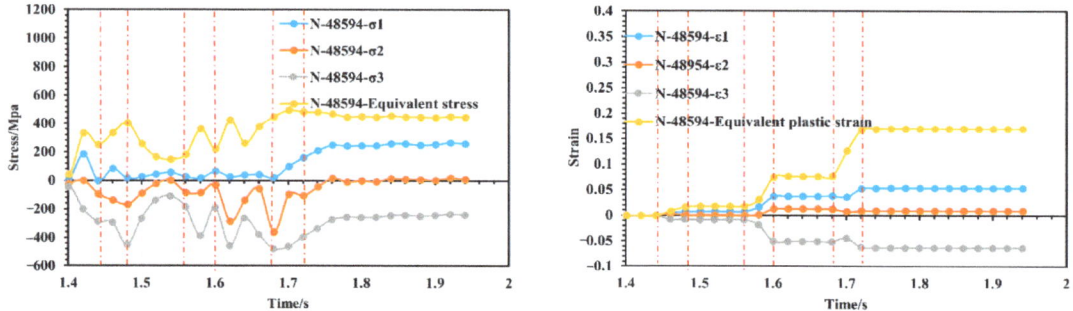

Figure 11. Stress–strain history variation of node 48594.

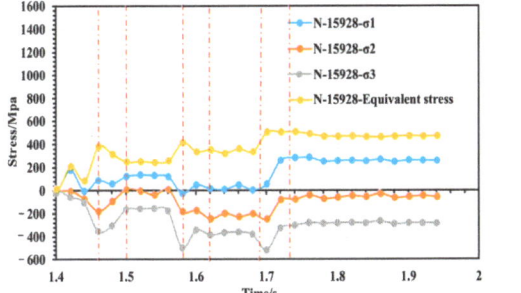 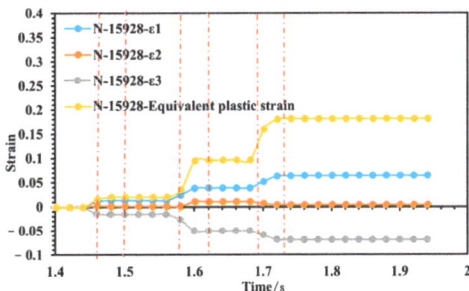

Figure 12. Stress–strain history variation of node 15928.

As depicted in Figure 10, the strip initiates entering the second set of rollers at t = 1.44 s and fully exits it at t = 1.48 s. During this period, a slight peak in the equivalent force is observed. The strip takes 1.48–1.58 s to travel between the second and third roller groups. During this time, the equivalent strain remains constant; however, tension between the two roller groups causes internal stress to fluctuate, with a predominance of elastic stress. At 1.58 s, the strip enters the third roll group. Due to the narrowing of the strip's cross section, it quickly experiences peak stress and further deformation. The equivalent stress continues to increase; however, as the increase in bending angle is small at this stage of deformation, the increase in equivalent plastic strain is not significant and only increases from 0.1 to 0.18. At 1.66 s into the process, the strip begins to enter the fourth roll group and as the cross section of the strip narrows, node 55786 at the inner edge of the flange triangle region quickly moves to the top of corner 2, resulting in a small peak of stress just before entering the fourth roller group. Upon entering the fourth roll group, the node experiences additional plastic tensile deformation at the top of corner 2. This results in a rapid increase in the equivalent plastic strain from 0.18 to 0.52. Although the bend angle increment is basically the same, the bend angle at this point is significantly smaller, and both sides of the triangle region have a large tensile effect on the top of corner 2 during deformation, which ultimately leads to greater strain at this point, making it the most susceptible to defects in the forming process. Overall, it appears that unit node 55786 has a step increase in strain during the cold roll forming process. It only has a strain surge just between the roller sets entering, after which it remains in a constant strain state for the duration of the movement of the two roller sets. Also, based on Figure 10, it can be concluded that whenever the strip passes through the rollers, node 55786 experiences a three-way stress state consisting of two tensile stresses and one compressive stress. The strain state is plane strain. Since node 55786 is situated on the outer surface of corner 2, this corresponds with the theoretical mechanical analysis at that particular location. As shown in Figures 11 and 12, the stress–strain histories of nodes 48594 and 15928 are similar, the two nodes are located in the inside of the bending corners, and in the process of deformation, with the increase in the number of forming passes, their equivalent stresses show an increasing trend during the action of each group of rollers. However, during the travel of the roller groups, node 48594 on the strip experiences more violent stress fluctuations; this is likely due to its location at the junction of the edge of the flange triangular region and the web plate, which causes it to undergo bidirectional bending deformation, and it is therefore subjected to more complex stresses than the right-angled edge of the flange, which is subjected to unilateral bending. In total, the maximum equivalent strain of both nodes is below 0.18, indicating a small strain value and relatively smooth deformation; the stress state of the two nodes also shows three-way stress, but its type is two compressions and one tension, and the strain state also basically corresponds to the characteristics of plane strain.

3.1.3. Influence of Different Roll Gaps on the Asymmetric Cross-Section Dimensional Accuracy of Strips

From Figure 13, it can be seen that the roll gap has little effect on the deformed dimensions of the right-angled part of the flange of the strip asymmetric cross section, but it has a significant effect on the triangular part of the flange. As the roll gap decreases, the triangular part approaches the desired size. However, as the number of forming passes increases, the cumulative strain of the triangular part becomes greater than that of the right-angled part. This results in deformation phenomena such as warping of the triangular part under stress after forming. When the roll gap is 0.3 mm, it becomes comparable to the thickness of the strip due to the rebound of the bent corner regions, which results in an excessive squeezing of the triangular part during the fourth pass forming. The result is an excessive elongation of the flange triangular region of the strip and a concave phenomenon in corner 4 connecting the flange triangular region and the web plate. The thinning phenomenon occurs in this region, indicating that the strain in this location ought to be greater, which can be verified from Figure 8. Therefore, taking into account factors such as the stress–strain of the strip cross section and the accuracy of the strip dimensions, the roll gap of 0.4 mm is relatively good.

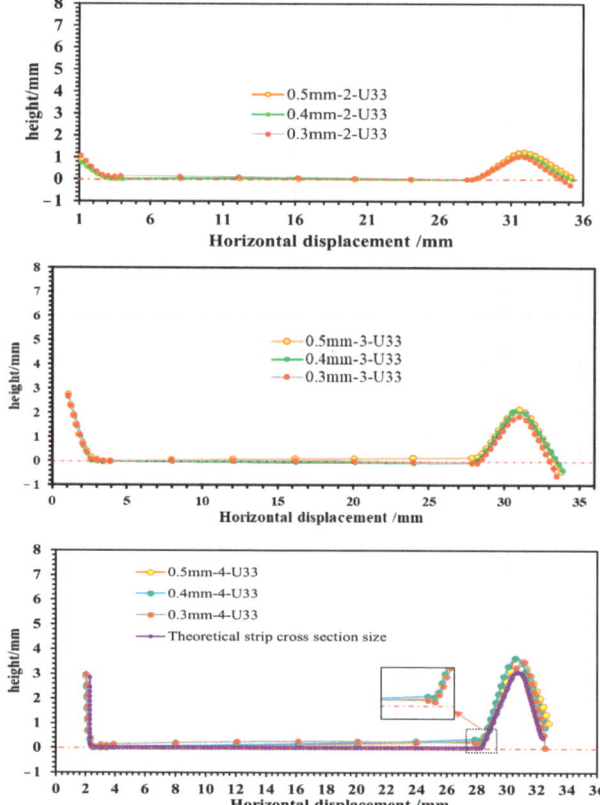

Figure 13. Deformation of strip asymmetric cross-sectional dimensions during the 2nd/3rd/4th passes under different roll gaps.

3.2. Effect of Roll Spacing on the Longitudinal Stress–Strain of Strips

When the roll gap was 0.4 mm, the changes in longitudinal stress and strain in the strip after the second, third and fourth forming passes during the strip forming process were investigated when the strip flange triangular part is 2.64 mm away from the outer edge, and the roll spacing is 100 mm, 120 mm and 140 mm, respectively. The established paths are shown in Figure 14, together with the stress and strain distributions along the paths shown in Figures 15 and 16. In addition, strip edge wave fluctuation diagrams at the edge of the flange triangular part (i.e., at 0 mm) were also determined, as shown in Figure 17.

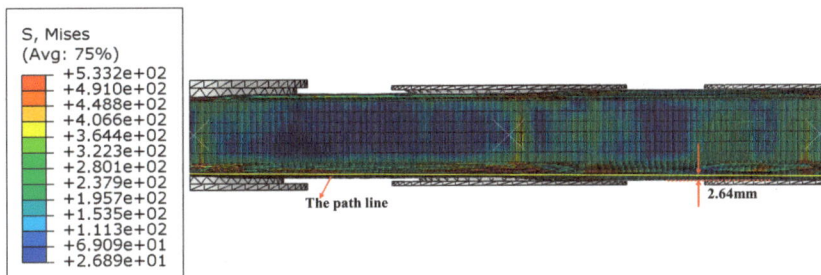

Figure 14. Schematic of the path in the strip at 2.64 mm from the outer edge of the flange triangular part.

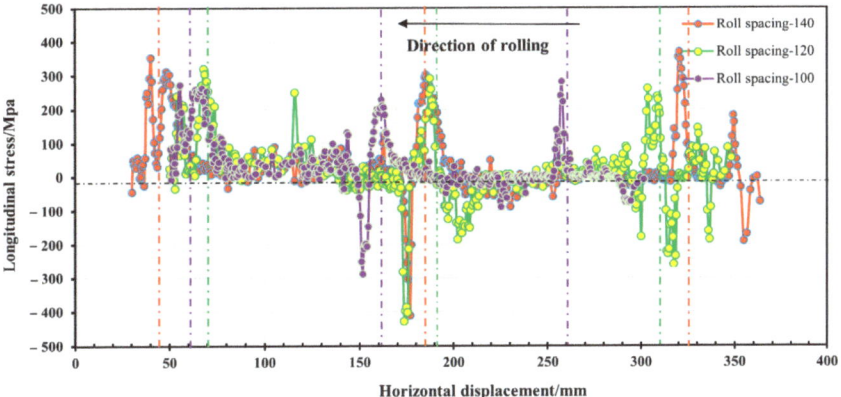

Figure 15. Longitudinal stress distribution in the strip at 2.64 mm from the outer edge of the flange triangular part.

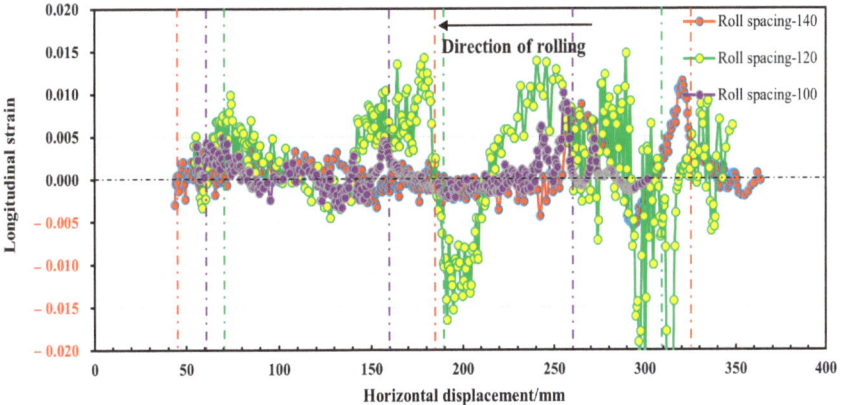

Figure 16. Longitudinal strain distribution in the strip at 2.64 mm from the outer edge of the flange triangular part.

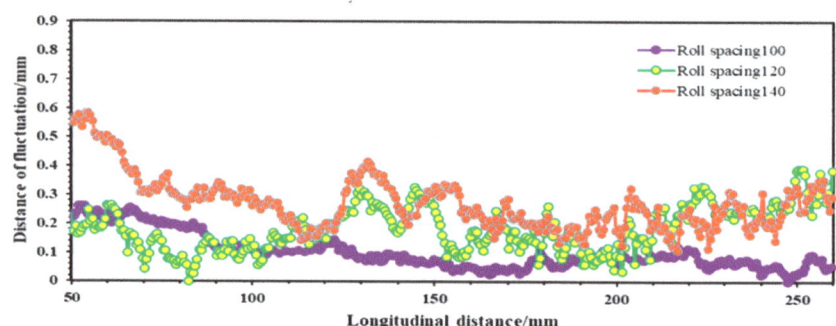

Figure 17. Fluctuations in the edge wave at the edge of the flange triangular part (i.e., at 0 mm).

It is evident from Figure 15 that the longitudinal stresses within the strip during the forming process are predominantly tensile stresses. However, a fleeting longitudinal compressive stress is observed when the strip departs from the roll group in the third pass. With an increase in roll spacing, there is a slight incremental trend observed in the peak stress of the strip while passing through each roller. The peak stress measures around 380 Mpa, suggesting that the longitudinal stress of the strip is predominantly elastic stress. It can be seen from Figure 16 that the peak longitudinal strain at different roll spacings shows a decreasing trend as the number of passes increases. However, the longitudinal strain fluctuation is greatest when the roll spacing is 120 mm. As the number of passes increases, especially after the fourth pass, the longitudinal strain value for a roll spacing of 140 mm (peak strain of 0.002) is smaller than the longitudinal strain value for a roll spacing of 100 mm (peak strain of 0.004). It is shown that when the roll spacing is sufficiently large, it is possible to effectively reduce longitudinal strain. However, as roll spacing increases, it also leads to greater longitudinal stress, and this increase can result in strip rebound and exacerbate the strip's fluctuation phenomenon, ultimately aggravating the edge wave phenomenon of the strip. Figure 17 illustrates the strip fluctuation of the edge wave at the edge of the flange triangular part t (i.e., at 0 mm). As observed, the edge wave phenomenon of the strip becomes more serious with an increase in roll spacing. At a roll spacing of 140 mm, the edge wave fluctuates between 0–0.5 mm, at 120 mm, it fluctuates between 0–0.35 mm, and at 100 mm, the edge wave of the strip is at its smallest, with a fluctuation of 0–0.26 mm. Furthermore, augmenting the roll spacing will result in increased volume of the forming roll set equipment and manufacturing costs. Taking into account multiple factors, the roll spacing of 100 mm appears to be the most suitable option.

3.3. Experimental Validation and Analysis of Defects in Strip Cold Roll Forming Process

According to the simulation results, the cold forming equipment was prepared with a roll base diameter of 70 mm, roll spacings of 100 mm and roll gaps with a self-correcting function. The physical equipment is shown in Figure 18.

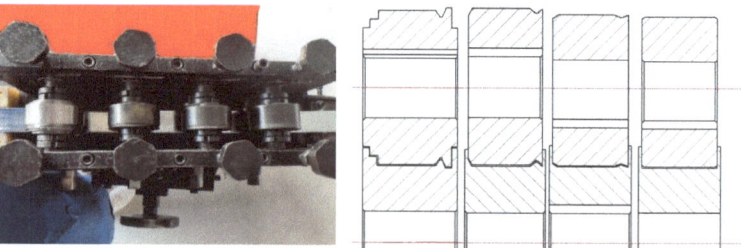

Figure 18. Preparation of cold forming roll set equipment with 100 mm roll spacing and designed roll set cross section.

During the simulation, the two main types of defects that appeared during strip forming were edge waves and intermittent crack lines that appeared at the top of corner 2, as shown in Figure 19. This was also verified in subsequent experiments. The appearance of these two types of defects is mainly due to the roll gap being too small, so that the strip in the passing roller groups surge in strain, and ultimately lead to defects; the above defects are generated in the roll gap of 0.3 mm when appearing. When the roller gap was increased to 0.35 mm, the experiment showed that the strip edge wave phenomenon almost disappeared, but there were still microcracks at the top of corner 2. When the roll gap increased to 0.4 mm, the surface quality of the strip was good, and the physical object of the strip formed in each pass is shown in Figure 20. By measuring the cross-sectional size of the strip, it was found that the test data of the strip cross section were in good agreement with the theoretical value. Comparison of experimental and theoretical values of strip cross-section dimensions is shown in Figure 21. When the roll gap was increased to 0.5 mm, the dimensions of the strip cross section were found to be slightly larger than the theoretical values, and the experimental values and simulation results showed consistency. This indicates that when the roll gap is large, the strain concentration phenomenon is effectively alleviated and defects such that cracks do not occur, but it has a greater effect on the forming accuracy of the strip cross section.

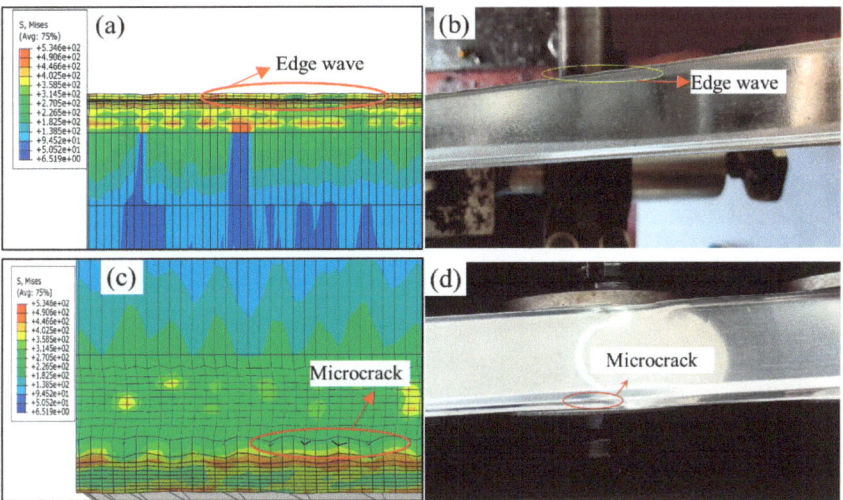

Figure 19. Simulation and experimental diagrams of the phenomenon of edge wave and tearing at the top of corner 2 during strip forming. (**a**) Edge wave phenomenon in simulation... (**b**) Edge wave phenomenon during strip forming (**c**) Tearing at the top of corner 2 in simulation (**d**) Tearing at the top of corner 2 during strip forming.

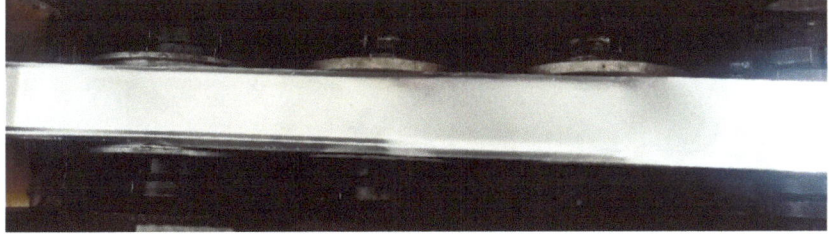

Figure 20. Surface quality topography of formed strip with roll gap of 0.4 mm.

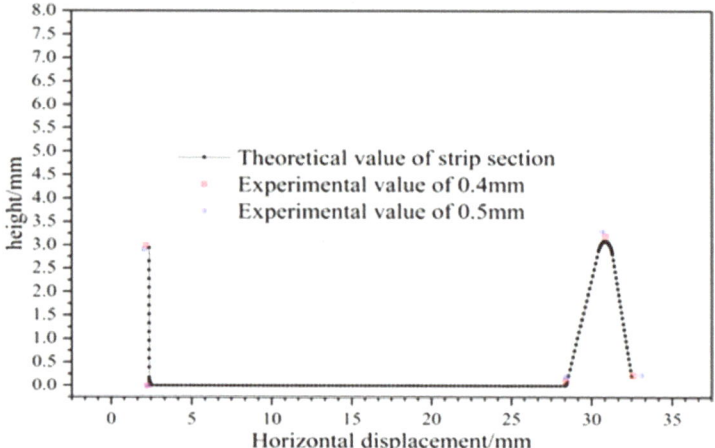

Figure 21. Test value of strip cross section at roll gap of 0.4 mm/0.5 mm.

The microstructure morphology of the strip after the second/third/fourth pass at corner 2 with the roll gap of 0.4 mm is shown in Figure 22, As the number of forming passes increased, the transverse tensile force on Corner 2 increased, causing the grains on its outer side to elongate gradually; by the fourth pass, the fibrillation of the grains became highly apparent. Nevertheless, the grains situated in the inside of corner 2 experienced a transformation from tension to compression while the forming angle decreased progressively, and this part of the grains showed extrusion characteristics (the inner side was mainly stressed by compressive stress): still isomorphic crystals, even finer crushed crystals, appeared on the edge of the inner region. The morphology of this feature corresponds to the stress analysis inside and outside the characteristic nodes in Section 3.1.2.

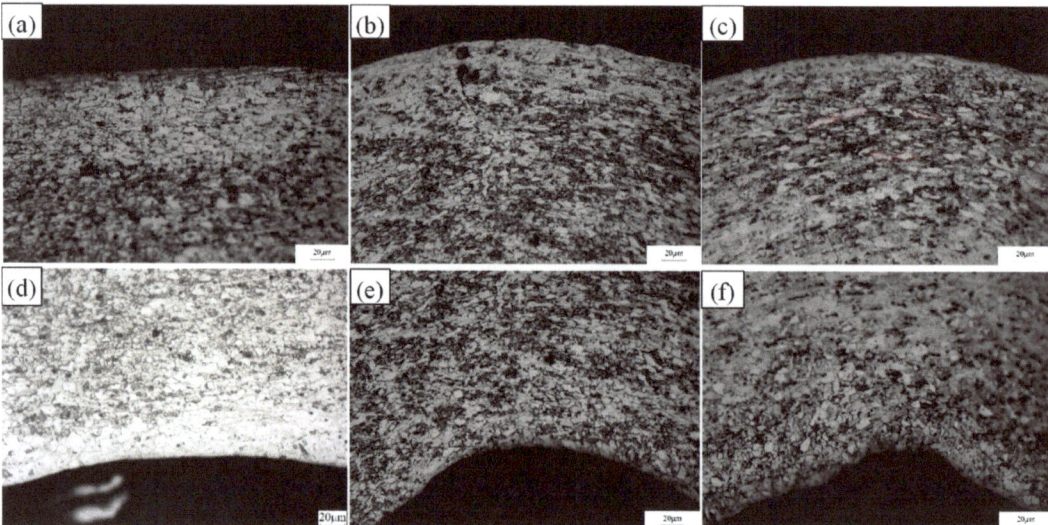

Figure 22. Microstructure variation of the inside and outside part of corner 2 for each pass at a roll gap of 0.4 mm. (**a**) Outside of corner 2 of pass 2. (**b**) Outside of corner 2 of pass 3. (**c**) Outside of corner 2 of pass 4. (**d**) Inside of corner 2 of pass 2. (**e**) Inside of corner 2 of pass 3. (**f**) Inside of corner 2 of pass 4.

The intermittent crack defects that occurred at the top of corner 2 were scanned and the microscopic morphology is shown in Figure 23. As can be seen from Figure 23a,b, once the crack was formed at the top of corner 2, the crack extension region showed a laminar tearing pattern and the layer-by-layer transition showed a step-like shape, which was very similar to a disintegrated fracture from the microscopic morphology, indicating that a relatively severe stress concentration phenomenon occurs at this location. At the early stage of crack initiation, the region of corner 2 was subjected to a large lateral stress from the roll gap, which provided the tangential stress of the laminar crack, resulting in stress concentration in this region, and strain was greater and crack propagation rate was extremely fast. Figure 23c shows that a typical tough nest morphology appeared when the laminar crack reached the mid-region, indicating that when the crack extended to a certain distance, its expansion started to slow gradually, and since the strip was a plastic material, a typical tough nest morphology started to appear in the slow region. A partial enlargement of its ligamentous fossa morphology is shown in Figure 23d. Thus, the mechanism of longitudinal crack propagation is mainly a composite form of cleavage fracture + ductile fracture, and once crack propagation occurs, it will rapidly form tearing propagation and produce destructive cracks as cold roll forming proceeds.

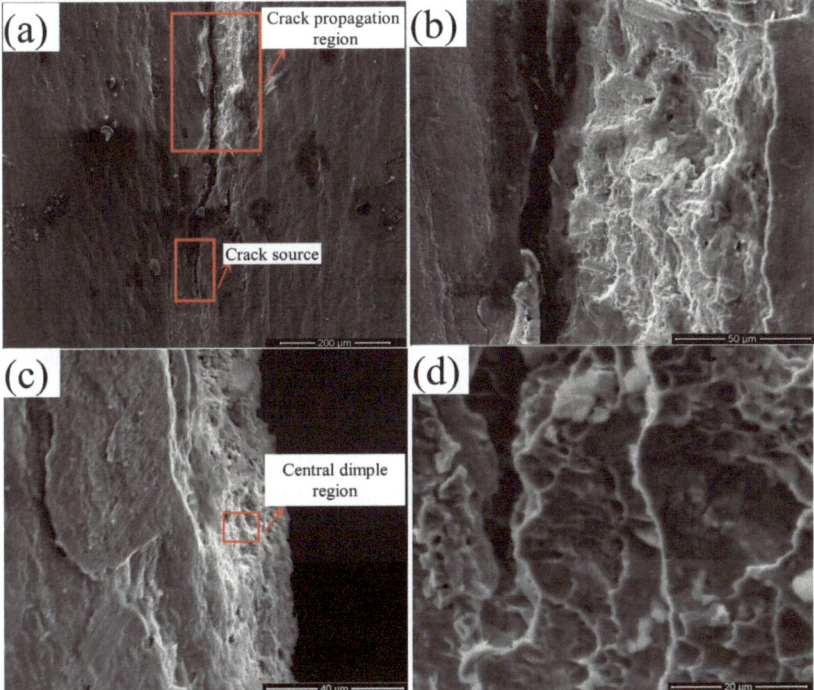

Figure 23. Morphological view of crack propagation at the top of corner 2. (**a**) Macroscopic morphology of crack source and crack extension region. (**b**) Enlarged view of the crack extension region. (**c**) Typical tough nest morphology of the mid-region (**d**) Enlarged view of the typical tough nest morphology.

4. Conclusions

(1) This paper proposes a new technology for the on-rail cold forming of space truss rods. It was found by studying the effect on cold roll forming of asymmetric strip cross section under different roll gaps that the stresses on the strip during the forming process are mainly concentrated in corners 2/4/6, with the largest strain values in corner 2. As the number of passes increases, the maximum equivalent strain values

are 0.09, 0.24 and 0.64 in succession when the roll gap is 0.3 mm. The equivalent roll gap variation of 0.4 mm/0.5 mm is relatively similar, and for corner 2, the equivalent variation is approximately 0.07/0.15/0.44 as the number of forming passes increases. The forces in the web plate region are dominated by elastic stresses, and little plastic deformation occurs. From the analysis of the stress–strain history of the characteristic nodes in corners 2/4/6, it can be seen that the stress and strain are mainly at the moment of action of the two roller wheel sets, the stress during the travel period is mainly dominated by the elastic stress, and there is no change in the strain. During the deformation process, the stress type of node 55786 shows two tensile types and one compressive type, and the stress types of nodes 48594 and 15928 show two compressive types and one tensile type, and the strains of all three nodes conform to the plane strain characteristics. Taking into account the stress–strain of the strip cross section and the precision of the strip dimensions, a roll gap of 0.4 mm is relatively good.

(2) As the roll spacing increases, the peak longitudinal stress of the strip passing through each roller group shows a small incremental trend, but the peak stress is around 380 MPa. When the roll spacing is large enough, it can effectively reduce the longitudinal strain, but it will lead to an increase in the longitudinal stress, which easily leads to the rebound of the strip and aggravate the phenomenon of strip edge wave. Combined with the fluctuation in strip edge at different roll spacings and factors such as manufacturing cost and volume, a roll spacing of 100 mm is appropriate.

(3) After experimental verification, the strip can be successfully prepared according to the cross-section requirements when the roll gap is 0.4 mm and the roll spacing is 100 mm. Observation of the microstructure of corner 2 shows that as the number of passes increases, the grains on the outside of corner 2 show obvious fibrosis characteristics, while the grains on the inside are extruded and still show an equiaxial shape. When the roll gap is reduced during strip formation, the strip edge is susceptible to edge waviness and the top of corner 2 is prone to intermittent cracking. Analysis of the cracking region of the strip has shown that the crack propagation mechanism is mainly based on cleavage fracture and ductile fracture.

Author Contributions: Conceptualization, X.Y. and J.H.; methodology, X.Y.; formal analysis, X.Y.; investigation, X.Y.; resources, X.Y. and J.H.; data curation, X.Y. and R.L.; writing—original draft preparation, X.Y.; writing—review and editing, X.Y. and R.L.; supervision, J.H. All authors have read and agreed to the published version of the manuscript.

Funding: This work was funded by the Beijing Space Vehicle General Design Department of the Chinese Academy of Space Technology (WY-YY/30508010701JY0011).

Institutional Review Board Statement: Not applicable.

Informed Consent Statement: Not applicable.

Data Availability Statement: The data presented in this study are available on request from the corresponding author.

Conflicts of Interest: Author Jingtao Han was employed by the company Guangzhou Sino Precision Steel Tube Industry Research Institute Co., Ltd. The remaining authors declare that the research was conducted in the absence of any commercial or financial relationships that could be construed as a potential conflict of interest. The authors declare no conflict of interest.

References

1. Yang, J.; Li, J.; Wu, W.J.; Yu, N. Research Status and Prospect of On-orbit Additive Manufacturing Technology for Large Space Truss. *Mater. Rep.* **2021**, *35*, 3159–3167.
2. Ding, J.F.; Gao, F.; Zhong, X.P.; Wang, G.; Liang, D.P.; Zhang, Y.C.; Li, W.J. The key mechanical problems of on-orbit construction. *Sci. Sin. -Phys. Mech. Astron.* **2019**, *49*, 50–57. [CrossRef]
3. Brown, M.A. A deployable mast for solar sails in the range of 100–1000 m. *Adv. Space Res.* **2011**, *48*, 1747–1753. [CrossRef]
4. Chai, Y.; Luo, J.; Wang, M. Bi-level game-based reconfigurable control for on-orbit assembly. *Aerosp. Sci. Technol.* **2022**, *124*, 107527. [CrossRef]

5. Xue, Z.H.; Liu, J.G.; Wu, C.C.; Tong, Y.C. Review of in-space assembly technologies. *Chin. J. Aeronaut.* **2021**, *34*, 21–47. [CrossRef]
6. Cao, K.; Li, S.; She, Y.; Biggs, J.D.; Liu, Y.; Bian, L. Dynamics and on-orbit assembly strategies for an orb-shaped solar array. *Acta Astronaut.* **2021**, *178*, 881–893. [CrossRef]
7. Fenci, G.E.; Currie, N.G. Deployable structures classification: A review. *Inter. J. Space Stru.* **2017**, *32*, 112–130. [CrossRef]
8. Tang, J.; Kwan, T.H.; Wu, X. Extrusion and thermal control design of an on-orbit 3D printing platform. *Adv. Space Res.* **2022**, *69*, 1645–1661. [CrossRef]
9. Taylor, S.L.; Jakus, A.E.; Koube, K.D.; Ibeh, A.J.; Geisendorfer, N.R.; Shah, R.N.; Dunand, D.C. Sintering of micro-trusses created by extrusion-3D-printing of lunar regolith inks. *Acta Astronaut.* **2018**, *143*, 1–8. [CrossRef]
10. Wang, T.; Wang, Y.F.; Gao, C.; Jiang, S.Y. Feasibility study on feeding wire electron beam brazing of pure titanium using an electron gun for space welding. *Vacuum* **2020**, *180*, 109575.
11. Paton, B.E.; Lobanov, L.M.; Naidich, Y.V.; Asnis, Y.A.; Zubchenko, Y.V.; Ternovyi, E.G.; Volkov, V.S.; Kostyuk, B.D.; Umanskii, V.P. New electron beam gun for welding in space. *Sci. Technol. Weld. Join.* **2019**, *24*, 320–326. [CrossRef]
12. Li, J.; Huang, X.; Ma, G.; Wang, J.; Pan, J.; Ruan, Q. Numerical Simulation and Parameter Design of Strip Cold Rolling Process of 301L Stainless Steel in 20-Roll Mill. *Metall. Mater. Trans. B* **2020**, *51*, 1370–1383. [CrossRef]
13. Zhang, L.; Ni, J.; Lai, X. Dimensional errors of rollers in the stream of variation modeling in cold roll forming process of quadrate steel tube. *Int. J. Adv. Manuf. Technol.* **2008**, *37*, 1082–1092. [CrossRef]
14. Qadir, S.; Nguyen, V.; Hajirasouliha, I.; Ceranic, B.; Tracada, E.; English, M. Shape optimisation of cold roll formed sections considering effects of cold working. *Thin-Walled Struct.* **2022**, *170*, 108576. [CrossRef]
15. Su, C.-J.; Yang, L.-Y.; Lou, S.-M.; Cao, G.-H.; Yuan, F.-R.; Wang, Q. Optimized bending angle distribution function of contour plate roll forming. *Int. J. Adv. Manuf. Technol.* **2018**, *97*, 1787–1799. [CrossRef]
16. Panton, S.; Duncan, J.; Zhu, S. Longitudinal and shear strain development in cold roll forming. *J. Mater. Process. Technol.* **1996**, *60*, 219–224. [CrossRef]
17. Panton, S.M.; Zhu, S.D.; Duncan, J.L. Geometric Constraints on the Forming Path in Roll Forming Channel Sections. *J. Mater. Process. Technol.* **1992**, *206*, 113–118. [CrossRef]
18. Han, Z.W.; Liu, C.; Lu, W.P.; Ren, L.Q. The effects of forming parameters in the roll-forming of a channel section with an outer edge. *J. Mater. Process. Technol.* **2001**, *116*, 205–210. [CrossRef]
19. Poursafar, A.; Saberi, S.; Tarkesh, R.; Vahabi, M.; Fesharaki, J.J. Experimental and mathematical analysis on spring-back and bowing defects in cold roll forming process. *Int. J. Interact. Des. Manuf. (IJIDeM)* **2022**, *16*, 531–543. [CrossRef]
20. Su, C.J.; Li, X.X.; Li, X.M.; Feng, Z.Y.; Naveen, B.S.; Wang, R.; Huang, W.M. Influence of bending angle function on local thinning defects of roll-formed sheets based on segmental boundary combination optimization. *Int. J. Adv. Manuf. Tech.* **2023**, *127*, 5803–5816. [CrossRef]
21. Cheng, J.J.; Cao, J.G.; Zhao, Q.F.; Liu, J.; Yu, N.; Zhao, R.G. A novel approach to springback control of high-strength steel in cold roll forming. *Int. J. Adv. Manuf. Tech.* **2020**, *107*, 1793–1804.
22. Xu, C.T.; Wei, Y.F.; Pan, T.; Chen, J.C. A study of cold roll forming technology and peak strain behavior of asymmetric corrugated channels. *Int. J. Adv. Manuf. Tech.* **2022**, *118*, 4213–4223. [CrossRef]
23. Shim, D.S.; Son, J.Y.; Lee, E.M.; Baek, G.Y. Improvement strategy for edge waviness in roll bending process of corrugated sheet metals. *Int. J. Mater. Form.* **2017**, *10*, 581–596. [CrossRef]
24. Safdarian, R.; Naeini, H.M. The effects of forming parameters on the cold roll forming of channel section. *Thin-Walled Struct.* **2015**, *92*, 130–136. [CrossRef]
25. Najafabadi, H.M.; Naeini, H.M.; Safdarian, R.; Kasaei, M.M.; Akbari, D.; Abbaszadeh, B. Effect of forming parameters on edge wrinkling in cold roll forming of wide profiles. *Int. J. Adv. Manuf. Technol.* **2019**, *101*, 181–194. [CrossRef]
26. Wang, J.; Liu, H.M.; Li, S.F.; Chen, W.J. Cold Roll Forming Process Design for Complex Stainless-Steel Section Based on COPRA and Orthogonal Experiment. *Materials* **2022**, *15*, 8023. [CrossRef]

Disclaimer/Publisher's Note: The statements, opinions and data contained in all publications are solely those of the individual author(s) and contributor(s) and not of MDPI and/or the editor(s). MDPI and/or the editor(s) disclaim responsibility for any injury to people or property resulting from any ideas, methods, instructions or products referred to in the content.

Article

A Molecular Dynamics Study on the Dislocation-Precipitate Interaction in a Nickel Based Superalloy During the Tensile Deformation

Chang-Feng Wan [1,†], Li-Gang Sun [1,†], Hai-Long Qin [2], Zhong-Nan Bi [2] and Dong-Feng Li [1,*]

[1] School of Science, Harbin Institute of Technology, Shenzhen 518055, China; 18B358027@stu.hit.edu.cn (C.-F.W.); sunligang@hit.edu.cn (L.G.S.)
[2] Beijing Key Laboratory of Advanced High Temperature Materials, Central Iron and Steel Research Institute, Beijing 100081, China; hailongqin@126.com (H.-L.Q.); bizhongnan21@aliyun.com (Z.-N.B.)
* Correspondence: lidongfeng@hit.edu.cn; Tel./Fax: +86-(0)755-86146861
† These authors contributed equally to this work.

Citation: Wan, C.-F.; Sun, L.-G.; Qin, H.-L.; Bi, Z.-N.; Li, D.-F. A Molecular Dynamics Study on the Dislocation-Precipitate Interaction in a Nickel Based Superalloy During the Tensile Deformation. *Materials* **2023**, *16*, 6140. https://doi.org/10.3390/ma16186140

Academic Editor: Wojciech Borek

Received: 7 July 2023
Revised: 31 August 2023
Accepted: 6 September 2023
Published: 9 September 2023

Copyright: © 2023 by the authors. Licensee MDPI, Basel, Switzerland. This article is an open access article distributed under the terms and conditions of the Creative Commons Attribution (CC BY) license (https://creativecommons.org/licenses/by/4.0/).

Abstract: In the present paper, the dislocation-precipitate interaction in the Inconel 718 superalloy is studied by means of molecular dynamics simulation. The atomistic model composed of the ellipsoidal Ni_3Nb precipitate (γ'' phase) and the Ni matrix is constructed, and tensile tests on the composite $Ni_3Nb@Ni$ system along different loading directions are simulated. The dislocation propagation behaviors in the precipitate interior and at the surface of the precipitate are characterized. The results indicate that the dislocation shearing and bypassing simultaneously occur during plastic deformation. The contact position of the dislocation on the surface of the precipitate could affect the penetration depth of the dislocation. The maximum obstacle size, allowing for the dislocation shearing on the slip planes, is found to be close to 20 nm. The investigation of anisotropic plastic deformation behavior shows that the composite system under the loading direction along the major axis of the precipitate experiences stronger shear strain localizations than that with the loading direction along the minor axis of the precipitate. The precipitate size effect is quantified, indicating that the larger the precipitate, the lower the elastic limit of the flow stress of the composite system. The dislocation accumulations in the precipitate are also examined with the dislocation densities given on specific slip systems. These findings provide atomistic insights into the mechanical behavior of nickel-based superalloys with nano-precipitates.

Keywords: precipitate strengthening; molecular dynamics; dislocation-precipitate interaction; anisotropy

1. Introduction

Inconel 718 (IN718), a precipitate-strengthened nickel-based superalloy, has been widely used in aerospace, petrochemical, and nuclear industries due to its exceptional mechanical properties [1–4]. With a yield strength of approximately 1050 MPa, IN718 exhibits high strength and excellent performance across a wide temperature range, from cryogenic temperatures up to 920 K [5]. Its remarkable combination of strength, toughness, and fatigue resistance makes it suitable for demanding applications. Furthermore, IN718 demonstrates excellent resistance to oxidation and corrosion, ensuring its longevity and reliability in harsh environments. The alloy's ability to maintain its mechanical properties under extreme conditions has solidified its reputation as a highly reliable and versatile material. The alloy has a quite complex nanostructure, which involves two kinds of nano-precipitates, namely γ' and γ'' phases, to affect the dislocation movement in the γ matrix phase [6–8]. The γ'' precipitate (ellipsoidal and disc-like) is the main strengthening precipitate and has a coherent DO_{22} structure with a volume fraction of approx. 13–15%. The spherical γ' precipitate has a volume fraction of approx. 3–5% and is less relevant than the γ'' precipitate from the strengthening point of view. There is also a small amount of needle-like δ phase [9], which can be transformed from the γ'' phase and is usually located at grain boundaries to stabilize the grain morphology. Knowledge of the underlying mechanisms of how dislocations interact with the complex precipitate system of IN718 is crucial to understanding the role that the nano-precipitates play on the mechanical properties of IN718.

There are two dislocation-precipitate interaction mechanisms in IN718, namely dislocation shearing and dislocation bypassing [7,8,10,11]. It has been known that the competition between the two mechanisms depends largely on the precipitate size [10–13] such that the mobile dislocations likely shear through the smaller precipitates and bypass the larger ones. The critical radius of the precipitates is often used to indicate the transition from the dislocation shearing to the dislocation bypassing [8,11]. However, the ellipsoidal geometry of γ'' precipitate has not been well accounted for. Molecular dynamics (MD) simulations are widely used to explore the lattice-scale physics of the dislocation-precipitate interaction [14–19]. Bacon, D.J. and Osetsky, Y.N. [14] studied the interaction between the edge dislocation and copper precipitates in α-iron alloy, quantitatively revealing the dependence of critical resolved shear stress (CRSS) on temperature, precipitate size, and the mismatch level. It is found that the smaller the size of the copper precipitates, the easier the dislocation shearing. Similar results have been obtained for other alloys, e.g., medium- and high-entropy alloys [15,16,18], Mg-Al alloy [17] and Mg-Zn alloy [19]. The MD method can also be used to study the dependence of the plastic flow and the dislocation motion on the size of nano-pore or nano-particle in copper [20]. Apart from the atomistic insights into the dislocation-precipitate interaction achieved, the MD modeling can also play a fundamental role in the multi-scale modeling, for example a combination of the MD method and discrete dislocation dynamics (DDD) model, to explore the evolution of dislocation patterns with the effect of precipitate morphology taken into account explicitly [21–26]. In these modeling studies, MD simulation is typically used to gain some material parameters at the atomic level and to support the higher-level DDD simulations. Although the modeling methods at atomic levels, such as MD, DDD, and their combination, have been illustrated as a powerful tool to fundamentally quantify the interplay between dislocations and precipitates in a wide range of alloys, explicit atomistic studies on the dislocation shearing and bypassing in IN718 are still limited.

In this work, MD studies are performed to investigate the interaction between the dislocations and the γ'' precipitates in the IN718 superalloy. The stress-driven motion of dislocations and its effect on the mechanical properties are focused on. The morphological features of the γ'' are explicitly considered in the MD model, and the uniaxial tensile tests are simulated along different loading directions. A series of snapshots on the microstructure evolution with the dislocation shearing and bypassing going through in the precipitate are given. Moreover, the anisotropic effect and the precipitate size effect are investigated. The paper is structured as follows: In Section 2, the details of the MD modeling are introduced. Section 3 shows the results of the MD simulation, followed by the discussion of the results in Section 4, and Section 5 presents the concluding remarks.

2. Molecular Dynamics Model

Figure 1 illustrates the molecular dynamics (MD) model employed in this study. A cubic unit cell, measuring $25 \times 25 \times 25$ nm^3, is constructed consisting of a face-centered cubic (FCC) Ni matrix. At the center of the cell, a disk-like single-crystalline Ni_3Nb particle with a DO_{22} structure is coherently embedded within the Ni matrix. The major axis of the ellipsoidal Ni_3Nb precipitate is set to 20 nm, while the minor axis is 10 nm. To investigate the interaction between dislocations and the Ni_3Nb precipitate, a pair of dislocations is preset within the pure Ni matrix. The total number of atoms in the unit cell is approximately 1,300,000. Due to the complex composition of IN718, and the lack of a reliable force field for simulating its atomic interactions, pure Ni is assumed to represent the matrix material in this study. Pure Ni and IN718 have similar stacking fault energies, which govern the evolution of crystalline defects such as dislocations and stacking faults during plastic deformation. The atomic interactions are described using the embedded-atom-method (EAM) potential developed by Zhang, Y. et al. [27]. Periodic boundary conditions are applied to the MD model, and the simulations are performed using the LAMMPS [28] package. The model is first relaxed at 300 K for 500 ps under a Nose-Hoover thermostat [29] and 0 bar external pressure. Subsequently, uniaxial tensile tests are simulated at a constant strain rate of 2.5×10^8 s^{-1} and a temperature of 300 K. Tensile tests are conducted with loading along the [010] direction (major axis of the precipitate) and the [001] direction (minor axis of the precipitate), which represent loading parallel and perpendicular to the disk plane, respectively. Local atomic shear strain is calculated to investigate plastic shearing during the tensile deformation [30]. To characterize the crystal structure, the common neighbor analysis (CNA) method is employed [31–33], with green atoms representing the FCC structure, red atoms representing the hexagonal close-packed (HCP) structure, and white representing grain boundaries, dislocations, and other disordered atoms. The dislocation extraction algorithm (DXA) [33,34], integrated into the OVITO 3.7.8 software [35], is utilized to determine the dislocation lines within the simulation.

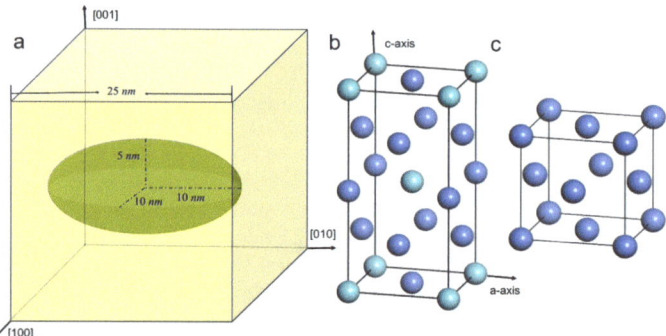

Figure 1. Illustration of the MD model with (**a**–**c**) showing the morphology of the unit cell, the atomic DO_{22} structure of the precipitate, and the fcc structure of the matrix, respectively. In (**b**,**c**), the cyan ball represents Niobium atoms and the blue represents Nickel atoms.

3. Results

3.1. Dislocation-Precipitate Interaction Simulations

When a single dislocation is initially introduced into the unit cell, it will undergo movement and evolution under the influence of the driving force during tensile deformation, leading to the emergence of additional new dislocations. Figure 2 shows the distribution of dislocations in the unit cell at a tensile strain of 6.25%. Here, the matrix is set transparent, and the dislocation lines and the atoms of the precipitate are visible. Figure 2a,b shows the views from different directions, with the precipitate viewed as an ellipsoid and a disk, respectively. It can be seen that a large number of dislocations surround the precipitate (marked by the black arrows). Quite a few dislocations can be seen on the surface of the precipitate (marked by the red arrows), which are penetrating into the precipitate. This could be the direct evidence of the dislocation shearing at this applied strain level. In order to examine whether the dislocation bypassing occurs or is ongoing, the unit cell and its periodic counterpart attached on the {010} plane are visualized in Figure 2c. Some bowed dislocation lines (marked by black arrows) can be seen between the two precipitates, probably indicating the dislocation bypassing based on the Orowan looping concept [36]. It is important to note that when the two unit cells are attached on the {001} plane, no dislocation lines can be detected. However, if the periodic unit cells are tiled throughout the entire space, it is possible to observe more curved dislocation lines. It should be noted that mobile dislocations tend to move along slip planes in the direction of shear stress under the influence of applied stress. Ideally, the Orowan bowing would be observed on the planes where the dislocation glides. However, since MD simulations are used here to capture the dynamic 3D configurations of dislocations, it is hard to visualize the dislocation lines on a specific plane.

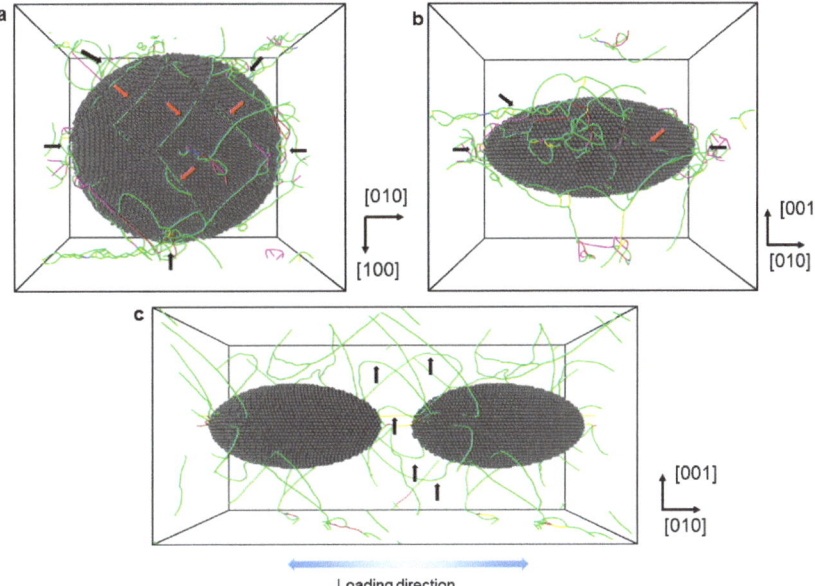

Figure 2. The distributions of dislocation lines in the unit cell at a tensile strain of 6.25%. The pure *Ni* matrix atoms are set transparent. The views of the unit cell, (**a**) from the [001] and (**b**) from the [100] directions are presented here. The view (**c**) from the [100] direction for the unit cell with its periodic duplication along the [010] direction is also presented. In these figures, the black arrows indicate the dislocations surrounding or looping the precipitates, while the red arrows mark the dislocations which are penetrating the precipitate.

In order to further examine the dislocation shearing phenomenon, the atomic configurations of the precipitate are focused on. Figure 3 shows the atoms of the precipitate at different applied strains. It can be observed that when a dislocation shears into the precipitate, a stacking fault is generated in the region where the dislocation sweeps through. This creates an easier path for subsequent dislocations to shear into the precipitate along the same path [6]. In addition, given in the figure are the identified slip systems (defined in Figure 3) for the dislocation shearing. Figure 3c,d shows that the dislocations shearing into the precipitate can be on the same slip plane but along different slip directions.

3.2. Atomic Shear Strain Distributions during the Tensile Deformation

In order to further probe the local deformation at the atomic level, a series of atomic configurations showing the local equivalent atomic shear strain in the precipitate under different applied strains are presented in Figure 4. In Figure 4a–c, most atoms experience almost no shear strain (blue), and only a few atoms experience local shear strain (with white and red colors). It can be seen in Figure 4a–c that atomic shear strain tends to occur locally on some lines on the surface of the precipitate. These lines represent the path of slip planes on the surface of the precipitate. As the applied strain increases, more trace lines become visible due to the activation of additional slip planes for the dislocations to glide on. Certain slip planes are preferred for dislocations, which can be identified by the relatively high atomic shear strains observed on their corresponding trace lines. To provide a clear view of the shear strain distribution within the precipitate, Figure 4d–f shows the shear strain distributions over the atoms in the precipitate, with most of the blue atoms neglected.

Four regions corresponding to four slip planes are marked in the figure. It can be seen that the atoms on the slip plane 'C' indicated by black arrows experience relatively low shear strain. This implies that the precipitate is hard to shear at the trace line position on this slip plane. For the atoms on the slip plane 'B', relatively high shear strain is experienced, implying that the dislocation shearing is prone to taking place. Note that the trace line of the slip plane 'C' is closer to the center of the precipitate than that of the slip plane 'B'. The atoms with shear strain experienced on the other two slip planes can also be seen. These results indicate that the dislocation shearing may occur all over the precipitate.

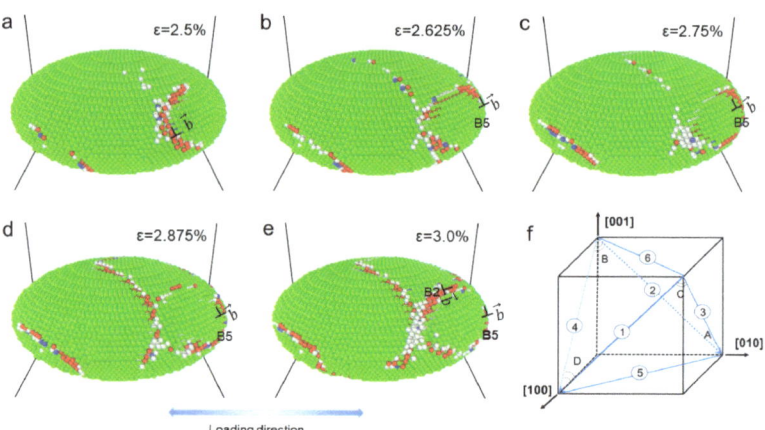

Figure 3. The snapshots of atomic configurations (CNA method) during the uniaxial tensile deformation of the Ni_3Nb system sample, at tensile strains of (**a**) 2.5% ; (**b**) 2.625%; (**c**) 2.75%; (**d**) 2.875%; (**e**) 3.0%. The ⊥ represents dislocations and $\vec{b}$ denotes burgers vector. The slip system illustration (**f**) for the FCC crystal is also presented. The green dots represent Ni_3Nb, while the red dots represent the HCP structure. The white dots represent boundaries and other defects.

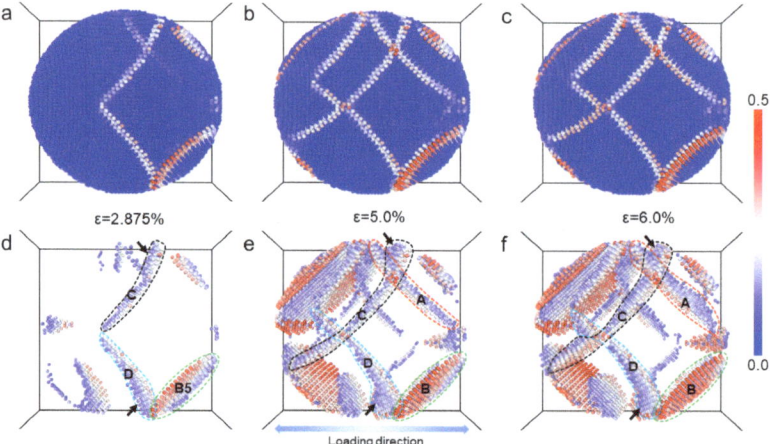

Figure 4. The snapshots of atomic configurations colored by atomic shear strain, at tensile strains of (**a**) $\epsilon = 2.875\%$, (**b**) $\epsilon = 5.0\%$ and (**c**) $\epsilon = 6.0\%$. In the according atomic configurations (**d**–**f**), the atoms with the an atomic shear strain of less than 0.15 are set invisible. The regions marked by the dashed lines highlight the atoms on some slip planes.

In order to compare the atomic shear strains in the Ni_3Nb precipitate with that in the pure Ni matrix, Figure 5 shows the shear strain on the two mid-planes vertical to the [001] and [100] directions, respectively, at 12.5% tensile strain. The boundary of the precipitate is outlined by the white lines. Uneven plastic shear can be observed in the figure, with the matrix undergoing more severe deformation compared to the precipitate. Furthermore, the dislocation is unable to easily cut through the central area of the precipitate, leading to a reduced ability of the precipitate to accommodate plastic deformation.

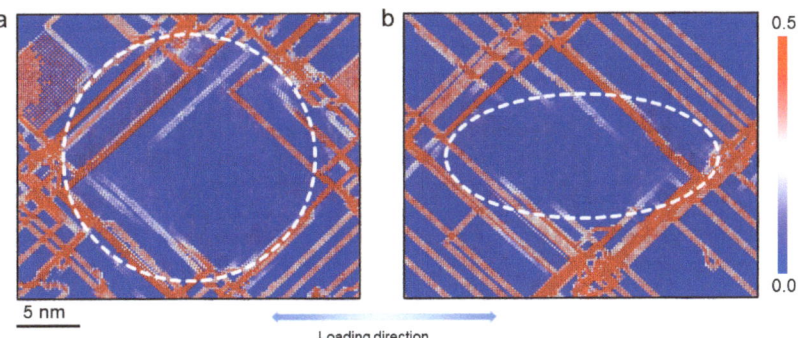

Figure 5. The atomic configurations of two mid-planes of the sample at a tensile strain of 12.5% (colored by atomic shear strain), (**a**) The mid-plane with the circular cross-section of the precipitate and (**b**) The mid-plane with the elliptical cross-section of the precipitate. The region of precipitate is highlighted by the white dashed lines.

3.3. Anisotropic Mechanical Responses

To examine the anisotropic behavior of the unit cell, the atomic shear strain distributions of the mid-plane at a tensile strain of 25% along two loading directions are investigated and shown in Figure 6, where the precipitates are indicated by the dashed lines. For both cases, strong localizations of the atomic shear strain can be seen in the matrix phase, with slight strain localizations in the precipitate. It can be observed from the figure that the precipitate with the loading direction parallel to the disk plane undergoes more plastic deformation compared to the one with the loading direction perpendicular to the disk plane. This suggests that dislocation shearing is more likely to occur when the loading direction aligns with the disk plane of the γ'' precipitate. This trend is not surprising, as the shape of the precipitate (ellipsoidal and disc-like) has a significant influence on dislocation shearing behavior. Additionally, Figure 6 presents the lengths of the precipitate axes and the corresponding half spacing between the two periodic precipitates in both the deformed configuration and the initial configuration. It demonstrates strong anisotropic deformation in both the matrix and the precipitate. Specifically, when the loading direction aligns with the major axis of the precipitate, the axis is stretched by 8.5%, while when the loading direction aligns with the minor axis of the precipitate, the axis is stretched by 3%.

3.4. The Effect of Precipitate Size on the Stress–Strain Response

In order to examine the precipitate size effect, the ratio between the major and the minor axes of the precipitate is fixed to 2. Additional four-unit cells are constructed with the half length of the major axis varying from 15 nm to 40 nm, and tensile tests along different loading directions are simulated. Figure 7a,b shows the uniaxial tensile stress–strain curves with different precipitate sizes and under different loading directions. When the unit cells undergo elastic deformation, the overall stresses along the tensile direction increase linearly with the applied strain. However, once the applied strain exceeds certain thresholds, strongly nonlinear stress–strain responses occur due to plastic deformation controlled by dislocation movements. After surpassing the elastic limits, the stress drops dramatically at first and then fluctuates. This nonlinear behavior is attributed to the dynamic nature of dislocations and the complex interaction between dislocations and precipitates. As expected, the size of the precipitate has a significant effect on the elastic limit of the flow stress, with larger precipitates leading to lower elastic limits of the flow stress. It can be seen in Figure 7a,b that the difference of the flow stresses at the elastic limit for 10 nm and 40 nm cases when the loading direction is parallel to the disk plane is approx. 7 GPa. When the loading direction is vertical to the disk plane, the difference is about 6 GPa. In addition, the amplitude of the stress fluctuations depends on the loading directions. In order to clearly see the precipitate size effect, Figure 7c shows the average flow stress vs. the half length of the major axis with error bars included. It is evident that as the size of the precipitate increases, the average flow stress decreases. Additionally, the unit cell with the loading direction perpendicular to the disk plane exhibits a higher average stress compared to the unit cell with the loading direction parallel to the disk plane. This difference becomes more prominent as the half length of the major axis increases from 10 nm to 30 nm. However, this trend does not hold for the case of a 40 nm precipitate.

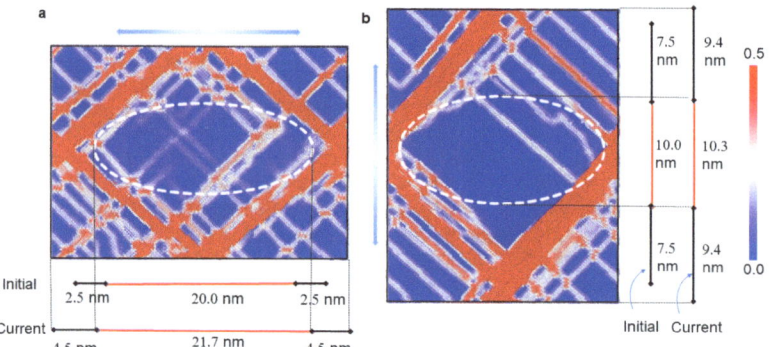

Figure 6. Atomic shear strain distributions in the mid-plane vertical to [100] at 25% tensile strain in (**a**) [010] and (**b**) [001] directions.

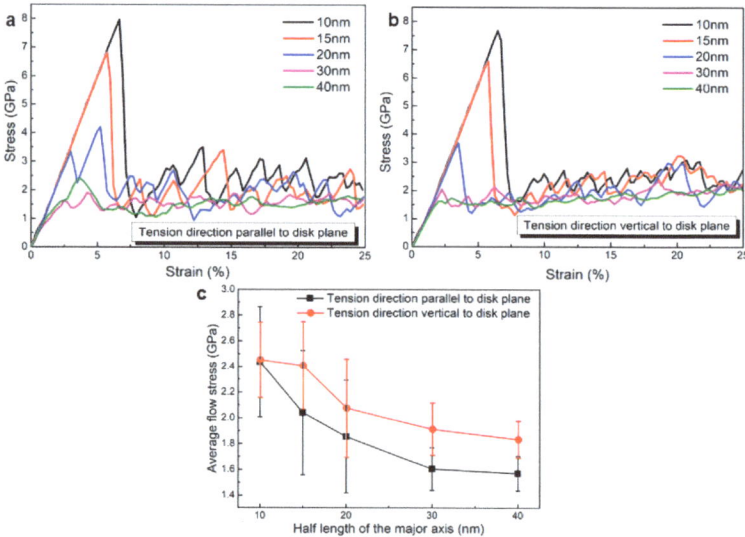

Figure 7. The size effect and loading direction dependent mechanical response in terms of (**a**,**b**) stress–strain curves with different precipitate sizes and loading directions and (**c**) the average flow stress obtained from (**a**,**b**).

4. Discussion
4.1. The Critical Precipitate Size for Dislocation Shearing

In the classical dislocation-precipitate interaction theory, the dislocation prefers to shear the precipitate if the precipitate size is small enough, otherwise, dislocation tends to bypass it with the so-called Orowan loop (see e.g., [37,38]) generated. The classic theory fits well to the cases with spherical precipitates [39,40]. To quantify the interaction mechanisms in the alloys containing non-spherical precipitates, the mean radius of the precipitate has been used [8,11]. The mean radius refers to the radius of a hypothetical spherical precipitate that has the same volume as the actual precipitate. However, the present paper shows that the interaction between the dislocation and the ellipsoidal γ'' precipitate in the IN718 superalloy is quite complex, and both dislocation shearing and bypassing may be induced simultaneously. The dislocation shearing and bypassing with regard to the γ'' precipitate can be illustrated as follows.

Figure 8 illustrates potential interactions between dislocations and precipitates. In Figure 8a, the dislocation line bypasses the precipitates aligned along the major axis. In Figure 8b, the dislocation line intends to shear into the precipitates aligned along the minor axis. The size of the obstacle on that plane can have a significant impact on the dislocation's behavior. In the case of dislocation bypassing, as shown in Figure 8a, the obstacle size refers to the length of the major axis of the precipitate. In the other case shown in Figure 8b, the obstacle size is defined by the length of the minor axis of the precipitate. Therefore, the maximum size of the γ'' phase for dislocation shearing can be estimated to be in the interval from 10 nm (the length of the minor axis) to 20 nm (the length of the major axis). This estimation is close to the prediction proposed in [11], where the critical size for dislocation shearing is approx. 18 nm. It should be noted that there exist other estimations of the critical precipitate size for dislocation shearing in the literature [8,41], ranging from 40 nm to 120 nm.

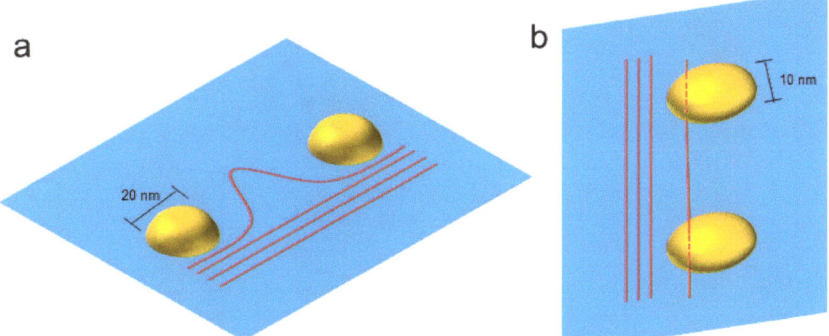

Figure 8. The illustrations of the dislocation-precipitate interaction with (**a**) the dislocation bypassing the two precipitates which are aligned along the major axis and (**b**) dislocation shearing into the two precipitates which are aligned along the minor axis.

To further verify the estimated critical precipitate size for dislocation shearing, the dislocation glide behavior in the precipitate with a major axis radius of 40 nm and a minor axis radius of 20 nm is investigated. Figure 9a,b shows the atomic shear strain distribution in the precipitate and the matrix under 12.5% tensile strain for the unit cell. It can be seen that, similar to the results in Figure 5, the high atomic shear strains are mainly located in the edge region of the precipitate. The estimated regions where dislocation shearing and bypassing could take place in the precipitate are given in Figure 9c,d. Here, the red regions are obtained based on the estimated critical obstacle size (20 nm here). It can be seen that the red regions in Figure 9c,d are qualitatively consistent with the high atomic shear strain areas in Figure 9a,b. Figure 9e,f shows the estimated dislocation shearing regions in the precipitate with a major axis radius of 20 nm and a minor axis radius of 10 nm. It can be seen that the portion of the red regions is bigger than that in Figure 9c,d, and it fits well with the simulated results shown in Figure 5.

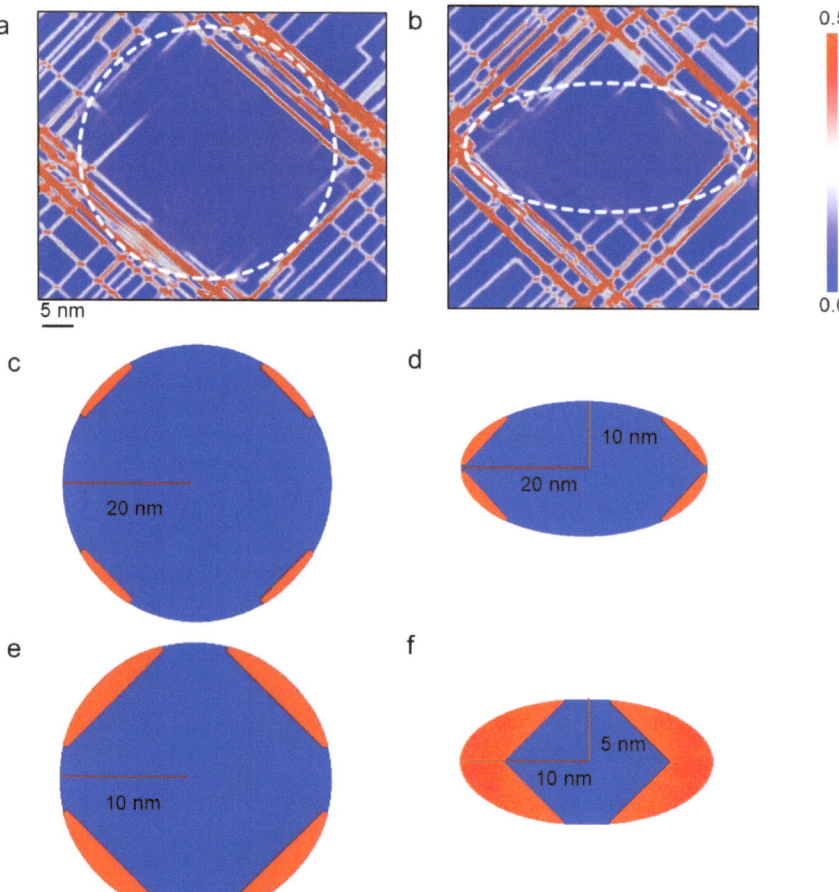

Figure 9. The atomic shear strain distributions of two mid-planes of the sample (the length of the major axis of the precipitate is 80 nm) at a tensile strain of 12.5% with (**a**) The mid-plane with the circular cross-section of the precipitate and (**b**) The mid-plane with the elliptical cross-section of the precipitate. The precipitate-occupied region is highlighted by the white dashed lines. The according illustrations of the dislocation shearing regions are also given based on the estimated critical obstacle size with (**c**,**d**) for large precipitate and (**e**,**f**) for standard precipitate.

4.2. Dislocation Accumulations in the Precipitate

From a mechanistic point of view, the dislocation density in the precipitate is a key state variable to reflect the precipitate strengthening [42,43]. However, accurately measuring or estimating dislocation density levels within the γ'' precipitate of IN718 superalloy can be challenging due to the small size of the precipitates. MD simulations offer a convenient and reliable way to investigate the evolution of dislocations within the precipitate. It can be observed in Figure 3 that when dislocations shear into the precipitate, certain slip systems are favored. To quantify the dislocation accumulations along different slip systems, the dislocation densities in the precipitate are statistically calculated for the unit cell with a standard precipitate as shown in Table 1. According to the symmetry of the slip directions, twelve slip systems are categorized into three groups. The slip systems in Group 1 are the slip systems with the slip direction perpendicular to the [100] direction (containing the slip systems, D1, C1, A2, and B2 as shown in Figure 3f). The slip systems D4, B4, A3, and C3 are in Group 2 with the slip direction perpendicular to [010]. The slip systems, A6, B5, C5, and D6 are in Group 3 with the slip direction perpendicular to [001]. Table 1 shows the dislocation densities at a tensile strain of 6.0% for the case with the loading direction along the major axis of the precipitate. It shows that the slip systems in Group 3 are most preferred by the dislocations to accumulate along the precipitate. The slip systems in Group 3 are particularly favored by dislocations in the precipitate. When dislocations attach to the precipitate along the slip systems in Group 3, they form relatively short trace lines. This arises because the slip directions are perpendicular to the minor axis, making the obstacle size on the slip plane relatively small. With a small obstacle size, the dislocations can easily cut through the precipitate along the slip systems where there is a higher density of dislocations. This phenomenon elucidates the anisotropy of atomic shear strain. In theory, the active direction of slip cannot be perpendicular to the direction of loading. Consequently, when loading runs parallel to the plane of the disk, it can activate the slip systems in Group 3 and Group 1, whereas loading vertically to the disk plane can activate the slip systems in Group 1 and Group 2. Given the preference for specific slip systems demonstrated in Table 1, loading parallel to the disk plane could induce more dislocation motion within the precipitate. Greater dislocation motion corresponds to a heightened atomic shear strain.

Table 1. Statistical dislocation density in the precipitate when $\varepsilon = 6.0\%$.

Group 1 (C1, D1, A2, and B2)	Group 2 (A3, C3, B4 and D4)	Group 3 (B5, C5, A6 and D6)
2.87×10^{11} mm^{-2}	3.30×10^{11} mm^{-2}	5.84×10^{11} mm^{-2}

5. Conclusions

In this work, the MD simulations are performed to investigate the dislocation-precipitate interaction in the IN718 superalloy. The main findings and conclusions are summarized as follows.

1. The present paper explicitly examined the dislocation movements within and around the ellipsoidal γ'' nano-precipitate. Both dislocation bypassing and shearing are identified for the γ'' precipitate. This underscores the significant influence of the dislocation's attachment point on the precipitate's surface on the depth to which the dislocation can penetrate. Specifically, dislocations tend to shear into the precipitate from positions where the cross-section of the precipitate intersected by the active slip planes is relatively small. The maximum value of such a size, permitting dislocation shearing, is estimated to fall within the range of 10 nm to 20 nm, with a close approximation to 20 nm. This estimation of the precipitate's critical size aligns well with the simulation outcomes.
2. The analysis of dislocation accumulation demonstrates that, in scenarios where dislocations shear into the precipitate with a loading direction parallel to the major axis of the precipitate, all twelve FCC slip systems contribute to accommodating a certain quantity of dislocations. Among these slip systems, the dislocation with a slip direction perpendicular to the disk plane is particularly favored by exhibiting the highest dislocation density. These dislocations encountered obstacles of relatively small size during their motion.
3. The atomic shear strain is uneven and anisotropic. It tends to localize around the interface between the matrix and the precipitate. The unit cell with the loading direction parallel to the major axis of the precipitate experiences stronger strain localizations than that with the loading along the minor axis of the precipitate. In addition, the flow stress fluctuations

for the unit cell with the loading direction parallel to the major axis are stronger than that under the loading along the minor axis of the precipitate. This anisotropy arises from the anisotropic dislocation slip in the ellipsoidal precipitate. More precisely, the dislocations with a slip direction perpendicular to the disk plane experience lower resistance as they move within the precipitate.

4. The investigation into the precipitate size effect reveals that, by maintaining a constant volume fraction of precipitate, larger precipitates correspond to reduced elastic limits of flow stress. In the case of sizable γ'' precipitates, plastic deformation within the precipitate is constrained to the limited edge regions. This limitation stems from the fact that slip resistance within the γ'' phase escalates from the edges to the center of the precipitate. In essence, shearing across the center of the precipitate necessitates a relatively substantial driving force that can overcome the heightened slip resistance posed by the larger obstacle size.

The present paper focuses on the dislocation-precipitate interaction and its implications on the mechanical behavior of the superalloy. A detailed analysis of the corresponding atomistic mechanisms is performed. The findings are expected to be beneficial for the development of high-performance superalloys by rationally utilizing nano-precipitation.

Author Contributions: C.-F.W.: writing—original draft, data curation, resources. L.-G.S.: modeling, review and editing. H.-L.Q. and Z.-N.B.: resources, review and editing. D.-F.L.: funding acquisition and project administration, review and editing. All authors have read and agreed to the published version of the manuscript.

Funding: This publication has emanated from research conducted with the financial support of National Natural Science Foundation of China under grant number 12272106, 12002108 and the Shenzhen Municipal Science and Technology program under grant numbers ZDSYS202106- 16110000001 and GXWD20220817151830003 and the Guangdong Basic and Applied Basic Research Foundation under grant number 2020A1515110236 and 2022A1515011402. Helpfull comments given by Prof. Esteban Busso are acknowledged.

Institutional Review Board Statement: Not applicable.

Informed Consent Statement: Not applicable.

Data Availability Statement: The data presented in this study are available on reasonable request from the corresponding author.

Acknowledgments: Helpful discussions with Esteban Busso are acknowledged.

Conflicts of Interest: The authors declare no conflict of interest. The authors declare that they have no known competing financial interests or personal relationships that could have appeared to influence the work reported in this paper.

Abbreviations

The following abbreviations are used in this manuscript:

IN718	Inconel 718
MD	molecular dynamics
DDD	Discrete dislocation dynamics
FCC	Face-centered cubic
EAM	Embedded-atom-method
CNA	common neighbor analysis
HCP	hexagonal close-packed
DXA	dislocation extraction algorithm

References

1. Hall, E. The deformation and ageing of mild steel: III discussion of results. *Proceedings of the Physical Society. Section B* **1951**, *64*, 747.
2. Decker, R. The evolution of wrought age-hardenable superalloys. *JOM* **2006**, *58*, 32–36. [CrossRef]
3. Reed, R.C. *The Superalloys: Fundamentals and Applications*; Cambridge University Press: Cambridge, UK, 2008.
4. Nageswara Rao, M. Factors influencing the notch rupture life of superalloy 718. *Trans. Indian Inst. Met.* **2010**, *63*, 363–367 [CrossRef]
5. Song, R.-H.; Qin, H.-L.; Li, D.-F.; Bi, Z.-N.; Busso, E.P.; Yu, H.-Y.; Liu, X.-L.; Du, J.-h.; Zhang, J. An Experimental and Numerical Study of Quenching-Induced Residual Stresses Under the Effect of Dynamic Strain Aging in an IN718 Superalloy Disc. *J. Eng. Mater. Technol.* **2021**, *144*, 011002. [CrossRef]
6. Oblak, J.; Paulonis, D.; Duvall, D. Coherency strengthening in Ni base alloys hardened by DO_{22} γ'' precipitates. *Metall. Mater. Trans. B* **1974**, *5*, 143–153. [CrossRef]
7. Han, Y.-f.; Deb, P.; Chaturvedi, M. Coarsening behaviour of γ''-and γ'-particles in Inconel alloy 718. *Met. Sci.* **1982**, *16*, 555–562. [CrossRef]
8. Sundararaman, M.; Mukhopadhyay, P.; Banerjee, S. Deformation behaviour of γ'' strengthened inconel 718. *Acta Metall.* **1988**, *36*, 847–864. [CrossRef]
9. Valle, L.; Araújo, L.; Gabriel, S.; Dille, J.; De Almeida, L. The effect of δ phase on the mechanical properties of an Inconel 718 superalloy. *J. Mater. Eng. Perform.* **2013**, *22*, 1512–1518. [CrossRef]
10. Merrick, H. The low cycle fatigue of three wrought nickel-base alloys. *Metall. Mater. Trans. B* **1974**, *5*, 891–897. [CrossRef]
11. Fisk, M.; Ion, J.C.; Lindgren, L.-E. Flow stress model for IN718 accounting for evolution of strengthening precipitates during thermal treatment. *Comput. Mater. Sci.* **2014**, *82*, 531–539. [CrossRef]
12. Lv, D.; McAllister, D.; Mills, M.; Wang, Y. Deformation mechanisms of D022 ordered intermetallic phase in superalloys. *Acta Mater.* **2016**, *118*, 350–361. [CrossRef]
13. Eghtesad, A.; Knezevic, M. A full-field crystal plasticity model including the effects of precipitates: Application to monotonic, load reversal, and low-cycle fatigue behavior of Inconel 718. *Mater. Sci. Eng. A* **2021**, *803*, 140478. [CrossRef]
14. Bacon, D.J.; Osetsky, Y.N. Hardening due to copper precipitates in α-iron studied by atomic-scale modelling. *J. Nucl. Mater.* **2004**, *329*, 1233–1237. [CrossRef]
15. Osetsky, Y.N.; Pharr, G.M.; Morris, J.R. Two modes of screw dislocation glide in fcc single-phase concentrated alloys. *Acta Mater.* **2019**, *164*, 741–748. [CrossRef]
16. Antillon, E.; Woodward, C.; Rao, S.; Akdim, B.; Parthasarathy, T. A molecular dynamics technique for determining energy landscapes as a dislocation percolates through a field of solutes. *Acta Mater.* **2019**, *166*, 658–676. [CrossRef]
17. Vaida, A.; Guénoléb, J.; Prakashc, A.; Korte-Kerzelb, S.; Bitzeka, E. Atomistic Simulations of Basal Dislocations Interacting with Mg17Al12 Precipitates in Mg. 2019. Available online: https://www.academia.edu/download/67161095/1902.pdf (accessed on 20 May 2019).
18. Li, J.; Chen, H.; Fang, Q.; Jiang, C.; Liu, Y.; Liaw, P.K. Unraveling the dislocation–precipitate interactions in high-entropy alloys. *Int. J. Plast.* **2020**, *133*, 102819. [CrossRef]
19. Esteban-Manzanares, G.; Alizadeh, R.; Papadimitriou, I.; Dickel, D.; Barrett, C.; LLorca, J. Atomistic simulations of the interaction of basal dislocations with MgZn2 precipitates in Mg alloys. *Mater. Sci. Eng. A* **2020**, *788*, 139555. [CrossRef]
20. Li, J.; Fang, Q.; Liu, B.; Liu, Y. The effects of pore and second-phase particle on the mechanical properties of machining copper matrix from molecular dynamic simulation. *Appl. Surf. Sci.* **2016**, *384*, 419–431. [CrossRef]
21. Lehtinen, A.; Granberg, F.; Laurson, L.; Nordlund, K.; Alava, M.J. Multiscale modeling of dislocation-precipitate interactions in Fe: From molecular dynamics to discrete dislocations. *Phys. Rev. E* **2016**, *93*, 013309. [CrossRef] [PubMed]
22. Santos-Güemes, R.; Esteban-Manzanares, G.; Papadimitriou, I.; Segurado, J.; Capolungo, L.; LLorca, J. Discrete dislocation dynamics simulations of dislocation-θ' precipitate interaction in Al-Cu alloys. *J. Mech. Phys. Solids* **2018**, *118*, 228–244. [CrossRef]
23. Krasnikov, V.; Mayer, A.; Pogorelko, V.; Latypov, F.; Ebel, A. Interaction of dislocation with GP zones or θ'' phase precipitates in aluminum: Atomistic simulations and dislocation dynamics. *Int. J. Plast.* **2020**, *125*, 169–190. [CrossRef]
24. Krasnikov, V.S.; Mayer, A.E.; Pogorelko, V.V. Prediction of the shear strength of aluminum with θ phase inclusions based on precipitate statistics, dislocation and molecular dynamics. *Int. J. Plast.* **2020**, *128*, 102672. [CrossRef]
25. Santos-Güemes, R.; Bellón, B.; Esteban-Manzanares, G.; Segurado, J.; Capolungo, L.; LLorca, J. Multiscale modelling of precipitation hardening in Al–Cu alloys: Dislocation dynamics simulations and experimental validation. *Acta Mater.* **2020**, *188*, 475–485. [CrossRef]
26. Santos-Güemes, R.; Capolungo, L.; Segurado, J.; LLorca, J. Dislocation dynamics prediction of the strength of Al–Cu alloys containing shearable θ'' precipitates. *J. Mech. Phys. Solids* **2021**, *151*, 104375. [CrossRef]
27. Zhang, Y.; Ashcraft, R.; Mendelev, M.; Wang, C.; Kelton, K. Experimental and molecular dynamics simulation study of structure of liquid and amorphous Ni62Nb38 alloy. *J. Chem. Phys.* **2016**, *145*, 204505. [CrossRef]
28. Plimpton, S. Fast parallel algorithms for short-range molecular dynamics. *J. Comput. Phys.* **1995**, *117*, 1–19. [CrossRef]
29. Evans, D.J.; Holian, B.L. The Nose–Hoover thermostat. *J. Chem. Phys.* **1985**, *83*, 4069–4074. [CrossRef]
30. Shimizu, F.; Ogata, S.; Li, J. Theory of shear banding in metallic glasses and molecular dynamics calculations. *Mater. Trans.* **2007**, *48*, 2923–2927. [CrossRef]
31. Honeycutt, J.D.; Andersen, H.C. Molecular dynamics study of melting and freezing of small Lennard-Jones clusters. *J. Phys. Chem.* **1987**, *91*, 4950–4963. [CrossRef]

32. Faken, D.; Jónsson, H. Systematic analysis of local atomic structure combined with 3D computer graphics. *Comput. Mater. Sci.* **1994**, *2*, 279–286. [CrossRef]
33. Stukowski, A.; Bulatov, V.V.; Arsenlis, A. Automated identification and indexing of dislocations in crystal interfaces. *Model. Simul. Mater. Sci. Eng.* **2012**, *20*, 085007. [CrossRef]
34. Stukowski, A.; Albe, K. Extracting dislocations and non-dislocation crystal defects from atomistic simulation data. *Model. Simul. Mater. Sci. Eng.* **2010**, *18*, 085001. [CrossRef]
35. Stukowski, A. Visualization and analysis of atomistic simulation data with OVITO–the Open Visualization Tool. *Model. Simul. Mater. Sci. Eng.* **2009**, *18*, 015012. [CrossRef]
36. Orowan, E. Fracture and strength of solids. *Rep. Prog. Phys.* **1949**, *12*, 185. [CrossRef]
37. Coutney, T.H. *Mechanical Behaviour of Materials*; McGrawHill Inc.: New York, NY, USA, 2000.
38. Maciejewski, K.; Jouiad, M.; Ghonem, H. Dislocation/precipitate interactions in IN100 at 650° C. *Mater. Sci. Eng. A* **2013**, *582*, 47–54. [CrossRef]
39. Chen, H.; Chen, Z.; Ji, G.; Zhong, S.; Wang, H.; Borbély, A.; Ke, Y.; Bréchet, Y. The influence of shearable and nonshearable precipitates on the Portevin-Le Chatelier behavior in precipitation hardening AlMgScZr alloys. *Int. J. Plast.* **2021**, *147*, 103120. [CrossRef]
40. Chatterjee, S.; Li, Y.; Po, G. A discrete dislocation dynamics study of precipitate bypass mechanisms in nickel-based superalloys. *Int. J. Plast.* **2021**, *145*, 103062. [CrossRef]
41. Chaturvedi, M.; Han, Y.-f. Strengthening mechanisms in Inconel 718 superalloy. *Met. Sci.* **1983**, *17*, 145–149. [CrossRef]
42. Busso, E.; Meissonnier, F.; O'dowd, N. Gradient-dependent deformation of two-phase single crystals. *J. Mech. Phys. Solids* **2000**, *48*, 2333–2361. [CrossRef]
43. Muiruri, A.; Maringa, M.; du Preez, W. Evaluation of dislocation densities in various microstructures of additively manufactured Ti6Al4V (ELI) by the method of x-ray diffraction. *Materials* **2020**, *13*, 5355. [CrossRef]

Disclaimer/Publisher's Note: The statements, opinions and data contained in all publications are solely those of the individual author(s) and contributor(s) and not of MDPI and/or the editor(s). MDPI and/or the editor(s) disclaim responsibility for any injury to people or property resulting from any ideas, methods, instructions or products referred to in the content.

Article

Influence of Manganese Content on Martensitic Transformation of Cu-Al-Mn-Ag Alloy

Lovro Liverić [1], Tamara Holjevac Grgurić [2,*], Vilko Mandić [3] and Robert Chulist [4]

1. Faculty of Engineering, University of Rijeka, Vukovarska 58, 51000 Rijeka, Croatia; lliveric@riteh.hr
2. School of Medicine, Catholic University of Croatia, Ilica 242, 10000 Zagreb, Croatia
3. Faculty of Chemical Engineering and Technology, University of Zagreb, Marulićev trg 19, 10000 Zagreb, Croatia; vmandic@fkit.hr
4. Institute of Metallurgy and Materials Science, Polish Academy of Sciences, 25 Reymont Str., 30-059 Krakow, Poland; r.chulist@imim.pl
* Correspondence: tamara.grguric@unicath.hr

Citation: Liverić, L.; Holjevac Grgurić, T.; Mandić, V.; Chulist, R. Influence of Manganese Content on Martensitic Transformation of Cu-Al-Mn-Ag Alloy. *Materials* **2023**, *16*, 5782. https://doi.org/10.3390/ma16175782

Academic Editor: Wojciech Borek

Received: 2 August 2023
Revised: 19 August 2023
Accepted: 22 August 2023
Published: 24 August 2023

Copyright: © 2023 by the authors. Licensee MDPI, Basel, Switzerland. This article is an open access article distributed under the terms and conditions of the Creative Commons Attribution (CC BY) license (https://creativecommons.org/licenses/by/4.0/).

Abstract: The influence of manganese content on the formation of martensite structure and the final properties of a quaternary Cu-Al-Mn-Ag shape memory alloy (SMA) was investigated. Two alloys with designed compositions, Cu- 9%wt. Al- 16%wt. Mn- 2%wt. Ag and Cu- 9%wt. Al- 7%wt. Mn- 2%wt. Ag, were prepared in an electric arc furnace by melting of high-purity metals. As-cast and quenched microstructures were determined by optical microscopy and scanning electron microscopy equipped with EDS. Phases were confirmed by high-energy synchrotron radiation and electron backscatter diffractions. Austenite and martensite transformations were followed by differential scanning calorimetry and hardness was determined using the Vickers hardness test. It was found that the addition of silver contributes to the formation of the martensite structure in the Cu-Al-Mn-SMA. In the alloy with 7%wt. of manganese, stable martensite is formed even in the as-cast state without additional heat treatment, while the alloy with 16%wt. of manganese martensite transforms only after thermal stabilization and quenching. Two types of martensite, $\beta_1{'}$ and $\gamma_1{'}$, are confirmed in the Cu-9Al-7Mn-2Ag specimen. The as-cast SMA with 7%wt. Mn showed significantly lower martensite transformation temperatures, M_s and M_f, in relation to the quenched alloy. With increasing manganese content, the M_s and M_f temperatures are shifted to higher values and the microhardness is lower.

Keywords: Cu-Al-Mn-Ag alloys; shape memory alloy; heat treatment; microstructure; phase transformations; martensitic transformation; thermal analysis

1. Introduction

Shape memory alloys (SMAs) possess two unique effects, the memory effect and pseudoelasticity [1,2]. The most commercially used and also one of the most expensive SMAs is Nitinol (NiTi). However, currently it is increasingly being replaced in non-medical applications by Cu-based SMAs, which have excellent functional and electrical properties, low cost, and easy and cheap processability [3].

Both fundamental Cu-SMAs, those grounded in Cu-Zn and Cu-Al, are capable of undergoing a thermoelastic martensitic transition from the high-temperature β-parent phase, often referred to as the austenite phase, facilitated by rapid cooling or immersion in water for quenching. Under equilibrium conditions or slow cooling, the β-phase undergoes eutectoid decomposition to form a combined α + γ₁-phase [4–7].

Cu-Al-Ni and Cu-Zn-Al are Cu-based SMAs and are the most commercially used SMAs, but they have limitations in some applications due to their high brittleness [8–10]. The addition of manganese to a Cu-Al alloy increases the ductility and cold workability of the material [11–15].

Previous research on Cu-Al-Mn SMAs has shown that the high-temperature β-phase, which is crucial for martensite formation, undergoes an order–disorder transformation, β (A2, BCC Cu) → $β_2$ (B2, CuAl) → $β_1$ ($L2_1$, Cu2MnAl), during cooling [16–18].

The β-phase decomposes the eutectoid into α- and $γ_2$-phases under equilibrium conditions. Similarly, the metastable martensite phase is formed from the β-phase by rapid cooling. Depending on the A2 or $L2_1$ (Heusler) parent phase, the 2M or 6M martensite structure is formed [19]. Depending on the specific chemical composition, an array of martensite structures, including 2H (a hexagonal structure) and 9R, 18R, 6R, and 3R (rhombohedral structures), can manifest in Cu-based SMAs [20–22].

Thus far, the influence of some elements such as microalloying components on the functional properties of Cu-Al-Mn alloys has been studied. The results of microalloying with nickel showed that there is a decrease in grain size and an improvement in the shape memory effect, but at the same time there is a decrease in the ductility of the alloy and a shift in the martensitic transformation to lower values [23–26]. Moreover, the nickel is completely soluble in the β-phase, as are the microalloying elements Zn, Sn, and Au, forming a single-phase system, in contrast to Cu-Al-Mn microalloys with Fe, Ti, Cr, V, Co, and Si, which exhibit lower solubility in the matrix and a tendency to precipitate [25,27].

Moreover, microalloying of the Cu-Al-Mn SMA with Au, Co, and Zn does not increase the ductility of the material or the tendency of cold deformation, while the addition of Sn to the ternary alloy significantly reduces ductility [17]. Of the elements that are not completely soluble in the β-phase, Fe shows the greatest microalloying effect on the properties of Cu-Al-Mn alloys and, depending on the content added, increases the temperature of the martensitic transformation. Studies have also shown that the addition of Ti, Co, Cr, and Si mainly contributes to the poorer mechanical properties of Cu-Al-Mn SMAs [28–30]. Microalloying with magnesium, on the other hand, does not significantly change the properties of the ternary alloy and does not affect the change in martensitic transformation temperatures [18].

There are limited publications available on the effect of the addition of silver on the functional properties of Cu-Al-Mn SMAs. Silva [29] pointed out that the hardness, corrosion resistance, and ageing properties are improved by the addition of Ag to Cu-Al-Mn alloys. It was also found that silver increased the fraction of the ferromagnetic $L2_1$-phase and the magnetic properties of the material, while Santos [30] reported an increase in the microhardness of the ternary alloy due to the formation of bainite [31].

The present work focuses on the effect of different compositions of a quaternary Cu-Al-Mn-Ag SMA on the formation of the martensite structure during casting and after thermal stabilization. The composition and microstructure were correlated with martensitic transformation temperatures and microhardness.

2. Materials and Methods

Cu-9Al-16Mn-2Ag and Cu-9Al-7Mn-2Ag alloys were prepared by melting raw metals (Mateck Material-Technologie & Kristalle, Jülich, Germany): copper 99.9%, aluminum 99.5%, manganese 99.8%, and silver 99.99%. The metals were melted in an electric arc furnace undergoing cycles of vacuuming and argon leaking and were re-melted four times for better homogenization.

The specimens were then cast in cylindrical molds with dimensions of 8 mm × 12 mm. Heat treatment was carried out at 900 °C for 30 min in chamber furnaces (OVER, Zagreb, Croatia) followed by quenching in water.

For metallographic analysis, the samples were cut, cold mounted, and ground with 600#, 800#, and 1200# SiC abrasives followed by final polishing with 3 μm and 1 μm diamond paste performed on Citopress-20 and Tegramin-30 (Struers, Willich, Germany). The prepared specimens were etched with a 2.5 g $FeCl_3$/48 mL CH_3OH/10 mL H_2O solution.

The microstructure was studied using an Axio Vert A1 optical microscope with the AxioCam ERc 5s microscope module (Carl Zeiss NTS GmbH, Oberkochen, Germany) and a scanning electron microscope (FEG QUANTA 250, FEI, Hillsboro, Oregon, USA) with an energy dispersion X-ray spectroscopy detector (EDS) (Oxford Instruments plc, Tubney Woods, Abingdon, Oxon, UK).

Electron backscatter diffraction (EBSD) analyses were conducted using a Supra 35 scanning electron microscope (SEM) (Carl Zeiss NTS GmbH, Oberkochen, Germany) operating at an acceleration voltage of 15 kV, a working distance of 17 mm, a tilt angle of 70°, and step sizes ranging from 0.4 to 0.06 µm. Samples for EBSD analysis were prepared according to standard metallographic techniques, which included mechanical grinding using SiC papers, polishing with diamond pastes, and a final polishing step for 1 h using 0.04 µm colloidal silica.

The crystal structure and global texture of samples in the cast and quenched state were examined by high-energy X-ray diffraction measurements at DESY, Hamburg, Germany, using the beamline P07B (87.1 keV, λ = 0.0142342 nm). For phase analysis, the diffraction patterns were recorded in the so-called continuous mode using a 2D Mar345 Image Plate detector. In order to obtain textureless measurements, all samples were rotated by 180° about the ω-axis when X-rayed. To ensure the Bragg condition for all satellite reflections, the samples were continuously rotated around the ω sample axis by a $\omega < \pm 10°$. The beam size was 1×1 mm^2. Subsequently, the obtained 2D patterns were integrated using the Fit2D Version 18 (beta) software and presented in a graph of relative intensity vs. 2Theta angle.

The atomic order was calculated as the intensity ratio of I_{hkl}/I_{220} using the reflection of the dominant phase, i.e., up to 900 °C from austenite reflections, and then from martensite reflections.

Transformation temperatures were determined using a Modulated Differential Scanning Calorimeter (MDSC) Mettler-Toledo 822e (Mettler-Toledo, Columbus, OH, USA). Dynamic measurements were performed by 2 heating/cooling measurement cycles from −100 °C to 350 °C in an inert atmosphere with a heating/cooling rate of 10 K/min.

The microhardness of the alloys was determined using the Future Tech FM-ARS-F-9000 with an FM-700 microhardness tester (FM-ARS 9000, Future-Tech, Kanagawa, Japan) using a load of HV 100 g and a dwell time of 15 s. The Vickers microhardness values were calculated as the average of five individual measurements taken from each sample.

3. Results

OM micrographs in the bright field and polarized light of the as-cast Cu-9Al-16Mn-2Ag SMA are shown in Figure 1. A two-phase morphology, (α + β), is observed in the microstructure, with some very thin needles of martensite forming at the grain boundaries. After solution treatment and quenching, the morphology changed, and a completely formed martensite structure with different orientations inside grains was observed (Figure 2). The grain size in the quenched Cu-9Al-16Mn-2Ag alloy is significantly smaller. The grain size is influenced by parameters of thermal stabilization, i.e., temperature, retention time, cooling medium, etc. [32]. In Cu-Al-Mn alloys during the heat-induced thermoelastic martensite transformation, two martensite structures, $\beta_1{}'$ (18R) and $\gamma_1{}'$ (2H), can co-exist, depending on chemical composition, the e/a ratio, and thermal stabilization routes [33].

Increasing structural order stabilizes more $\gamma_1{}'$-martensite than $\beta_1{}'$-martensite, while the driving force for the nucleation of martensite sites is higher for the 2H-type martensite. The two types of martensites are very similar; they only show differences in morphology caused by different modes of inhomogeneous shear. The $\beta_1{}'$ (18R)-martensite is formed from the L2$_1$ (Heusler) phase, and its stacking sequence is AB'CB'CA'CA'BA'BC'BC'AC'AB'. The $\gamma_1{}'$-martensite is usually formed at higher aluminum compositions, specifically more than 13 at%.

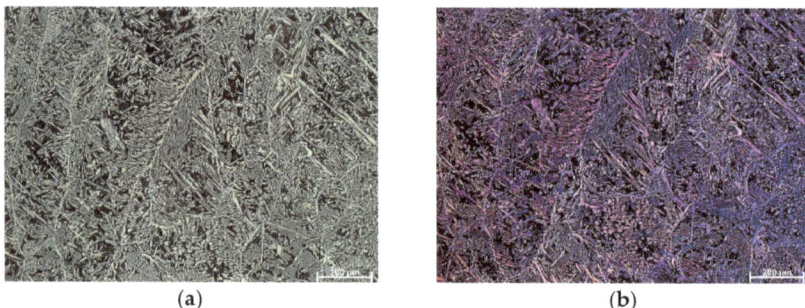

Figure 1. OM micrographs of as-cast Cu-9Al-16Mn-2Ag SMA: (**a**) BF, mag. 100×, (**b**) POL, mag 100×.

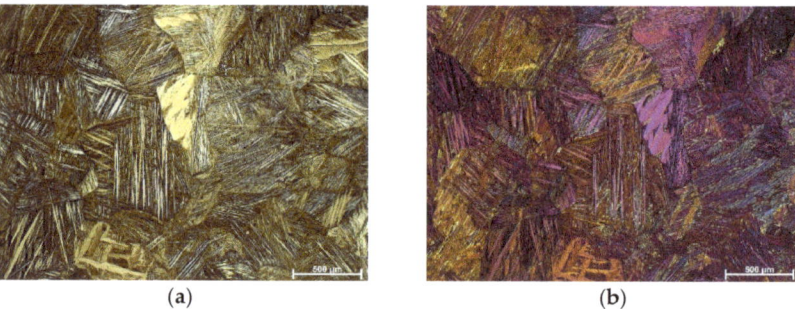

Figure 2. OM micrographs of quenched Cu-9Al-16Mn-2Ag SMA: (**a**) BF, mag. 50×, (**b**) POL, mag 50×.

The SEM analysis of the investigated samples confirmed the complete transformation from austenitic to martensitic structure after quenching without the precipitation of the α-phase (Figure 3b,d). The mostly spear-like shape of martensite can be observed in the quenched alloy, which refers to the β_1'-type of martensite, with a monoclinic structure. At some parts a zig-zag morphology of the β_1'-martensite is also detected (Figure 3d). Some coarse shape martensite plates can be also observed in the quenched alloy with 16%wt. of manganese (Figure 3d). Most likely, this can point to the existence of small amounts of another martensite type, γ_1', with an orthorhombic structure, but it should be confirmed by XRD analysis. The martensite pattern depends on the nucleation process and growth-type kinetics. It is well known that during thermoelastic martensitic transformation the growth of martensite plates takes place progressively and the formation of new plates occurs only when existing plates cannot grow further due to grain boundaries [33]. Needles of the β_1'-martensite type exhibit high thermoelastic behavior attributed to controlled growth in self-accommodating groups.

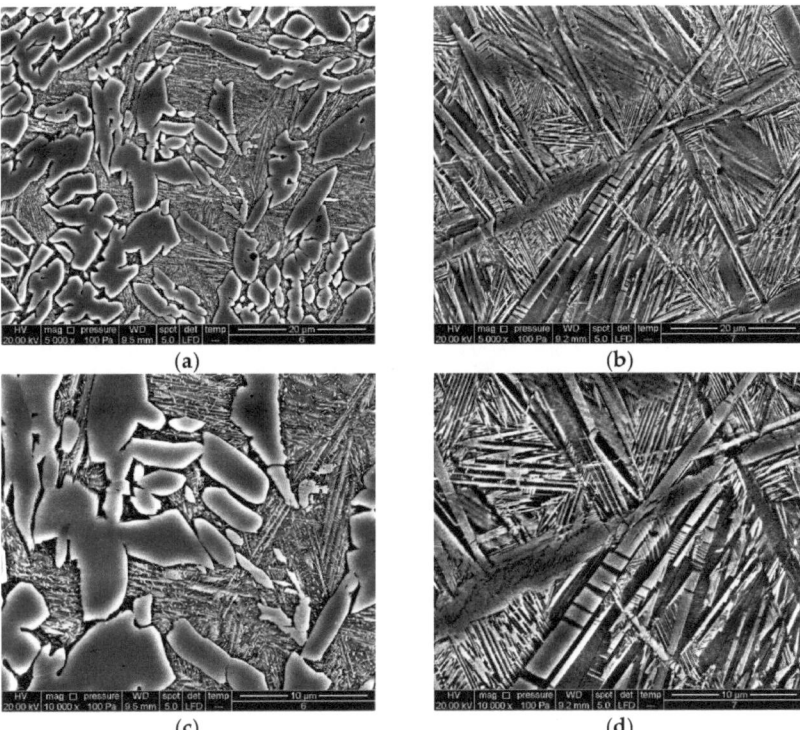

Figure 3. SEM micrograph of Cu-9Al-16Mn-2Ag SMA, mag. 5000×: (**a**) as-cast state, (**b**) quenched state, mag. 10,000×, (**c**) as-cast state, (**d**) quenched state.

Martensite plates nucleate and grow at different sites, as can be observed in the SEM micrographs (Figure 3). The SEM micrographs reveal the initial stages of needle formation between α-precipitates in the as-cast alloy (Figure 3a,c).

Figures 4 and 5 display the fully formed martensite structure in the alloy with lower manganese content in both the as-cast and quenched states. The quenched alloy exhibits more intense and thicker martensite formed by twinning, as depicted in Figure 6. The EDS analysis reveals a similar composition in various positions of the martensite matrix for both the as-cast (Figure 7) and quenched (Figure 8) states of the Cu-9Al-7Mn-2Ag alloy.

Figure 9 shows the BS images, band contrast, and phase maps for the Cu-9Al-16Mn-2Ag alloy in the as-cast state, with the precipitates of the fcc-Cu-phase (α-phase) colored green. Figure 9e,f primarily depicts spear-shaped morphologies in the β_1'-martensite plate group in the quenched alloy. The EBSD analysis reveals varying types of patterns in microareas associated with habit variants of the 18R martensite.

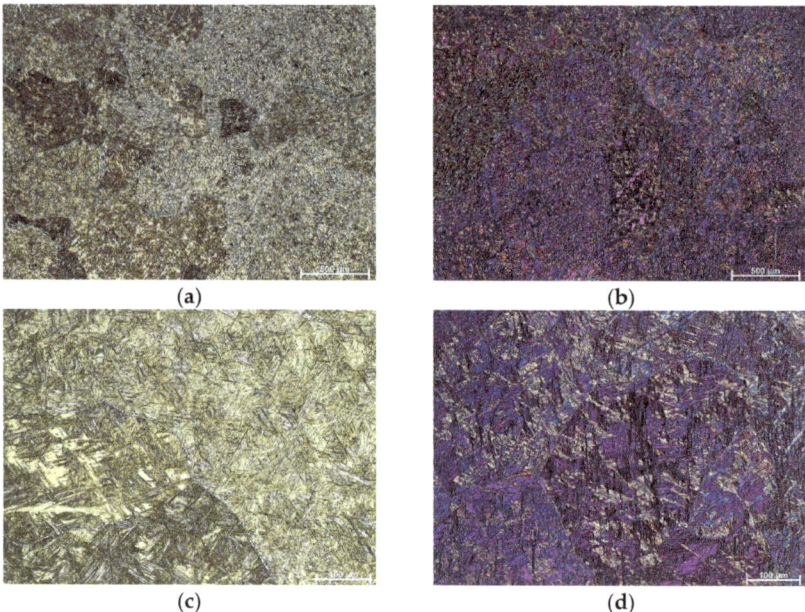

Figure 4. OM micrographs of as-cast Cu-9Al-7Mn-2Ag SMA: (**a**) BF, mag. 50×, (**b**) POL, mag 50×, (**c**) BF, mag. 200×, (**d**) POL, mag. 200×.

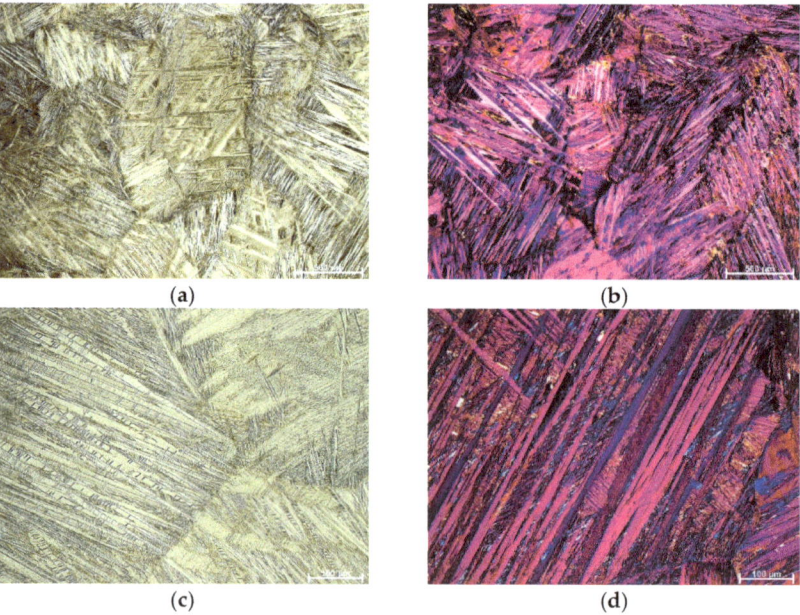

Figure 5. OM micrographs of quenched Cu-9Al-7Mn-2Ag SMA: (**a**) BF, mag. 50×, (**b**) POL, mag 50×, (**c**) BF, mag. 200×, (**d**) POL, mag. 200×.

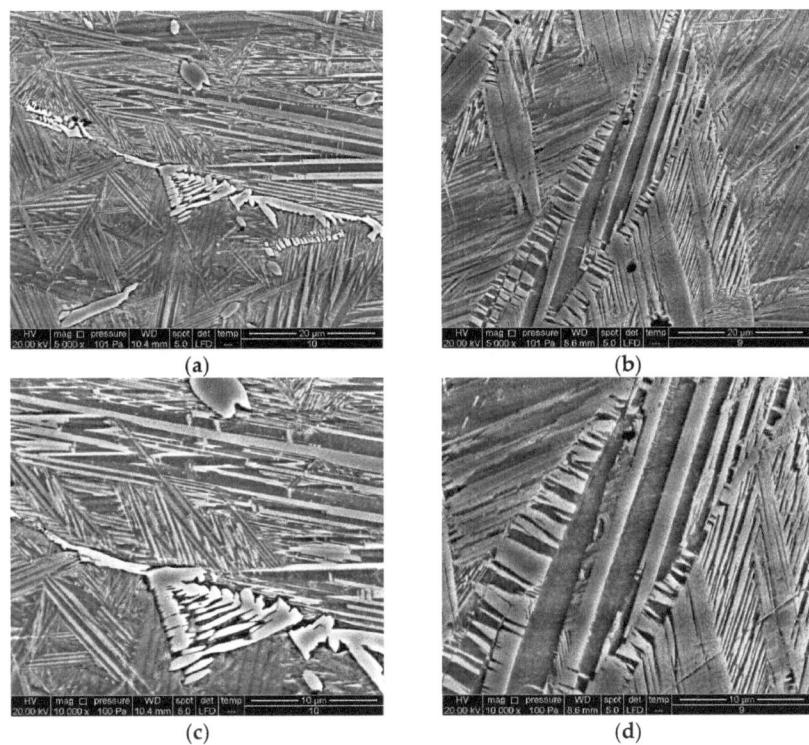

Figure 6. SEM micrograph of Cu-9Al-7Mn-2Ag SMA, mag. 5000×: (**a**) as-cast state, (**b**) quenched state, mag. 10,000×, (**c**) as-cast state, (**d**) quenched state.

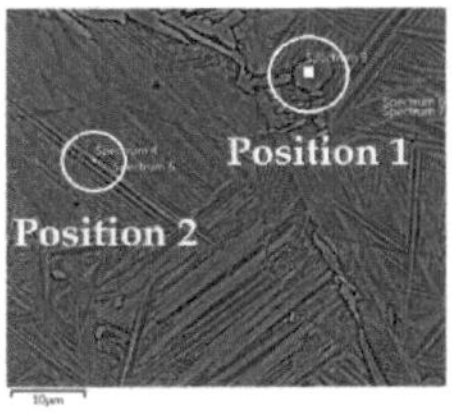

	Position 1		Position 2	
	Element	Wt%	Element	Wt%
	Al	8.33 ± 0.20	Al	9.75 ± 0.30
	Mn	6.56 ± 0.18	Mn	6.89 ± 0.25
	Cu	81.29 ± 0.55	Cu	79.81 ± 0.78
	Ag	1.86 ± 0.20	Ag	1.96 ± 0.29

Figure 7. EDS analysis of as-cast Cu-9Al-7Mn-2Ag alloy.

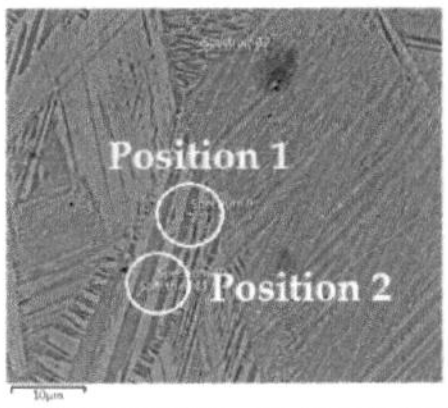

	Position 1		Position 2	
	Element	Wt%	Element	Wt%
	Al	8.36 ± 0.39	Al	8.27 ± 0.62
	Mn	6.38 ± 0.34	Mn	6.27 ± 0.57
	Cu	80.10 ± 1.14	Cu	82.49 ± 0.97
	Ag	1.80 ± 0.39	Ag	2.98 ± 0.61

Figure 8. EDS analysis of quenched Cu-9Al-7Mn-2Ag alloy.

Figure 9. Results of EBSD measurements: (**a**) band contrast image, (**b**) phase map for Cu9Al16Mn2Ag alloy in the as-cast state; (**c**) band contrast image, (**d**) phase map for Cu9Al16Mn2Ag alloy in the quenched state; (**e**) band contrast image, (**f**) phase map for the Cu9Al7Mn2Ag alloy in the quenched state.

The XRD analysis confirmed the existence of 18R (β_1')-martensite in both the as-cast and quenched Cu-9Al-7Mn-2Ag alloy (Figure 10). In the as-cast alloy with 16 wt.% of manganese, intensive peaks for fcc Cu (α-phase) are detectable, indicating that a very low cooling rate gives rise to Cu precipitation. On the other hand, quenching the alloy leads to the formation of 18R-martensite, similar to that observed in the Cu-9Al-7Mn-2Ag

SMA (Figure 11). The XRD diffractogram for the Cu-9Al-16Mn-2Ag alloy confirmed the co-existence of two martensitic phases, β_1' (18R) and orthorhombic γ_1' (2H).

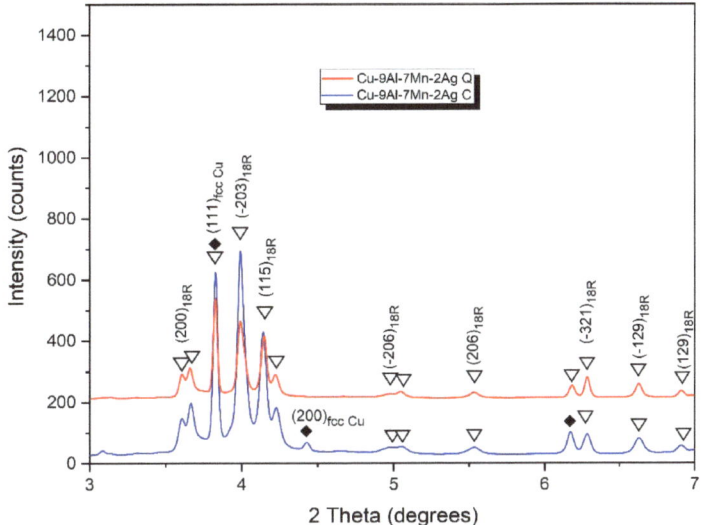

Figure 10. XRD diffractogram for Cu-9Al-7Mn-2Ag alloy.

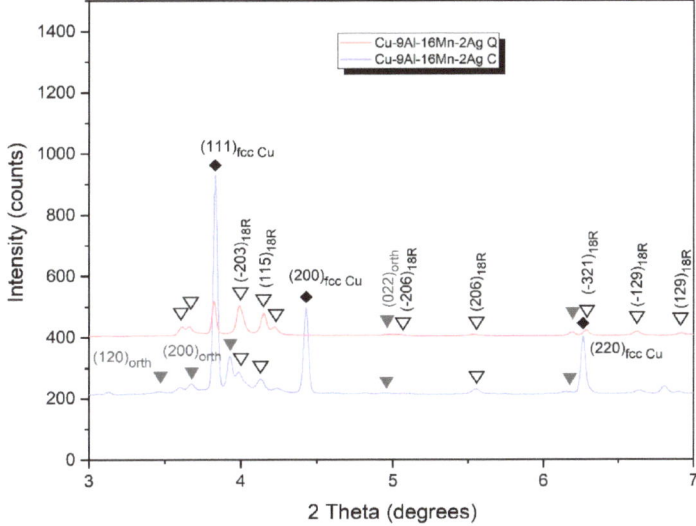

Figure 11. XRD diffractogram for Cu-9Al-16Mn-2Ag alloy.

The DSC results and transitions are presented in Table 1 and in Figures 12 and 13. A martensite transformation is a first-order transition, and it is not solely related to the change in specific heat capacity (cp) but is also accompanied by the emission of fusion heat during the transformation. The martensite transformation in the quenched Cu-9Al-16Mn-2Ag alloy shows that the start of the martensitic transformation was at M_s = 65 °C in both cooling cycles, and the finish temperatures were at M_f = 1 °C (1st cooling cycle) and M_f = −15 °C (2nd cooling cycle), respectively (Figure 12). With a lower content of manganese, 7 wt.%, the as-cast sample shows a significantly lower martensitic start temperature, $M_s = 22$ °C, in the first cooling cycle, and $M_f = -56$ °C (Table 1). After quenching, transition temperatures were shifted to higher values, $M_s = 55$ °C and $M_f = -22$ °C (1st cycle) and $M_s = 63$ °C and $M_f = -44$ °C (2nd cycle). Quenched samples with 7 wt.% and 16 wt.% of manganese exhibited similar martensitic transformation temperatures, but the enthalpy of transformation was significantly higher in the Cu-9Al-7Mn-2Ag alloy due to the more intense formation of martensite layers (Figure 13). The multiple exothermic peaks observable in the DSC thermograms are related to re-orientations and different martensitic structures.

Table 1. DSC results of martensitic transformation temperatures and fusion enthalpy.

Sample	$M_s/°C$ (1st Cycle)	$M_f/°C$ (1st Cycle)	$M_s/°C$ (2nd Cycle)	$M_f/°C$ (2nd Cycle)	ΔH (J/g)
as-cast Cu-9Al-7Mn-2Ag	22	−56	14	−63	2.25
quenched Cu-9Al-7Mn-2Ag	54	−22	64	−44	6.8
quenched Cu-9Al-16Mn-2Ag	65	1	65	−15	1.27

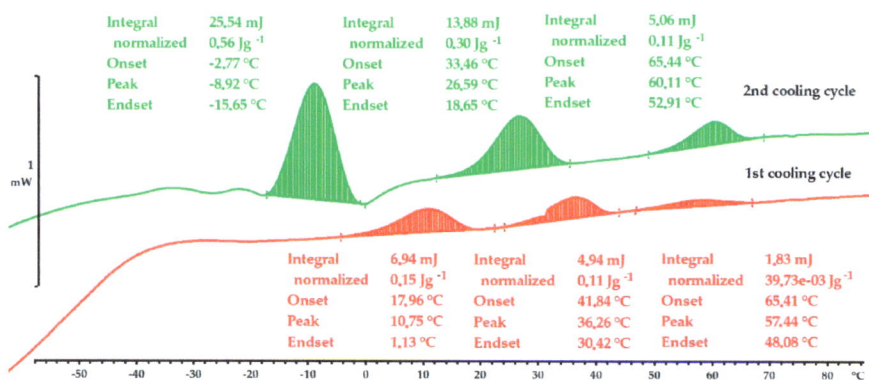

Figure 12. DSC cooling curves for quenched Cu-9Al-16Mn-2Ag alloy.

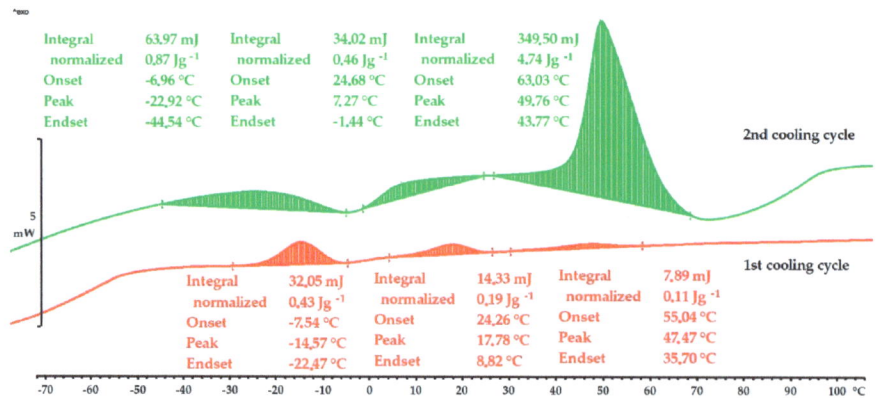

Figure 13. DSC cooling curves for quenched Cu-9Al-7Mn-2Ag alloy.

The microhardness of the studied Cu-Al-Mn-Ag shape memory alloy (SMA) is presented in Table 2. The quenched Cu-9Al-7Mn-2Ag alloy exhibits slightly lower microhardness compared to the as-cast condition, a finding that aligns with previous SMA investigations [16,29–31,34,35]. Increasing the manganese content to 16 wt.% results in a decrease in the hardness of the Cu-SMA. In contrast to the Cu-9Al-7Mn-2Ag alloy, the Cu-SMA with 16 wt.% manganese displays higher microhardness in the quenched state, which is atypical for shape memory materials and could be linked to the formation of the brittle γ_1'-phase in the quenched alloy.

Table 2. Microhardness of Cu–Al–Mn-Ag SMAs.

Hardness		HV					
		1	2	3	4	5	Average
Cu9Al16Mn2Ag	As-cast state	167.27	192.51	201.81	170.12	180.97	182.54
	Quenched state	244.93	235.50	237.01	238.24	241.69	239.47
Cu9Al7Mn2Ag	As-cast state	241.72	248.14	246.48	244.68	246.12	245.42
	Quenched state	241.72	248.14	238.52	243.67	239.84	242.38

According to Silva [29], the addition of Mn to the Cu-11%Al alloy significantly increases its microhardness value, while the addition of 3% Ag leads to a slight decrease in the microhardness of the Cu-11%Al alloy. Silva [29] also notes that Mn alters the range of phase stability and that the eutectoid reaction is no longer detectable in the annealed Cu-11%Al and Cu-11%Al-3%Ag alloys. The presence of Ag does not significantly influence the phase transformation sequence or microhardness but increases the magnetic moment of the Cu-11%Al-10%Mn alloy by about 2.7 times and decreases the rates of eutectoid and peritectoid reactions in the annealed Cu-11%Al alloy.

Furthermore, according to Jain [16], increasing the Al:Mn ratio in as-cast samples of the alloy system leads to an increase in hardness. However, in quenched samples, the hardness decreases, and this decrease is largely consistent with the Al:Mn ratio, which is attributed to the formation of softer martensitic phases at higher ratios.

4. Conclusions

In this work, we investigated the microstructures and phase transitions of Cu- 9wt.% Al- 16wt.% Mn- 2wt.% Ag and Cu- 9wt.% Al- 7wt.% Mn- 2wt.% Ag shape memory alloys (SMA), examining the differences between the as-cast and quenched states. We found that the Cu- 9wt.% Al- 7wt.% Mn- 2wt.% Ag SMA exhibits a stable martensite morphology in both the as-cast and quenched states, while the Cu- 9wt.% Al- 16wt.% Mn- 2wt.% Ag SMA develops a martensitic structure only after heat treatment and quenching in water. In both Cu alloys, the existence of 18R (β_1')-martensite is confirmed, but the γ_1'-phase is present only in the Cu-9Al-16Mn-2Ag alloy. The Ms temperatures for the quenched samples are notably similar, at 63 °C or 65 °C, but the Mf value shifts to lower values at a manganese content of 7 wt%. Furthermore, our study shows that increasing the manganese content leads to a decrease in microhardness, with the quenched Cu-9Al-7Mn-2Ag alloy exhibiting lower microhardness compared to its as-cast counterpart.

Author Contributions: Conceptualization, L.L. and T.H.G.; methodology L.L. and T.H.G., investigation, L.L., T.H.G., V.M. and R.C.; resources L.L. and T.H.G.; data curation, L.L. and T.H.G.; writing—original draft preparation, L.L. and T.H.G.; writing—review and editing, L.L., T.H.G., V.M. and R.C.; visualization, L.L. and T.H.G.; supervision, T.H.G. All authors have read and agreed to the published version of the manuscript.

Funding: This research received no external funding.

Institutional Review Board Statement: Not applicable.

Informed Consent Statement: Not applicable.

Data Availability Statement: Not applicable.

Conflicts of Interest: The authors declare no conflict of interest.

References

1. Wilberforce, T.; Alaswad, A.; Baroutaji, A.; Abdelkareem, M.; Ramadan, M.; Olabi, A.; Sayed, E.; Elsaid, K.; Maghrabie, H. Future Directions for Shape Memory Alloy Development. In *Encyclopedia of Smart Materials: Smart Biomaterials and Bioinspired Smart Materials and Systems*; Elsevier: Amsterdam, The Netherlands, 2022; Volume 1, pp. 231–242. [CrossRef]
2. Rodriguez, J.; Zhu, C.; Duoss, E.; Wilson, T.; Spadaccini, C.; Lewicki, J. Shape-morphing composites with designed microarchitectures. *Sci. Rep.* **2016**, *6*, 27933. [CrossRef]
3. Dobrzański, L.; Dobrzański, L.; Dobrzańska-Danikiewicz, A.; Dobrzańska, J. Nitinol Type Alloys General Characteristics and Applications in Endodontics. *Processes* **2022**, *10*, 101. [CrossRef]
4. Šimšić, Z.S.; Živković, D.; Manasijević, D.; Grgurić, T.H.; Du, Y.; Gojić, M.; Todorović, R. Thermal analysis and microstructural investigation of Cu-rich alloys in the Cu–Al–Ag system. *J. Alloys Compd.* **2014**, *612*, 486. [CrossRef]
5. Šimšić, Z.S.; Manasijević, D.; Živković, D.; Grgurić, T.H.; Kostov, A.; Minić, D.; Živković, Ž. Experimental investigation and characterization of selected as-cast alloys in vertical $Cu_{0.5}Ag_{0.5}$–Al section in ternary Cu–Al–Ag system. *J Therm Anal Calorim.* **2015**, *120*, 149. [CrossRef]
6. Huang, H.Y.; Zhu, Y.Z.; Chang, W.S. Comparison of Bending Fatigue of NiTi and CuAlMn Shape Memory Alloy Bars. *Adv. Mater. Sci. Eng.* **2020**, *2020*, 8024803. [CrossRef]
7. Stošić, Z.; Manasijević, D.; Balanović, L.J.; Grgurić, T.H. Effects of Composition and Thermal Treatment of Cu-Al-Zn Alloys with Low Content of Al on their Shape-memory Properties. *Mater. Res.* **2017**, *20*, 1425. [CrossRef]
8. Abolhasani, D.; Han, S.; VanTyne, C.; Kang, N.; Moon, Y. Enhancing the shape memory effect of Cu–Al–Ni alloys via partial reinforcement by alumina through selective laser melting. *J. Mater. Res. Technol.* **2021**, *15*, 4032–4047. [CrossRef]
9. Alaneme, K.; Okotete, E. Reconciling viability and cost-effective shape memory alloy options—A review of copper and iron based shape memory metallic systems. Engineering Science And Technology. *Int. J. Res.* **2016**, *19*, 1582–1592. [CrossRef]
10. Gustmann, T.; dos Santos, J.; Gargarella, P.; Kühn, U.; Van Humbeeck, J.; Pauly, S. Properties of Cu-Based Shape-Memory Alloys Prepared by Selective Laser Melting. *Shape Mem. Superelast.* **2016**, *3*, 24–36. [CrossRef]
11. Ren, C.; Wang, Q.; Hou, J.; Zhang, Z.; Yang, H.; Zhang, Z. Exploring the strength and ductility improvement of Cu–Al alloys. *Mater. Sci. Eng. A* **2020**, *786*, 139441. [CrossRef]
12. Konen, R.; Fintov, S. Copper and Copper Alloys: Casting, Classification and Characteristic Microstructures. In *Copper Alloys—Early Applications and Current Performance—Enhancing Processes*; IntechOpen: London, UK, 2012; pp. 3–30. [CrossRef]
13. Van Humbeeck, J. High Temperature Shape Memory Alloys. *J. Eng. Mater. Technol.* **1999**, *121*, 98–101. [CrossRef]
14. Najah Saud Al-Humairi, S. Cu-Based Shape Memory Alloys: Modified Structures and Their Related Properties. In *Recent Advancements in the Metallurgical Engineering and Electrodeposition*; IntechOpen: London, UK, 2020. [CrossRef]
15. Carl, M.; Smith, J.; Van Doren, B.; Young, M. Effect of Ni-Content on the Transformation Temperatures in NiTi-20 at. % Zr High Temperature Shape Memory Alloys. *Metals* **2017**, *7*, 511. [CrossRef]
16. Jain, A.; Hussain, S.; Kumar, P.; Pandey, A.; Dasgupta, R. Effect of Varying Al/Mn Ratio on Phase Transformation in Cu–Al–Mn Shape Memory Alloys. *Trans. Indian Inst.* **2015**, *69*, 1289–1295. [CrossRef]
17. Mallik, U.; Sampath, V. Effect of composition and ageing on damping characteristics of Cu–Al–Mn shape memory alloys. *Mater. Sci. Eng. A* **2008**, *478*, 48–55. [CrossRef]
18. Canbay, C.; Ozgen, S.; Genc, Z. Thermal and microstructural investigation of Cu-Al-Mn-Mg shape memory alloys. *Appl. Phys. A* **2014**, *117*, 767–771. [CrossRef]
19. Kainuma, R.; Satoh, N.; Liu, X.; Ohnuma, I.; Ishida, K. Phase equilibria and Heusler phase stability in the Cu-rich portion of the Cu–Al–Mn system. *J. Alloys Compd.* **1998**, *266*, 191–200. [CrossRef]
20. Suru, M.G.; Lohan, N.M.; Pricop, B.; Mihalache, M.; Mocanu, M.; Bujoreanu, L.G. Precipitation Effects on the Martensitic Transformation in a Cu-Al-Ni Shape Memory Alloy. *J. Mater. Eng. Perform.* **2016**, *25*, 1562. [CrossRef]
21. Zárubová, N.; Novák, V. Phase stability of CuAlMn shape memory alloys. *Mater. Sci. Eng. A* **2004**, *378*, 216. [CrossRef]
22. Qian, S.; Geng, Y.; Wang, Y.; Pillsbury, T.E.; Hada, Y.; Yamaguchi, Y.; Fujimoto, K.; Hwang, Y.; Radermacher, R.; Cui, J.; et al. Elastocaloric effect in CuAlZn and CuAlMn shape memory alloys under compression. *Philos. Trans. Royal Soc. A* **2016**, *374*, 20150309. [CrossRef] [PubMed]
23. Sedmák, P.; Šittner, P.; Pilch, J.; Curfs, C. Instability of cyclic superelastic deformation of NiTi investigated by synchrotron X-ray diffraction. *Acta Mater.* **2015**, *94*, 257–270. [CrossRef]
24. Yang, Q.; Wang, S.; Chen, J.; Zhou, T.; Peng, H.; Wen, Y. Strong heating rate-dependent deterioration of shape memory effect in up/step quenched Cu-based alloys: A ductile Cu Al Mn alloy as an example. *Acta Mater.* **2016**, *111*, 348–356. [CrossRef]
25. Sutou, Y.; Omori, T.; Koeda, N.; Kainuma, R.; Ishida, K. Effects of grain size and texture on damping properties of Cu–Al–Mn-based shape memory alloys. *Mater. Sci. Eng. A* **2006**, *438–440*, 743–746. [CrossRef]
26. Zhao, X.; Huang, H.; Wen, C.; Su, Y.; Qian, P. Accelerating the development of multi-component Cu-Al-based shape memory alloys with high elastocaloric property by machine learning. *Comput. Mater. Sci.* **2020**, *176*, 109521. [CrossRef]

27. Yang, Q.; Yin, D.; Ge, J.; Chen, J.; Wang, S.; Peng, H.; Wen, Y. Suppressing heating rate-dependent martensitic stabilization in ductile Cu-Al-Mn shape memory alloys by Ni addition: An experimental and first-principles study. *Mater. Charact.* **2018**, *145*, 381–388. [CrossRef]
28. Silva, R.A.G.; Adorno, A.T.; Carvalho, T.M.; Magdalena, A.G.; Santos, C.M.A. Precipitation Reaction in alpha-Cu-Al-Ag Alloys. *Mater. Rev.* **2012**, *16*, 747–753. [CrossRef]
29. Silva, R.; Paganotti, A.; Gama, S.; Adorno, A.; Carvalho, T.; Santos, C. Investigation of thermal, mechanical and magnetic behaviors of the Cu-11%Al alloy with Ag and Mn additions. *Mater. Charact.* **2013**, *75*, 194–199. [CrossRef]
30. Santos, C.; Adorno, A.; Paganotti, A.; Silva, C.; Oliveira, A.; Silva, R. Phase stability in the Cu-9 wt%Al-10 wt%Mn-3 wt%Ag alloy. *J. Phys. Chem. Solids* **2017**, *104*, 145–151. [CrossRef]
31. Santos, C.M.A.; Adorno, A.T.; Stipcich, M.; Cuniberti, A.; Souza, J.S.; Bessa, C.V.X.; Silva, R.A.G. Effects of Ag Presence on Phases Separation and order-disorder Transitions in Cu-xAl-Mn Alloys. *Mater. Chem. Phys.* **2019**, *227*, 184–190. [CrossRef]
32. Ivanić, I.; Kožuh, S.; Grgurić, T.H.; Vrsalović, L.; Gojić, M. The Effect of Heat Treatment on Damping Capacity and Mechanical Properties of CuAlNi Shape Memory Alloy. *Materials* **2022**, *15*, 1825. [CrossRef]
33. Ziółkowski, A. *Pseudoelasticity of Shape Memory Alloys Theory and Experimental Studies*; Butterworth-Heinemann: Oxford, UK, 2015.
34. Gholami-Kermanshahi, M.; Wu, Y.Y.; Lange, G.; Chang, S.H. Effect of alloying elements (Nb, Ag) on the damping performance of Cu–Al–Mn shape memory alloys. *J. Alloys Compd.* **2023**, *930*, 167438. [CrossRef]
35. Silva, R.A.G.; Gama, S.; Paganotti, A.; Adorno, A.T.; Carvalho, T.M.; Santos, C.M.A. Effect of Ag addition on phase transitions of the Cu–22.26 at.% Al–9.93 at.% Mn alloy. *Thermochim. Acta* **2013**, *554*, 71–75. [CrossRef]

Disclaimer/Publisher's Note: The statements, opinions and data contained in all publications are solely those of the individual author(s) and contributor(s) and not of MDPI and/or the editor(s). MDPI and/or the editor(s) disclaim responsibility for any injury to people or property resulting from any ideas, methods, instructions or products referred to in the content.

Article

Effect of Lankford Coefficients on Springback Behavior during Deep Drawing of Stainless Steel Cylinders

Fei Wu [1], Yihao Hong [1], Zhengrong Zhang [1,2,*], Chun Huang [1] and Zhenrong Huang [1]

[1] School of Material and Energy, Guangdong University of Technology, Guangzhou 510006, China; julie.wu@gdut.edu.cn (F.W.); hyhlztt@163.com (Y.H.)
[2] Guangdong Provincial Key Laboratory of Metal Forming and Forging Equipment Technology, Foshan 528300, China
* Correspondence: zzr@gdut.edu.cn

Abstract: Accurate prediction of springback is increasingly required during deep-drawing formation of anisotropic stainless steel sheets. The anisotropy of sheet thickness direction is very important for predicting the springback and final shape of a workpiece. The effect of Lankford coefficients (r_{00}, r_{45}, r_{90}) with different angles on springback was investigated using numerical simulation and experiments. The results show that the Lankford coefficients with different angles each have a different influence on springback. The diameter of the straight wall of the cylinder along the 45-degree direction decreased after springback, and showed a concave valley shape. The Lankford coefficient r_{90} had the greatest effect on the bottom ground springback, followed by r_{45} and then r_{00}. A correlation was established between the springback of workpiece and Lankford coefficients. The experimental springback values were obtained by using a coordinate-measuring machine and showed good agreement with the numerical simulation results.

Keywords: lankford coefficient; deep drawing; springback; stainless steel cylinder

Citation: Wu, F.; Hong, Y.; Zhang, Z.; Huang, C.; Huang, Z. Effect of Lankford Coefficients on Springback Behavior during Deep Drawing of Stainless Steel Cylinders. *Materials* 2023, *16*, 4321. https://doi.org/10.3390/ma16124321

Academic Editor: Wojciech Borek

Received: 9 May 2023
Revised: 5 June 2023
Accepted: 6 June 2023
Published: 11 June 2023

Copyright: © 2023 by the authors. Licensee MDPI, Basel, Switzerland. This article is an open access article distributed under the terms and conditions of the Creative Commons Attribution (CC BY) license (https://creativecommons.org/licenses/by/4.0/).

1. Introduction

As an important metal-forming process, sheet metal stamping is widely applied in the modern industry [1,2]. Springback is an inevitable physical phenomenon during the metal sheet-forming process [3–5]. The influence of springback on the accuracy and tolerance of a dimension is remarkable. The traditional trial-and-error and empirical methods for weakening springback and obtaining height-precision parts are time-consuming and expensive. The occurrence of defects, such as wrinkling, cracking, and springback, during sheet formation can be predicted with numerical simulations [6–9]. However, the predictions of springback and the final shape of the workpieces have a low accuracy rate because of the strong plastic anisotropy in the thickness direction.

A lot of research has been carried out in order to understand the influence of material properties and process parameters on springback behavior. Huang [10] analyzed the effects of different process parameters on springback during the stamping process using finite element numerical simulations. Minh [11] also used finite element simulation to analyze the effects of various factors—such as the blank holder force, friction coefficient, and blank thickness—on the springback of high-strength steel. Based on the numerical simulation results, it was evident that the blank holder force and blank thickness were the main factors affecting springback. Hashem and Roohi [12] utilized a numerical simulation to determine the effect of die and punch profile radii, as well as blank holder force on the springback and thinning percentage in the deep-drawing process of the cylindrical parts. The results show that an increased springback is observed due to an increased punch radius, and punch corner radius has been identified as the most significant effect on springback. Lajarin [13] found the blank holder force to be the most influential parameter for the springback of high-strength steel, followed by the die radius and friction conditions. Starman [14] proposed a numerical method to optimize the blank shape and tool geometry in a 3D sheet-metal-forming operation, with the effects of sheet-metal edge geometry and springback after forming and trimming being considered throughout the optimization process. Huang et al. [15] studied the defect behavior during the stamping of thin-walled semicircular shells with a bending angle via an analytical model, experiments, and a finite element (FE) simulation. The springback decreased with the increasing of blank holder force and decreasing of stamping speed. Aydın et al. [16] investigated the formability and springback behavior of dualphase (DP600) and high-strength low-alloy (HSLA) sheets bonded with laser-beam welding. The results show that springback behavior changes depending on the die angle and holding time. Where the die angle is up to $45°$, the springback angle increased by 10.4%. When holding time was increased by 10-s, the springback angle decreased by 21.2%, on average. Saito et al. [17] carried out a springback experiment of 980 MPa on high-strength steel sheets with V-shaped and U-shaped bending at temperatures ranging from room temperature to 973 K. The amount of springback decreased with the temperature rise, especially at temperatures above 573 K. Springback was much reduced at lower forming speeds. The influence of stress relaxation on springback was investigated using the V-shaped bending springback test and viscoplastic stress analysis. Chang et al. [18] studied the bending springback of medium-Mn steel, a third-generation automobile steel, under different working conditions through experiment and simulation, and they also analyzed the influences of rolling direction, bending angle, and punch fillet radius on springback. The effect of the rolling direction on the springback angle was negligible. The bending angle had a positive effect on the springback angle, while the punch fillet radius had a negative effect.

During recent decades, many researchers have explored the influence of anisotropy on springback and its prediction. Ragai et al. [19] provided an experimental and computational study of springback during draw-bending of stainless steel 410. The effect of several parameters such as blank holding force, lubrication, and anisotropy on springback were discussed. Parsa et al. [20] studied the springback of hyperboloid sheet metal formation theoretically, numerically, and experimentally. Emphatically, they analyzed the influence of thickness and curvature radius on springback. The experimental results showed that the influence of material anisotropy on the forming springback of hyperboloid sheet metal is related to material parameters. Gomes et al. [21] investigated the variation in springback in high-strength steels due to material anisotropy. They analyzed and compared material models based on different yield criteria using the geometry of a standard U-shape. The results showed a discrepancy between springback predicted by the various material models and the variations in springback from 0- and 90-degree material orientation. Leu and Zhuang [22] developed a simplified approach by considering the thickness ratio, normal anisotropy, and the strain-hardening exponent to estimate the springback angle in the vee bending process for high-strength steel sheets. The numerical simulation showed that the springback ratio increased as normal anisotropy increased or as the thickness ratio and the strain-hardening exponent decreased. Verma and Haldar [23] investigated the effect of anisotropy on springback for the benchmark problem in Numisheet-2005. An analytical model was developed to cross-check the prediction from the finite element analysis. Both of the models predicted that higher anisotropy leads to more springback. Lee et al. [24] investigated the influences of the anisotropy and friction models for high-strength steel sheets during U-draw/bending and suggested the optimum selection of the models for springback simulations. Jung et al. [25] developed an elastoplastic material constitutive model with anisotropic evolution, which they then applied to a U-bending process. By comparing the measured springback angle with the predicted springback angle, they showed that the model could accurately predict the springback angle in different directions.

Stainless steel sheets are widely used in modern industry, and springback during the stamping process is a common problem. Various process parameters can affect springback during the deep-drawing process of anisotropic stainless steel sheets. Although many scholars have conducted investigations on the springback problem during sheet metal formation using the finite element numerical simulation method and experimental methods, the research on springback of anisotropic sheet metal is still mostly confined to the forming of V/U-shaped parts, and there is little research on other common forming parts, for example, cylindrical cups. The influence of Lankford coefficients with different angles on springback during the cylinder deep-drawing process has not been clearly researched.

In this paper, a cylinder deep-drawing process with anisotropic stainless steel sheets was simulated based on the Barlat–Lian 1989 anisotropy yield criterion [26] by using Dynaform 5.9 software, and was used to predict springback. The Taguchi and ANOVA techniques were utilized to establish the correlation between springback at different angles from the rolling direction and Lankford coefficients (r_{00}, r_{45}, r_{90}) of 304 stainless steel. The ANOVA showed that the Lankford coefficient had a significant effect on springback. This research shows that each Lankford coefficient has an obvious influence on springback in diffident angles from the rolling direction by using the experimental and numerical simulation.

2. Finite Element Simulation (FEM) Analysis
2.1. FEM Simulation Procedure

In this paper, the finite element numerical simulation was carried out on Dynaform. Dynaform software is a special piece of software jointly developed by ETA and LSTC for numerical simulation of sheet metal formation. It is a combination of LS-DYNA solver and ETA/FEMB front and back processor, and it is one of the most popular CAE tools for sheet metal formation and die design. Figure 1 shows the cylinder deep-drawing die; the actual object of the model is thecooking pot. The dimensions of the blank, die, punch, and blank holder are given in Table 1. One quarter of the 3D numerical model can be applied to the FEM model. The simulations require a large amount of computational time if they are not simplified, but they can provide a greater degree of precision. In this paper, a complete 3D numerical model was used. The 3D numerical model is shown in Figure 2. Table 2 shows the Lankford coefficients of metal sheets in different rolling directions with two horizontal factors set. The other mechanical properties of the materials were imported from the materials library in Dynaform. The punch and die were set as rigid, and the velocity of the punch was set at 2000 mm/s. The friction coefficient between the tools and the blank were set to 0.125. The contact-one-way surface-to-surface mode was employed to determine the friction type, and the adaptive meshing method was adopted to mesh the geometry model [27]. The full integrated planar shell was used, and the element type was defined as the time-efficient full-order integral Belytschko–Tsay shell element. This allowed for the adoption of four-point integration in order to avoid the appearance of "hourglassing" mode. A dynamic explicit algorithm was used to calculate the forming process. The implicit algorithm was applied to calculate the springback process.

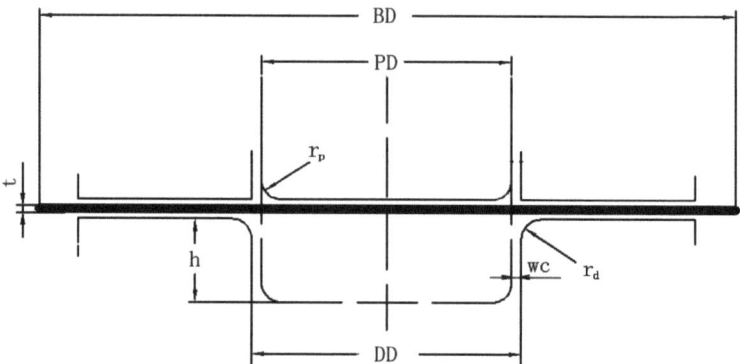

Figure 1. Geometry of drawing die.

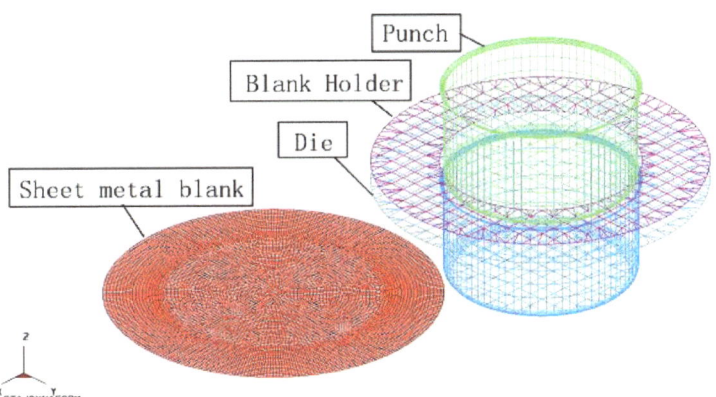

Figure 2. Model in FEM.

Table 1. Basic geometrical parameters.

Parameter	Dimension in mm
blank size diameter (BD)	315
blank thickness (t)	0.6
punch diameter (PD)	180
punch nose radius (rp)	8
die shoulder radius (rd)	4
die diameter (DD)	181.32
radial clearance between punch and die (wc)	0.66
height of drawing (h)	80

The material was modeled as an elastic–plastic material. The anisotropic characteristic was described by the Barlat–Lian 1989 anisotropic yield criterion [28]. The Barlat–Lian 1989 anisotropic yield criterion and the Hosford series' yield criterion were used to analyze the plastic flow law of the drawing process [29–31]. Three stress–strain curves were obtained from the tensile test for the model material, as shown in Figure 3. The different curves were determined according to the ratio of the Lankford coefficients in each direction of the actual material.

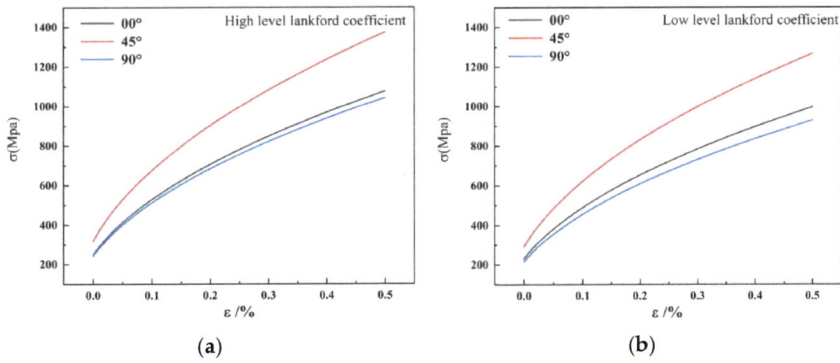

Figure 3. The stress–strain curves in different angles to the rolling direction: (**a**) high-level Lankford coefficient material hardening curve and (**b**) low-level Lankford coefficient material hardening curve.

Table 2. Test factors and their levels.

Level	Factors	r		
		r_{00}	r_{45}	r_{90}
Low-level		0.99	1.26	0.92
High-level		1.07	1.36	1.03

2.2. Taguchi Technique

The Taguchi technique was applied to the design scheme of the numerical simulation [32]. The two levels of the three-parameter orthogonal design, considering interactions (2^7), are presented in Table 3. The springback of different angles from the rolling direction was the process response. In order to understand the influence of Lankford coefficients, the ANOVA technique was applied to illustrate the degree of significance of each Lankford coefficient, including interactions.

Table 3. Experimental design of orthogonal considering interactions (2^7) for FEM simulation.

Simulation No.	Factors						
	r_{90}	r_{45}	$r_{90} \times r_{45}$	r_{00}	$r_{90} \times r_{00}$	$r_{45} \times r_{00}$	Error
1	Low	Low	Low	Low	Low	Low	Low
2	Low	Low	Low	High	High	High	High
3	Low	High	High	Low	Low	High	High
4	Low	High	High	High	High	Low	Low
5	High	Low	High	Low	High	Low	High
6	High	Low	High	High	Low	High	Low
7	High	High	Low	Low	High	High	Low
8	High	High	Low	High	Low	Low	High

2.3. Measurement Set-Up

Figure 4 shows the typical shape characteristics and measurement locations of the cylindrical cup. After formation, the workpiece was measured using CMM. Angles (α) were measured every 45 degrees from the rolling direction, and diameters were measured every 15 mm in the five sections along the height. A diagrammatic sketch of angles from the rolling direction is shown in Figure 4.

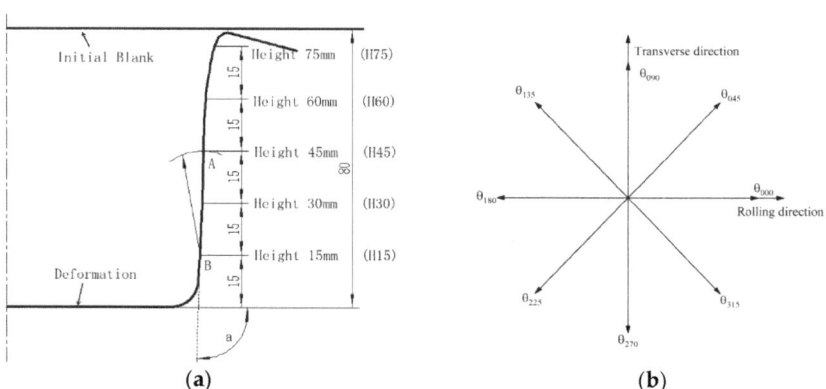

Figure 4. The Schematic diagram of the measurement position: (**a**) deformation and measurement locations and (**b**) diagrammatic sketch of angles from the rolling direction.

2.4. Formation Analysis

The forming limit diagram (FLD) and thickness change diagram can intuitively show the dynamic drawing process of the sheet metal and predict the formation of defects, such as cracking and wrinkling, and the thickness distribution of the sheet metal [33]. Figure 5a shows the forming limit diagram of the cylindrical cup after deep-drawing formation, Figure 5b shows the forming limit diagram of the cylindrical cup after springback, and Figure 5c is the cloud diagram of springback change in the cylindrical cup after springback calculation. It can be seen that the cylindrical cup fluctuates after springback with different degrees in the flange. The springback is apparent at 0°, 45°, and 90° positions, which shows a cyclical trend of first decreasing and then increasing along the rolling direction. The straight wall of the cylinder also showed uneven springback.

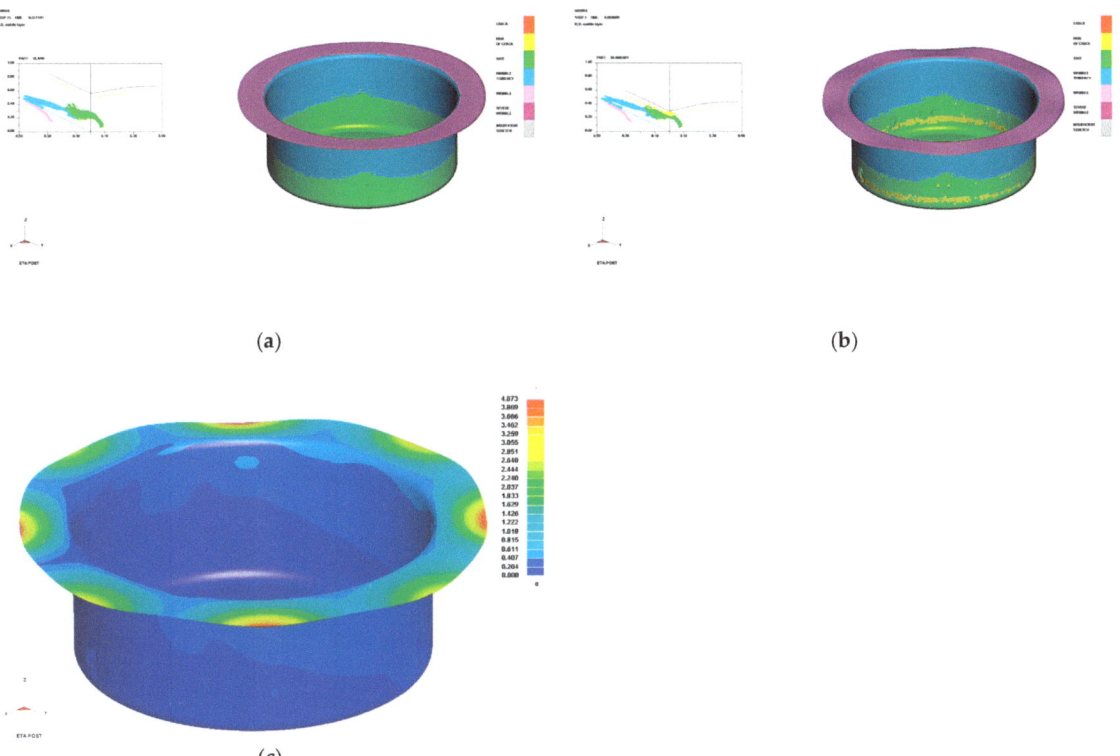

Figure 5. Forming limit diagram of the cylindrical cup and the cloud diagram of springback: (**a**) limit diagram of the cylindrical cup, (**b**) Limit diagram of the cylindrical cup after springback, and (**c**) cloud map of the cylindrical cup after springback.

Due to the uneven springback deformation of the straight wall of the cylindrical cup, the sections with heights of 45 mm and 60 mm were selected for measurement, and 120 coordinate points were measured for each section. The difference between coordinate values of data points before and after springback was calculated. The cross-section difference point cloud diagrams are shown in Figure 6. The co-ordinates only represent the position of data points on the section of the cylindrical drawing section. The distance between each point and the origin represents the springback value. It can be seen that the springback difference between the two heights is similar, and is in the range of 0.150–25 mm. Within the angle of 0–45° from the rolling direction, the springback difference firstly decreases, and then it increases. At the position of the maximum plastic strain value of r_{45}, that is, at the positions 45°, 135°, 225°, and 315° from the rolling direction, the springback difference of the cylinder drawing part reaches its maximum value.

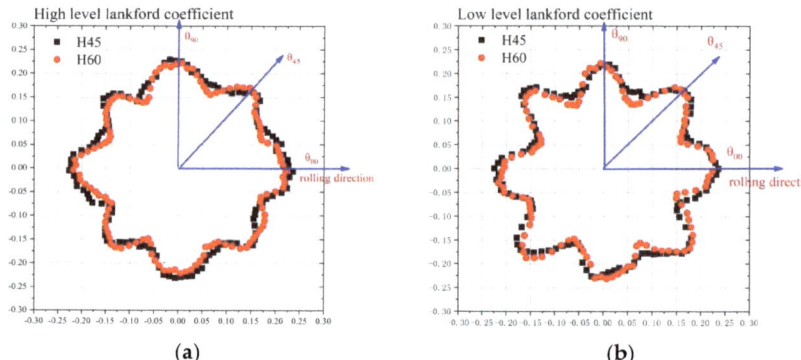

Figure 6. The springback difference point cloud of the cross-section of the cylindrical cup: (**a**) high–level Lankford coefficient material and (**b**) low–level Lankford coefficient material.

2.5. Stress–Strain Analysis

The straight wall of the cylindrical cup is an area of force transmission during deep drawing, and no more plastic deformation occurs. The straight wall experiences a single axial tensile stress. There is a small amount of axial elongation and deformation. The state of stress and strain during deep drawing is shown in Figure 7. The first principal stress and strain of the model of the straight wall model's middle layer was extracted to analyze the reasons for uneven springback.

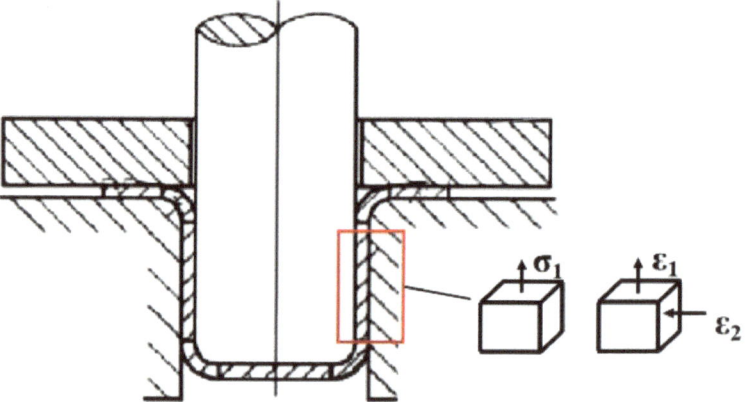

Figure 7. The stress and strain state of straight wall area.

The stress–strain analysis diagrams of cylindrical deep drawing at the heights of 45 mm and 60 mm are shown in Figure 8. The stress–strain data of 60 points on the circumference of the straight wall were extracted, and the red circle represents the average stress–strain value of all points. It can be seen that, at the height of 45 and 60 mm, the first principal stress was greater than the other two directions at the position of 45° from the rolling direction, while the first principal strain was smaller than the other two directions.

The stress–strain values for the three rolling directions of 0°, 45°, and 90° were compared and analyzed, and the results are shown in Table 4. The stress at 45° at the height of 45 mm is 23% higher, and the stress at 45° at the height of 60 mm is 19.37% higher than that of the other rolling directions. This is because the hardening curves are for different rolling directions. The value of the hardening curve at 45° from the rolling direction was larger, and the stress value required during deep drawing was larger. The strain in the 45° direction was smaller and contained more elastic stress in the deformation process, resulting in greater springback deformation after unloading.

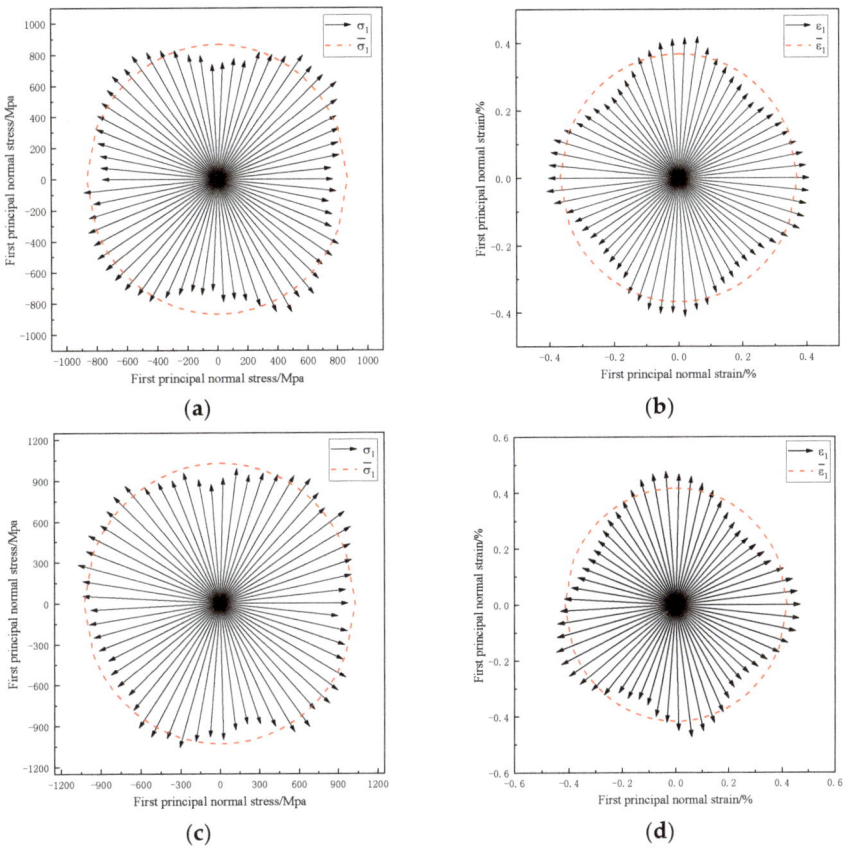

Figure 8. Stress and strain states on sections with different heights: (**a**) stress values at the 45 mm height position, (**b**) strain values at the 45 mm height position, (**c**) stress values at the 60 mm height position, and (**d**) strain values at the 60 mm height position.

Table 4. The stress and strain in three directions at the heights of 45 mm and 60 mm of high-level Lankford coefficient material.

Height	Direction	0°	45°	90°	Difference (Max−Min)
H45					
	Stress/Mpa	770.475	947.854	782.773	177.379 (23%)
	Strain	0.407	0.349	0.396	0.058
H60					
	Stress/Mpa	973.37	1162	997.662	188.63 (19.37%)
	Strain	0.468	0.368	0.463	0.1

2.6. Boundary Inflow Analysis

The diagram of inflow of the cylindrical cup boundary material is shown in Figure 9. It shows a cyclical trend of first increasing and then decreasing between 0° and 90° from the rolling direction. At the positions 45°, 135°, 215°, and 315°, there was a larger inflow, and the maximum value was 40.76 mm. The Lankford coefficients in the 45° direction were greater than those for the 0° and 90° directions. When the Lankford coefficients were large, the deformation resistance of the flange of the metal sheet was reduced, and the material flowed more easily. The flow stress value in the 45° direction was large, so the inflow of material was larger. The flow stress in the 0° and 90° directions was smaller, elongation deformation was easier, and the inflow was smaller. This may be one of the reasons for the greater springback difference in this direction.

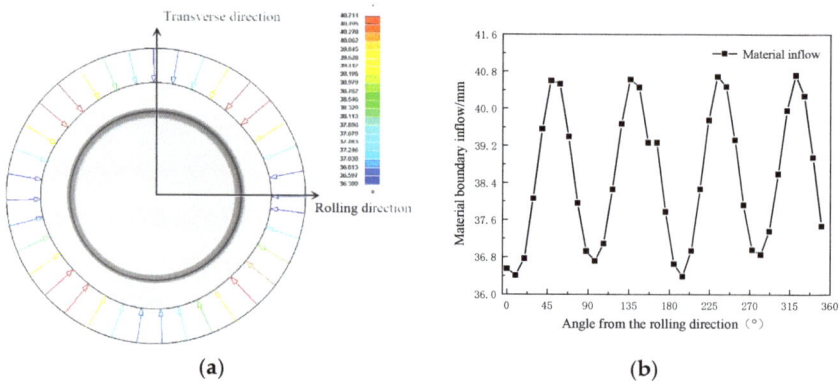

Figure 9. The boundary inflow of material and the trend of material inflow: (**a**) schematic diagram of the material boundary inflow and (**b**) numerical change diagram.

The sheet firstly underwent elastic deformation, and then plastic deformation occurred after the stress exceeded the flow stress during the deformation process. After unloading, the internal stress was redistributed, and then springback occurred. The deformation and plastic deformation of the cylindrical cup drawing in the 45° direction was less severe than the other two directions. The circumferential stress in the 45° direction was relatively large, resulting in larger springback deformation at this position. The diameter of the cylinder after springback in the 45° direction was smaller, showing a greater springback difference.

3. Experimental Procedures

3.1. Experimental Set-Up

Two different stainless steel sheets with the same thickness were selected for the experimental test. Based on the previous experimental tests [34,35], strong anisotropic properties were present in the two materials. Lankford coefficients of r_{00}, r_{45}, and r_{90} are listed in Table 2. For cylinder deep drawing, circular blanks with a diameter of 315 mm and a thickness of 0.6 mm were prepared. Figure 10 shows the drawing die designed and fabricated based on the simulation model.

Figure 10. Die assembly for experimental drawing.

3.2. Experimental Results

After calculating the weighted average of the co-ordinates of two types of stainless steel workpieces at different heights, the radius values at different angles were obtained, as shown in Figure 11. The average values of the cross-section point cloud can be compared at the height of 30 mm, 45 mm, and 60 mm. The valley shape of the depression was clearer and more obvious in the material with the larger Lankford coefficients at the positions of 45°, 135°, 225°, and 315°. At the height of 15 mm near the bottom of the cylinder, the low-Lankford-coefficient material showed a more rounded cross-section. The high-Lankford-coefficient material showed a less rounded cross-section after drawing, and it was close to an ellipse along the long axis of the rolling direction and the short axis perpendicular to the rolling direction. At a height of 75 mm near the flange, the sections of both kinds of stainless steel showed an elliptical shape after deep drawing and springback. The experimental results are in good agreement with the FEM simulation results.

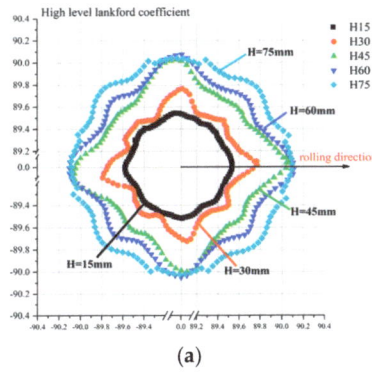

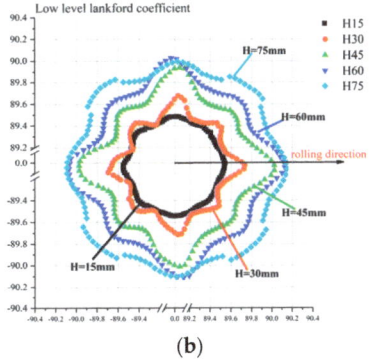

(a) (b)

Figure 11. *Cont.*

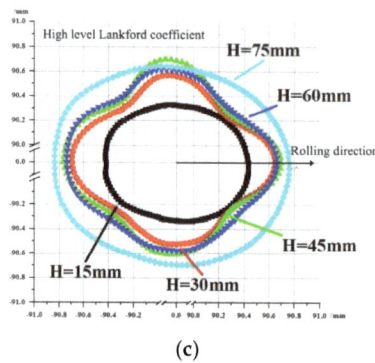
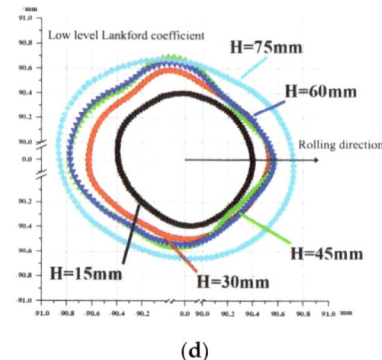

(c) (d)

Figure 11. Average vertical wall shell's dimension of the experiments and simulations: (**a**) high-level Lankford coefficient material simulation results, (**b**) low-level Lankford coefficient material simulation results, (**c**) high-level Lankford coefficient material experimental results, and (**d**) low-level Lankford coefficient material experimental results. Note: Black: H = 15 mm Red: H = 30 mm Green: H = 45 mm Blue: H = 60 mm Light Blue: H = 75 mm.

Tables 5–8 compare the radius values along the three rolling directions at five section heights obtained from numerical simulation and experiments. It can be seen that the difference in radius between different rolling directions is more obvious in the simulation, and it was the largest at the heights of 45 mm and 60 mm. In the material with high-level Lankford coefficients, the differences reached 0.433 mm and 0.318 mm, respectively. In the material with low-level Lankford coefficients, the differences reached 0.387 mm and 0.32 mm, respectively. The experimental difference between radii in different rolling directions was close to the simulation result in the material with high-level Lankford coefficients. The difference at the heights of 30 mm and 45 mm reached 0.158 mm and 0.204 mm, respectively. The experimental radius difference between different rolling directions in material with low-level Lankford coefficients was small, reaching 0.127 mm and 0.08 mm at the heights of 45 mm and 60 mm, respectively.

Table 5. The radius values of the simulations with high-level Lankford coefficients on the three directions.

Height/mm	Direction 0°/mm	45°/mm	90°/mm	Difference /mm
H15	89.561	89.492	89.537	0.068
H30	89.987	89.740	89.983	0.246
H45	90.506	90.072	90.486	0.434
H60	90.617	90.299	90.592	0.318
H75	90.599	90.575	90.517	0.024

Table 6. The radius values of the simulations with low-level Lankford coefficients on the three directions.

Height/mm	Direction 0°/mm	45°/mm	90°/mm	Difference /mm
H15	89.541	89.490	89.495	0.051
H30	89.884	89.710	89.883	0.174
H45	90.428	90.041	90.404	0.387
H60	90.603	90.283	90.557	0.320
H75	90.652	90.585	90.515	0.067

Table 7. The radius values of the experiments with high-level Lankford coefficients on the three directions.

Height/mm	Direction	0°/mm	45°/mm	90°/mm	Difference /mm
H15		90.546	90.506	90.452	−0.007
H30		90.781	90.563	90.660	0.158
H45		90.841	90.597	90.760	0.204
H60		90.811	90.642	90.708	0.118
H75		90.905	90.837	90.775	0.003

Table 8. The radius values of the experiments with low-level Lankford coefficients on the three directions.

Height/mm	Direction	0°/mm	45°/mm	90°/mm	Difference /mm
H15		90.517	90.498	90.516	0.019
H30		90.694	90.610	90.645	0.060
H45		90.781	90.628	90.729	0.127
H60		90.780	90.654	90.697	0.085
H75		90.895	90.826	90.767	0.005

Figure 12 shows experimental measurements of diameter at five sections along the height. They have a similar trend. The section at the height of 15 mm was close to the radius of the punch nose. The section at the height of 75 mm was similar to the radius of the die shoulder. The fillet radius had a great influence on the workpiece diameter. The cross-section of the cylinder after deep drawing showed an oval shape after springback. The diameters at the height of 30, 45, and 60 mm showed a similar trend. The results showed that the r_{45} Lankford coefficient is the maximum value. In addition, as the Lankford coefficient increased, the diameter decreased.

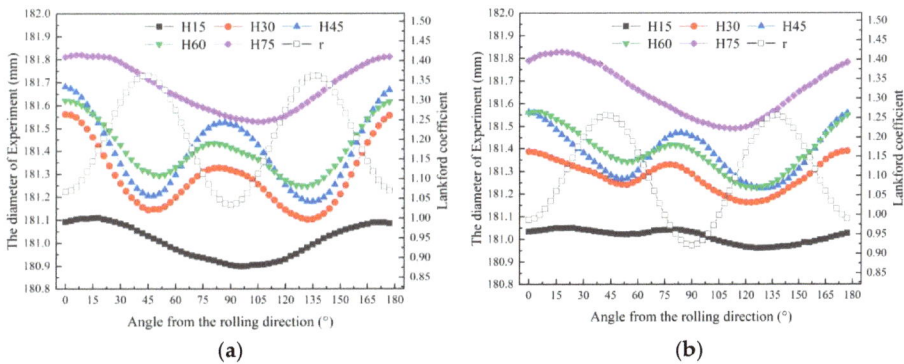

Figure 12. Diameter of different sections of experimental measurements: (**a**) the material with high-level Lankford coefficients and (**b**) the material with low-level Lankford coefficients.

4. Results and Discussions

4.1. Application of ANOVA

Table 9 shows the results of springback prediction by FEM simulation, which shows that the springback of every angle from the rolling direction is without symmetrical characteristics.

Table 9. The results of springback prediction by FEM simulation.

	The Amount of Springback/(°)								
	θ_{000}	θ_{045}	θ_{090}	θ_{135}	θ_{180}	θ_{225}	θ_{270}	θ_{315}	Difference (Max–Min)
1	90.205	90.052	90.313	90.623	90.629	90.645	89.999	90.303	0.44
2	90.294	90.748	90.645	90.838	90.617	90.729	90.363	90.315	0.544
3	90.255	90.566	90.779	91.021	90.445	90.931	90.601	90.407	0.766
4	90.217	90.242	90.027	90.342	90.654	90.412	90.268	90.112	0.627
5	90.205	90.518	90.379	90.529	90.626	90.514	90.338	90.577	0.421
6	90.295	90.507	90.652	90.723	90.175	90.481	90.418	90.447	0.428
7	90.027	90.14	90.246	90.006	90.069	89.968	90.148	89.884	0.362
8	90.225	89.968	90.303	90.188	90.249	90.161	90.279	90.058	0.257
average	90.215	90.375	90.385	90.534	90.433	90.480	90.302	90.263	0.319

To investigate the degree of significance of the Lankford coefficients, the ANOVA technique was used to analyze the springback. The mean overall value $S/N\left(\overline{S/N}\right)$ is expressed as Equation (1), where k is the number of simulations. The range of two levels (SR_j) is shown in Equation (2). The sum of squares owing to the variations of the overall mean (SS) and the mean of the Lankford coefficients with interactions (SS_j) are expressed as Equations (3) and (4), respectively. The percentage values ($\%p\text{-}Value_j$) were calculated using Equation (5), which is generally applied when measuring the degree of significance of each Lankford coefficient [36].

$$\overline{S/N} = \frac{1}{8}\sum_{k=1}^{8}(S/N)_k \qquad (1)$$

$$SR_j = \sum_{j=1}^{7}\left((S/N)_{1j} - (S/N)_{2j}\right) \qquad (2)$$

$$SS = \sum_{i=1}^{8}\left((S/N)_{ij} - \overline{S/N}\right)^2 \qquad (3)$$

$$SS_j = \sum_{j=1}^{7}\left((S/N)_{ij} - \overline{S/N}\right)^2 \qquad (4)$$

$$\%p - Value_j = \left(\frac{SS_j}{SS} \times 100\right) \qquad (5)$$

4.2. Effects of Process Parameters on Springback

The results of the range analysis and variance analysis are shown in Table 10. It is revealed that the influence of Lankford coefficients on springback is different at different angles. The source of r_{00} had a critical effect on the springback at θ_{000} from the rolling direction, the r_{45} is the key factor causing springback at θ_{045} and θ_{315}, and r_{90} is key at θ_{135}, θ_{180}, and θ_{225}. Furthermore, $r_{45} \times r_{00}$ has significant values for springback at θ_{090} because of interactions with Lankford coefficients. Meanwhile, $r_{90} \times r_{45}$ is the key factor of springback at θ_{270}. The measurement error also easily affected the range analyses, and so the ANOVA technique was used to analyze the springback. The ANOVA results shown in Table 9 correspond well with the range analysis results. Based on these analysis results, it has been found that the interactions of Lankford coefficients at different angles from the rolling direction have a clear effect on the springback.

Table 10. Results of range analysis and variance analysis.

	Range Analysis	Variance Analysis (The Key Factor and Contribution Value)
θ_{000}	r_{00}	$r_{00}(28.79\%)$, $r_{45}(18.94\%)$
θ_{045}	r_{45}	$r_{45}(37.53\%)$, $r_{45} \times r_{00}(23.21\%)$
θ_{090}	$r_{45} \times r_{00}$	$r_{45} \times r_{00}(54.46\%)$, $r_{45}(17.11\%)$
θ_{135}	r_{90}	$r_{90}(29.35\%)$, $r_{45}(20.66\%)$
θ_{180}	r_{90}	$r_{90}(48.00\%)$, $r_{45} \times r_{00}(23.13\%)$
θ_{225}	r_{90}	$r_{90}(47.89\%)$, $r_{45}(15.19\%)$
θ_{270}	$r_{90} \times r_{45}$	$r_{90} \times r_{45}(38.81\%)$, $r_{45} \times r_{00}(23.17\%)$
θ_{315}	r_{45}	$r_{45}(47.64\%)$, $r_{90} \times r_{45}(33.00\%)$

Figure 13a–h shows the sensitivity analysis of the effect of Lankford coefficients on springback. When the interactions of the Lankford coefficients were not considered, the amount of springback decreased with r_{90} and r_{45} at all angles, and the amount of springback decreased with increasing r_{00}, except at the angle of θ_{000}. When considering the interactions, the amount of springback increased with increasing $r_{90} \times r_{45}$. When the r_{45} and r_{90} increased simultaneously, the interaction of $r_{90} \times r_{45}$ hindered springback and caused it to decrease. The amount of springback increased with increasing $r_{45} \times r_{00}$, except at the angle of θ_{180}. The amount of springback decreased with increasing $r_{90} \times r_{00}$, except at the angles of θ_{045} and θ_{180}. The results show that the influences of the Lankford coefficient on springback at different angles are interrelated and interact with each other.

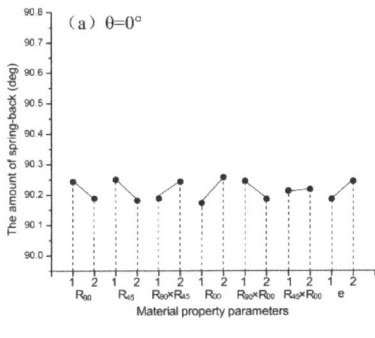
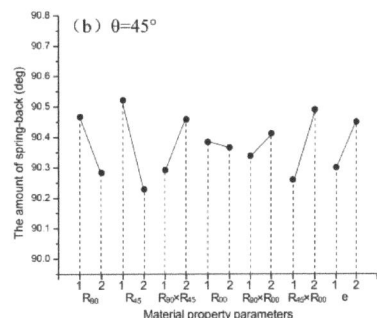
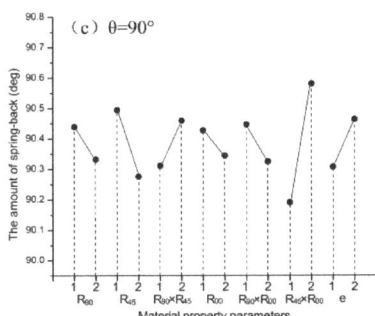
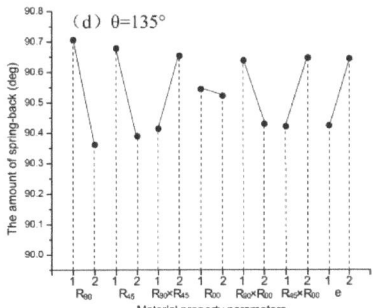

Figure 13. Cont.

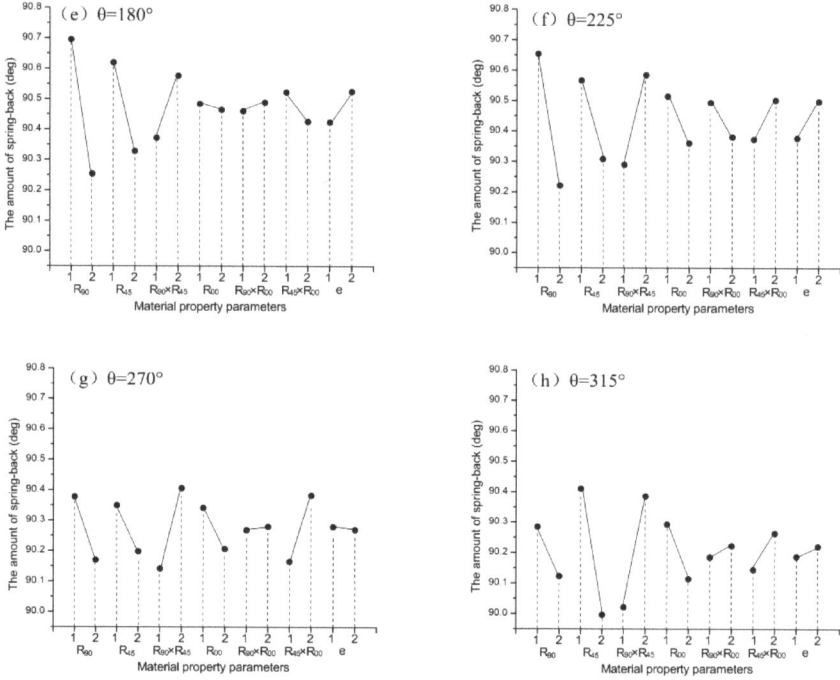

Figure 13. The sensitivity analysis of the effect of Lankford coefficients on springback.

4.3. Comparing the FEM Simulation and Experimental Results

The experimental results of two kinds of 304 stainless steel were measured using CMM, as shown in Table 11. The No. 1 and No. 8 FEM simulation results are shown in Table 12. Figure 14 shows the comparison of an average bottom fillet of the FEM simulation and experimental results at different angles from the rolling direction. The material with high-level Lankford coefficients had a larger amount of springback at 0, 90, and 270 degrees from the rolling direction. The experimental results have good agreement with the FEM simulation results, with the bottom fillet showing a similar trend. The springback of the cylinder bottom fillet occurred along the rolling direction, and there was an increasing trend with every 45° decrease.

Table 11. Average experimental results of two kinds of stainless steel.

Material	The Amount of Springback/(°)								Difference (Max–Min)
	θ_{000}	θ_{045}	θ_{090}	θ_{135}	θ_{180}	θ_{225}	θ_{270}	θ_{315}	
High–level Lankford coefficient	90.823	90.401	90.962	90.504	90.942	90.620	90.683	90.460	0.561
Low–level Lankford coefficient	90.756	90.529	90.845	90.700	91.009	90.792	90.671	90.565	0.48
Difference (Max–Min)	0.067	−0.128	0.117	−0.196	−0.389	−0.172	0.012	−0.105	

Table 12. FEM simulation results of springback.

Simulation No.	The Amount of Springback/(°)								Difference (Max–Min)
	θ_{000}	θ_{045}	θ_{090}	θ_{135}	θ_{180}	θ_{225}	θ_{270}	θ_{315}	
1	90.205	90.052	90.313	90.623	90.629	90.645	89.999	90.303	0.44
8	90.225	89.968	90.303	90.188	90.249	90.161	90.279	90.058	0.257
Difference (Max–Min)	0.02	−0.084	−0.01	−0.435	−0.38	−0.484	0.28	−0.245	

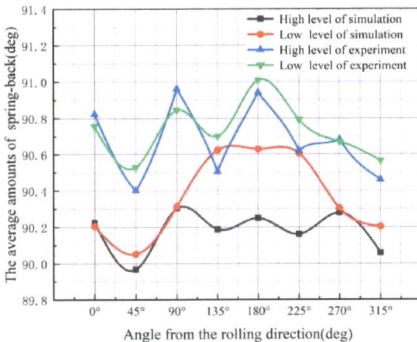

Figure 14. Comparison of the average of experimental results and the FEM simulation results.

The comparison between the bottom fillets of materials with high- and low-level Lankford coefficients is shown in Figure 15. The experimental results have good agreement with the FEM simulation results, showing a similar trend in springback. The amount of the bottom fillet decreased with an increase in the Lankford coefficient at all locations, except for 0, 90, and 270 degrees. The trend is more apparent in the FEM simulation results.

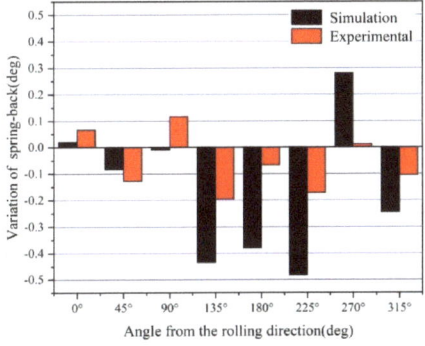

Figure 15. Comparison of the variation of angles between the FEM simulation results and the experimental results.

5. Conclusions

The influence of Lankford coefficients on stainless steel cylindrical cups was investigated using both experiments and numerical simulation. The conclusions are as follows:

The simulation and experimental results show that the Lankford coefficients had an obvious effect on the cross-section diameter. The flow velocity of the blank was different because of the anisotropy of the metal sheet, which makes the stress–strain values accumulate in different directions during the deep-drawing process, and finally causes a clear difference in springback. The simulated springback value for the straight wall was between 0.15 mm and 0.25 mm. The maximum springback value was at the position of

45° from the rolling direction. Specifically, the diameters at different height sections were related to the Lankford coefficients at different angles from the rolling direction, which were characterized by a concave valley in the 45 degree direction of the straight wall. The radius difference between the 45 degree rolling direction and the other two directions at each section height was between 0.1 mm and 0.3 mm.

The ANOVA results illustrated the influence of Lankford coefficients on the springback of the bottom fillet. The Lankford coefficient has different levels of effects on springback depending on the angle from the rolling direction. The 90-degree angle had the greatest influence, followed by the 45-degree, with 0 degrees having the least influence. The experimental results showed a similar trend to the simulation results. In addition, the springback of the bottom fillet decreased with the increasing overall Lankford coefficients.

The combination of the FEM simulation, the ANOVA technique, and the experimental study of the cylinder deep-drawing process is an effective method for studying the influence of Lankford coefficients on springback and predicting the final shape with high precision.

In this paper, the study of the effect of Lankford coefficients on springback of a cylindrical cup during the deep-drawing process remained at the macroscopic stage, and further analysis on the microscopic aspects was not carried out. The mechanism by which metal anisotropy influences the springback of a cylinder needs to be explored further. Although some characteristic rules regarding springback and cylindrical cup properties during deep drawing were obtained in this study, the analysis of the specific degree of influence of Lankford coefficients on springback properties is still in the preliminary stage. Therefore, the quantitative analysis of the influence of Lankford coefficients on the springback of the cylindrical cup during deep drawing will remain the focus of future research. Based on the previous research on the cylindrical cup, the springback of large complex thin-walled parts in deep drawing will be further explored.

Author Contributions: Conceptualization, F.W.; Methodology, Software, Visualization, and Writing, Y.H. and C.H.; Validation, Z.H.; Supervision, Project Administration, and Funding Acquisition, Z.Z. and F.W. All authors have read and agreed to the published version of the manuscript.

Funding: This research was funded by the National Natural Science Foundation of China (Grant No. 52175294 and No. 51705085), the State Engineering Research Center for Metallic Materials Net-shape Processing, South China University of Technology (Grant No. 2020011), and the Core Technology Research Foundation of Foshan City (Grant No. 1920001001369).

Institutional Review Board Statement: Not applicable.

Informed Consent Statement: Not applicable.

Data Availability Statement: The data presented in this study are available on request from the corresponding author. The data are not publicly available due to privacy.

Acknowledgments: The authors gratefully acknowledge the financial support from the National Natural Science Foundation of China (Grant No. 52175294 and No. 51705085), the State Engineering Research Center for Metallic Materials Net-shape Processing, South China University of Technology (Grant No. 2020011), and the Core Technology Research Foundation of Foshan City (Grant No. 1920001001369).

Conflicts of Interest: The authors declare no conflict of interest.

References

1. Trzepieciński, T. Recent Developments and Trends in Sheet Metal Forming. *Metals* **2020**, *10*, 779. [CrossRef]
2. Centeno, G.; Silva, M.B. Tube and Sheet Metal Forming Processes and Applications. *Metals* **2022**, *12*, 553. [CrossRef]
3. Guo, S.; Tian, C.; Pan, H.; Tang, X.; Han, L.; Wang, J. Research on the Springback Behavior of 316Ln Stainless Steel in Micro-Scale Bending Processes. *Materials* **2022**, *15*, 6373. [CrossRef] [PubMed]
4. Gia Hai, V.; Thi Hong Minh, N.; Nguyen, D.T. A Study on Experiment and Simulation to Predict the Spring-Back of Ss400 Steel Sheet in Large Radius of V-Bending Process. *Mater. Res. Express* **2020**, *7*, 16562. [CrossRef]
5. Kut, S.; Pasowicz, G.; Stachowicz, F. On the Springback and Load in Three-Point Air Bending of the Aw-2024 Aluminium Alloy Sheet with Aw-1050a Aluminium Cladding. *Materials* **2023**, *16*, 2945. [CrossRef]

6. Solfronk, P.; Sobotka, J.; Kolnerova, M.; Zuzanek, L. Spring-Back Prediction for Stampings from the Thin Stainless Sheets. *MM Sci. J.* **2016**, *2016*, 1090–1094. [CrossRef]
7. Ha, G.X.; Oliveira, M.G.; Andrade-Campos, A.; Manach, P.Y.; Thuillier, S. Prediction of Coupled 2D and 3D Effects in Springback of Copper Alloys After Deep Drawing. *Int. J. Mater. Form.* **2021**, *14*, 1171–1187. [CrossRef]
8. Ul Hassan, H.; Maqbool, F.; Güner, A.; Hartmaier, A.; Ben Khalifa, N.; Tekkaya, A.E. Springback Prediction and Reduction in Deep Drawing Under Influence of Unloading Modulus Degradation. *Int. J. Mater. Form.* **2016**, *9*, 619–633. [CrossRef]
9. Trzepiecinski, T.; Lemu, H.G. Effect of Computational Parameters on Springback Prediction by Numerical Simulation. *Metals* **2017**, *7*, 380. [CrossRef]
10. Huang, Y.; Chou, I.; Jiang, C.; Wu, Y.; Lee, S. Finite Element Analysis of Dental Implant Neck Effects on Primary Stability and Osseointegration in a Type Iv Bone Mandible. *Bio-Med. Mater. Eng.* **2014**, *24*, 1407–1415. [CrossRef]
11. Minh, N. Effect of Forming Parameters on Springback of Advanced High Strength Steel DP800. *Appl. Mech. Mater.* **2014**, *703*, 182–186. [CrossRef]
12. Hashemi, S.J.; Roohi, A.H. Minimizing Spring-Back and Thinning in Deep Drawing Process of St14 Steel Sheets. *Int. J. Interact. Des. Manuf.* **2022**, *16*, 381–388. [CrossRef]
13. Lajarin, S.F.; Marcondes, P.V. Influence of Process and Tool Parameters on Springback of High-Strength Steels. Proceedings of the Institution of Mechanical Engineers, Part B. *J. Eng. Manuf.* **2014**, *229*, 295–305. [CrossRef]
14. Starman, B.; Cafuta, G.; Mole, N. A Method for Simultaneous Optimization of Blank Shape and Forming Tool Geometry in Sheet Metal Forming Simulations. *Metals* **2021**, *11*, 544. [CrossRef]
15. Huang, X.; Guan, B.; Zang, Y.; Wang, B. Investigation of Defect Behavior During the Stamping of a Thin-Walled Semicircular Shell with Bending Angle. *J. Manuf. Process.* **2023**, *87*, 231–244. [CrossRef]
16. Aydın, K.; Karaağaç, İ.; Uluer, O. The Formability and Springback Characterization of Laser-Welded Dp–Hsla Sheets. *Appl. Phys. A* **2019**, *125*, 525. [CrossRef]
17. Saito, N.; Fukahori, M.; Hisano, D.; Hamasaki, H.; Yoshida, F. Effects of Temperature, Forming Speed and Stress Relaxation on Springback in Warm Forming of High Strength Steel Sheet. *Procedia Eng.* **2017**, *207*, 2394–2398. [CrossRef]
18. Chang, Y.; Wang, N.; Wang, B.T.; Li, X.D.; Wang, C.Y.; Zhao, K.M.; Dong, H. Prediction of Bending Springback of the Medium-Mn Steel Considering Elastic Modulus Attenuation. *J. Manuf. Process.* **2021**, *67*, 345–355. [CrossRef]
19. Ragai, I.; Lazim, D.; Nemes, J.A. Anisotropy and Springback in Draw-Bending of Stainless Steel 410: Experimental and Numerical Study. *J. Mater. Process. Technol.* **2005**, *166*, 116–127. [CrossRef]
20. Parsa, M.H.; Nasher Al Ahkami, S.; Pishbin, H.; Kazemi, M. Investigating Spring Back Phenomena in Double Curved Sheet Metals Forming. *Mater. Des.* **2012**, *41*, 326–337. [CrossRef]
21. Gomes, C.; Onipede, O.; Lovell, M. Investigation of Springback in High Strength Anisotropic Steels. *J. Mater. Process. Technol.* **2005**, *159*, 91–98. [CrossRef]
22. Leu, D.; Zhuang, Z. Springback Prediction of the Vee Bending Process for High-Strength Steel Sheets. *J. Mech. Sci. Technol.* **2016**, *30*, 1077–1084. [CrossRef]
23. Verma, R.K.; Haldar, A. Effect of Normal Anisotropy on Springback. *J. Mater. Process. Technol.* **2007**, *190*, 300–304. [CrossRef]
24. Lee, J.; Barlat, F.; Lee, M. Constitutive and Friction Modeling for Accurate Springback Analysis of Advanced High Strength Steel Sheets. *Int. J. Plast.* **2015**, *71*, 113–135. [CrossRef]
25. Jung, J.; Jun, S.; Lee, H.; Kim, B.; Lee, M.; Kim, J.H. Anisotropic Hardening Behaviour and Springback of Advanced High-Strength Steels. *Metals* **2017**, *7*, 480. [CrossRef]
26. Barlat, F.; Lian, J. Plastic behavior and stretch-ability of sheet metals. Part I A yield function for orthotropic sheet under plane stress conditions. *Int. J. Plast.* **1989**, *5*, 51–66. [CrossRef]
27. Zhang, D.; Bai, D.; Liu, J.; Guo, Z.; Guo, C. Formability Behaviors of 2a12 Thin-Wall Part Based on Dynaform and Stamping Experiment. *Compos. Part B Eng.* **2013**, *55*, 591–598. [CrossRef]
28. Panich, S.; Uthaisangsuk, V.; Suranuntchai, S.; Jirathearanat, S. Investigation of Anisotropic Plastic Deformation of Advanced High Strength Steel. *Mater. Sci. Eng. A* **2014**, *592*, 207–220. [CrossRef]
29. Zhang, F.F.; Chen, J.S.; Chen, J.; Huang, X.Z.; Lu, J. Review on development and experimental validation for anisotropic yield criterions. *Adv. Mech.* **2012**, *42*, 68–80.
30. Hosford, W.F. A Generalized Isotropic Yield Criterion. *J. Appl. Mech.* **1972**, *39*, 607–609. [CrossRef]
31. Huang, M.; Man, C. A Generalized Hosford Yield Function for Weakly-Textured Sheets of Cubic Metals. *Int. J. Plast.* **2013**, *41*, 97–123. [CrossRef]
32. Eklarkar, S.V.; Nandedkar, V.M. Fe Simulation and Optimization of Drawbead Parameters Using Taguchi Method for Hemispherical Cup. *Mater. Today Proc.* **2021**, *44*, 4709–4716. [CrossRef]
33. Shinge, V.R.; Dabade, U.A. Experimental Investigation on Forming Limit Diagram of Mild Carbon Steel Sheet. *Procedia Manuf.* **2018**, *20*, 141–146. [CrossRef]
34. Ailinei, I.; Galatanu, S.; Marsavina, L. Influence of Anisotropy on the Cold Bending of S600Mc Sheet Metal. *Eng. Fail. Anal.* **2022**, *137*, 106206. [CrossRef]

35. Džoja, M.; Cvitanić, V.; Safaei, M.; Krstulović-Opara, L. Modelling the Plastic Anisotropy Evolution of Aa5754-H22 Sheet and Implementation in Predicting Cylindrical Cup Drawing Process. *Eur. J. Mech. A Solids* **2019**, *77*, 103806. [CrossRef]
36. Thipprakmas, S. Application of Taguchi Technique to Investigation of Geometry and Position of V-Ring Indenter in Fine-Blanking Process. *Mater. Des.* **2010**, *31*, 2496–2500. [CrossRef]

Disclaimer/Publisher's Note: The statements, opinions and data contained in all publications are solely those of the individual author(s) and contributor(s) and not of MDPI and/or the editor(s). MDPI and/or the editor(s) disclaim responsibility for any injury to people or property resulting from any ideas, methods, instructions or products referred to in the content.

Article
Hot Deformation Behaviour of Additively Manufactured 18Ni-300 Maraging Steel

Błażej Tomiczek [1,*], Przemysław Snopiński [2], Wojciech Borek [2], Mariusz Król [2], Ana Romero Gutiérrez [3] and Grzegorz Matula [1]

1. Scientific and Didactic Laboratory of Nanotechnology and Material Technologies, Faculty of Mechanical Engineering, Silesian University of Technology, 44-100 Gliwice, Poland
2. Department of Engineering Materials and Biomaterials, Silesian University of Technology, 18A Konarskiego Street, 44-100 Gliwice, Poland
3. Escuela de Ingeniería Industrial y Aeroespacial, Institute of Applied Aeronautical Industry Research, Universidad de Castilla-La Mancha (UCLM), 45071 Toledo, Spain
* Correspondence: blazej.tomiczek@polsl.pl

Abstract: In this article, hot compression tests on the additively produced 18Ni-300 maraging steel 18Ni-300 were carried out on the Gleeble thermomechanical simulator in a wide temperature range (900–1200 °C) and at strain rates of 0.001 10 s^{-1}. The samples were microstructurally analysed by light microscopy and scanning electron microscopy with electron backscatter diffraction (EBSD). This showed that dynamic recrystallization (DRX) was predominant in the samples tested at high strain rates and high deformation temperatures. In contrast, dynamic recovery (DRV) dominated at lower deformation temperatures and strain rates. Subsequently, the material constants were evaluated in a constitutive relationship using the experimental flow stress data. The results confirmed that the specimens are well hot workable and, compared with the literature data, have similar activation energy for hot working as the conventionally fabricated specimens. The findings presented in this research article can be used to develop novel hybrid postprocessing technologies that enable single-stage net shape forging/forming of AM maraging steel parts at reduced forming forces and with improved density and mechanical properties.

Keywords: additive manufacturing; hot deformation; 18Ni-300 maraging steel; microstructure

1. Introduction

Maraging steels (MS) are attractive alloys for engineers because of their superior mechanical properties, corrosion resistance, and excellent fracture toughness. They are usually classified into groups according to their 0.2% proof stress or yield strength ranging from 200 to 350 in ksi (1400 to 2400 MPa), e.g., M200, M250, M300, and M350, respectively. Maraging steels are produced by a two-stage vacuum melting process in which the chemical composition and impurity content are strictly controlled [1]. With a proper ageing treatment, ultra-high strength MS can be produced, suitable for the aerospace-, tool- and die-making industries [2].

Recently, maraging steels have become one of the most interesting materials in rapid prototyping due to their good printability and excellent response to LPBF [3–5]. High-quality, nearly fully dense parts have been successfully produced. Furthermore, studies have shown that the very high cooling rates and layer-by-layer deposition during the selective laser melting process enable the formation of a unique and very fine cellular structure in AM maraging steels [6]. As a result, they generally exhibit higher yield strength, tensile strength and hardness than conventionally manufactured parts [7].

Thermomechanical postprocessing of additively manufactured parts has recently been considered a promising way to exploit the advantages of both processes [8,9]. Higher performance levels have been shown to be achieved by combining AM with conventional postprocessing methods. For example, Arconic has introduced the AmpliforgeTM process [10]. This is a hybrid technology that combines the benefits of additive manufacturing and advanced forging processes to produce stronger, tougher parts with less time, cost, and material waste. On the basis of the positive results of this technology, more attention needs to be paid to leveraging the benefits of both processes to incorporate them into new process chains to achieve a more favourable cost structure and better-performing materials.

Compared to traditional manufacturing processes that require multiple steps (for example, forging or rolling), metal 3D printing is more straightforward. It enables the production of near-net shape components with complex geometries that require little additional machining effort [11,12]. However, it also has multiple disadvantages such as slow build rates and limited component size [13]. Hybrid postprocessing, which combines plastic forming and additive manufacturing, makes it possible to partially eliminate the disadvantages of both technologies and make the forming process more energy- and cost-efficient (for low-volume production) [14–16].

Considering that for some metals, the energy required to process the scrap is greater than that required for the 3D printing process [17], it is clear that it is more cost-effective to forge an additively manufactured part than a cast or wrought alloy. However, to apply the hybrid method, the deformation behaviour and mechanism at high temperatures must first be understood. Until now, the deformation behaviour and mechanism of conventionally produced M250 [18], M300 [19], and M350 [20] at high temperatures have been studied by compression tests. Despite extensive studies on the annealing process [21], selective laser melting parameters [22], and plastic deformation under quasi-static loading conditions of LPBF-MS [23], the intrinsic workability and microstructure control by high-temperature deformation of 18Ni-300 alloys produced by the LPBF technique have not been studied in detail.

Due to the uniqueness of the rapid prototyping processes and to exploit the full potential of the AM 18Ni-300 alloy, it is of great importance to evaluate the specific microstructural features resulting from post-fabrication thermomechanical processing [24]. This is conducted mainly by isothermal compression tests on solid cylindrical specimens in the laboratory to select the process parameters and establish the main material constants in the constitutive relations. Employing the experimental data derived from hot compression tests and performing numerical simulations on the worksamples, it is possible to estimate the strain, strain rate, and temperature under nonisothermal conditions while using constitutive equations to accurately predict the flow stress over a wide range of hot deformation conditions.

The current work is an attempt to characterise the hot deformation behaviour of the AM 18Ni300 alloy by studying the effects of the conditions of the hot compression process on the evolution of the microstructure. This work is intended to provide a template that can be followed during hot working of additively manufactured 18Ni-300 maraging steel. With this in mind, the present study was carried out with the following specific objectives:

- To analyse the stress–strain and work-hardening behaviour;
- To provide constitutive equations allowing prediction of flow stress behaviour of AM 18Ni-300 maraging steel;
- To determine the dominant microstructural restoration mechanism.

The originality of this work lies in the determination of the constitutive equation describing the hot working behaviour (hot flow stress–strain relationship) of additively manufactured 18Ni300 maraging steel, and in the analysis of the influence of hot working parameters on softening mechanisms such as DRV/DRX and other microstructural features, such as grain size and morphology, which can then be used to establish a correlation between hot working parameters and the resulting mechanical properties.

2. Materials and Methods

In this article, 18Ni-300 maraging steel powder produced by BÖHLER was deposited using Renishaw's AM125 system to produce specimens. The process parameters given in Table 1 were used. The elemental composition of the spherical powder is given in Table 2.

A meander scanning strategy was used with a rotation of 67° after the deposition of each layer to fabricate samples with a diameter of 6 mm and a height of 10 mm.

Table 1. Selective laser melting process parameters used for the production of 18Ni-300 steel samples.

Power (W)	Layer Thickness, μm	Laser Speed, mm/s	Hatch Distance, mm
200	30	340	0.12

Table 2. Elemental composition of 18Ni-300 maraging steel powder.

Element	Fe	Ni	Co	Mo	Ti	Al	Cr	Cu	C	Mn	Si	P	S
Max %	Bal.	19.00	9.50	5.20	0.80	0.15	0.50	0.50	0.03	0.10	0.10	0.01	0.01
Min %	Bal.	17.00	8.5	4.50	0.60	0.05	–	–					

Isothermal compression tests were performed using a DSI (Dynamic System Inc.) Gleeble 3800 thermomechanical simulator on cylindrical specimens in the temperature range of 900 to 1200 °C with strain rates of 0.001, 0.01, 0.1, 1.0, and 10 s^{-1}. Prior to deformation, the specimens were rapidly heated to the desired temperature and then held for 5 min to remove the temperature gradient. All specimens were compressed to a true strain of 0.7 and then to preserve the deformed microstructures, cooled to room temperature for 60 s with compressed air. Figure 1 shows the comparison of the external surfaces of the hot compressed samples.

Figure 1. Comparison of the external surfaces of the hot compressed samples.

The hot compressed specimens were cut along the compression axis and then prepared using a standard metallographic technique. Etching with a 10% Nital solution (10 mL of HNO_3 + 90 mL C_2H_5OH) was used to reveal the details of the microstructure. Several sections of the specimens were polished and then viewed with an Axio Observer Z1 optical microscope (Carl Zeiss NTS GmbH, Oberkochen, Germany).

A field-emission Zeiss Supra 35 SEM microscope equipped with an EDAX EBSD system was used to analyse grain structure and orientation mapping. EBSD investigations were performed in a 200 × 200 μm^2 area with an accelerating voltage of 20 kV, a step size of 0.5–1 μm, and a tilt angle of 72°. ATEX 4.02 software was used to generate IPF maps, estimate GND density, and postprocess the EBSD data.

3. Results

3.1. Microstructure before Deformation

Figure 2 shows the typical microstructures of the 18Ni-300 maraging steel in the as-built (SLM) condition. Chemical etching shows distinct process-related features, such as melt pool boundary (melt line) crossing a previously melted area (see white dashed line) and pores, which is typical for additively produced maraging steel. On the basis of the microstructural analysis, it can be determined that the starting microstructure is composed of columnar martensite lamellae and residual austenite phases.

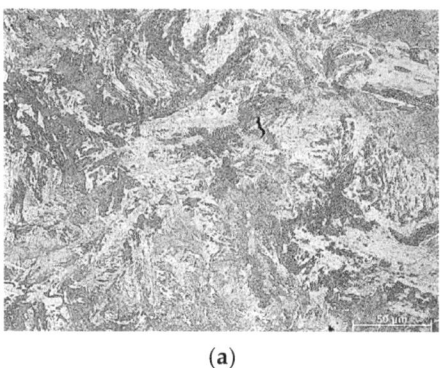
(a)

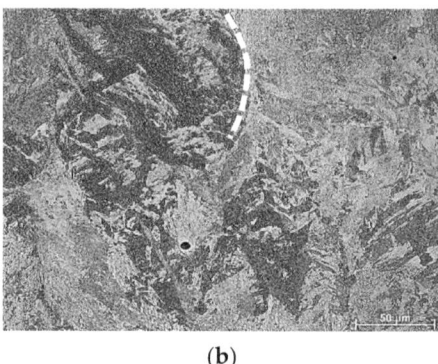
(b)

Figure 2. Microstructure of the 18Ni-300 maraging steel before deformation (**a**) cross-section (built) plane and (**b**) side plane.

To investigate only the effect of heating on the deformation temperature on the microstructure of AM 18Ni300 steel, the samples were annealed for 5 min at 900 and 1200 °C degrees and then immediately quenched with water. Figure 3a,b show typical optical micrographs of AM 18Ni300 samples subjected to the abovementioned procedure. To determine the effect of this annealing on the microstructure, the samples were alternatively etched in two reagents. As can be seen, using Nital solution etching revealed a typical martensitic microstructure almost without visible melt pool boundaries. However, the second reagent, a picric acid solution, revealed the microstructure characteristic of materials produced by the SLM technique (Figure 3c,d). The boundaries of the laser scan traces can be seen where cellular/columnar subgrains are present. It is worth noting that even higher annealing temperatures do not allow significant microstructural homogenization, because with longer heat treatment time, the typical features such as "laser scan trace boundaries" should disappear. Despite the high temperature, the heating time was insufficient for complete homogenisation and recrystallization of the microstructure. Significantly, given that lath martensite transformed to austenite microstructure during heating to hot deformation temperature, identification of prior-austenite grains (PAGS) are much more important [5,25].

The issue of austenitization temperature has already been analysed by two of the co-authors of this paper. In investigated steel, the austenite formation begins around 622–642 °C and austenite transformation finishes at 818–825 °C, depending on the applied heating rate. More details about all transformations taking place in analysed material during the heating and cooling cycle can be found in work [26]. Revealing the prior-austenite grain boundaries can be based on EBSD analysis. It has been found that the angles, primarily in the range of 20°–40° in the case of martensitic structures, correspond to the boundaries of the prior-austenite grains. This range of misorientation angles was also used for our microstructure, Figure 3e,f. It should be highlighted that the misorientation angles in the mentioned range did not reveal a continuous envelope at the prior-austenite grain boundaries (Figure 3g,h). However, the obtained results enable manual PAGS reconstruction based on both 20°–40° misorientation angles. The EBSD analysis reveals a difference in PAGS in the range of 8–10 μm. Apparently, the short annealing time and the inhomogeneity of the structure, typical for additively produced samples, did not allow for the significant growth of the austenite grain.

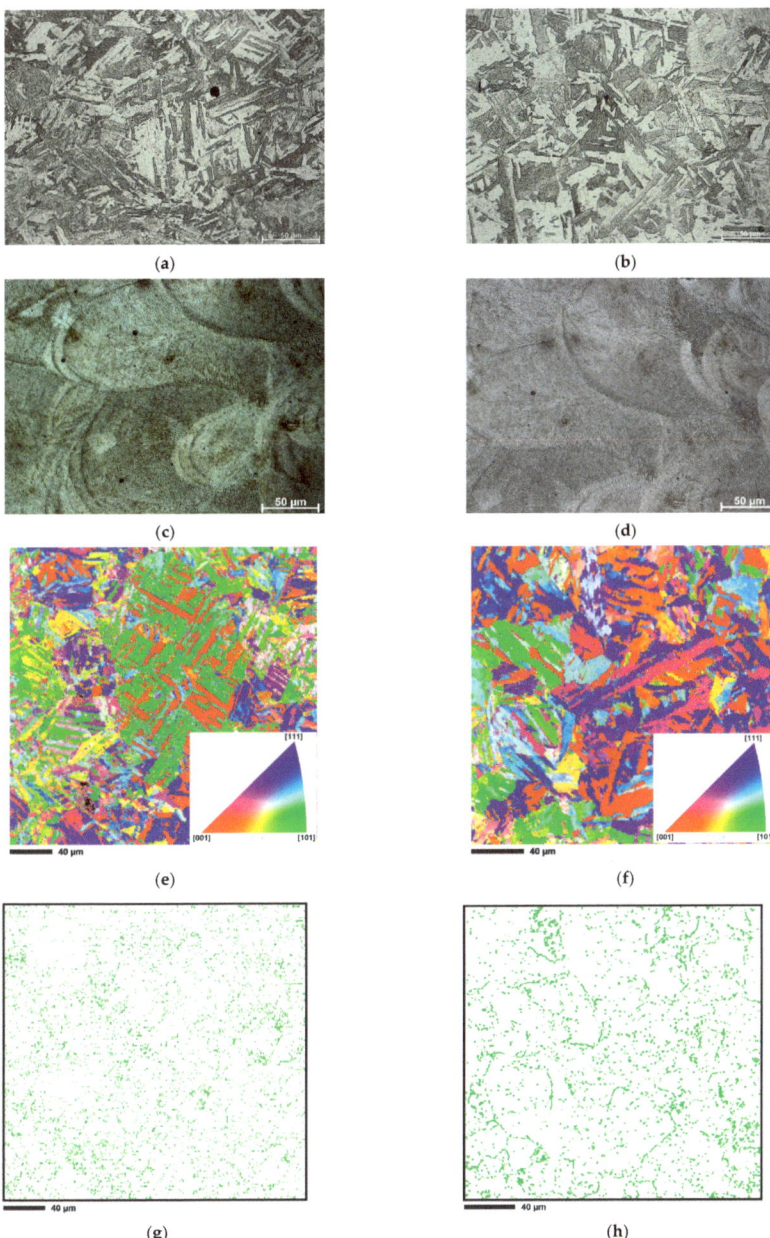

Figure 3. Microstructure of the 18Ni-300 maraging steel before deformation (**a**,**c**) after isothermal annealing at 900 °C, (**b**,**d**) at 1200 °C (**a**,**b**) etched in Nital and (**c**,**d**) saturated picric acid solution. Results of EBSD analysis and IPF-Z image of samples annealed at (**e**) 900 °C and (**f**) 1200 °C, IPF coloring triangles are shown in the lower right corners of figures, red [001], blue [111] and green [101], (**g**) grain boundary map (20–40°) of sample annealed at 900 °C, (**h**) grain boundary map (20–40°) of sample annealed at 1200 °C.

3.2. Flow Stress Behaviour

Figure 4 presents a series of typical stress–strain curves of AM 18Ni-300 maraging steel samples obtained under different deformation conditions during hot compression tests. It can be seen that the stress–strain curves at the strain rate of $10\ \text{s}^{-1}$ show significant stress drops at all deformation temperatures, which may be related to the low sampling frequency of the Gleeble thermomechanical simulator during the experiments, as well as insufficient movement control capability for high strain rate tests. A significant problem may also be interference related to the constant temperature control test during deformation, and as is known, high-speed deformation causes an increase in energy and thus an increase in temperature so that the device tries to react in a very short time in order to stabilise the temperature at a certain level. In general, deformation at high speeds using the Gleeble simulator is burdened with various disturbances that we are not always able to eliminate directly on the device, and only reverse analysis allows us to partially eliminate some forms of disorders recorded during the tests. For this reason, the data from these curves were not included in the next analysis.

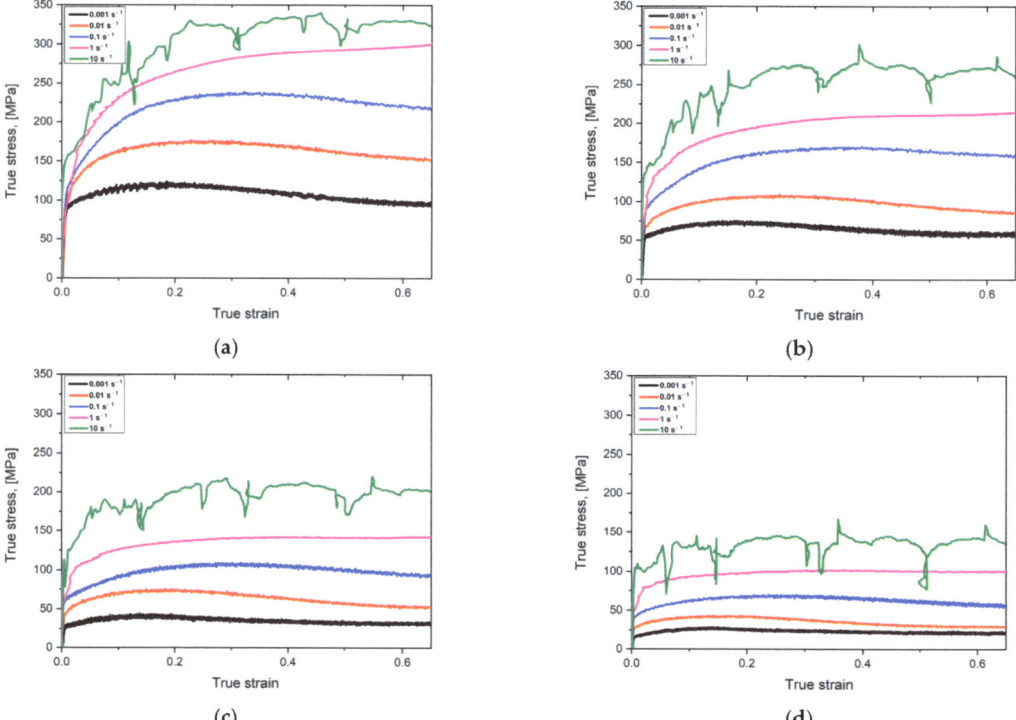

Figure 4. Typical stress–strain curves for the SLM M300 maraging steel obtained from uniaxial compression tests at various strain rates: (**a**) 900 °C, (**b**) 1000 °C, (**c**) 1100 °C, and (**d**) 1200 °C.

As expected, the decrease in the strain rate leads to a decrease in the flow stress. The main reason for this phenomenon is DRV (dynamic recovery) and DRX (dynamic recrystallization), which slow down the hardening at lower strain rates. Furthermore, the effect of the hot compression temperature on the flow stress is evident at any given strain rate. It can be clearly seen that the flow stress decreases significantly with increasing deformation temperature. This is due to the fact that the thermal activation and diffusion processes are more pronounced at high temperatures.

The next startling phenomenon is observed at a strain rate of 1 s^{-1}. Here, the flow stress increases rapidly during the initial deformation phase, then a transition to a relative equilibrium state is observed, and finally, the flow stress increases slowly with the actual strain increase, Figure 4a,b. This behaviour indicates an unstable phenomenon due to competition between the dynamic softening mechanism and the strain hardening. Other stress–strain curves show a classical dynamic recrystallisation (DRX) with a single stress peak followed by a gradual downward trend toward equilibrium.

According to Figure 4c,d, deformation at 1100 and 1200 °C at a strain rate of 1 s^{-1} results in largely flat flow stress without significant strain hardening or softening, indicating that DRV is the dominant softening mechanism. The remaining stress–strain curves show classic DRX behaviour with a single stress peak followed by a gradual decrease in flow stress with increasing strain.

3.3. Constitutive Analysis

Constitutive equations are usually derived to determine the relationship between hot deformation parameters such as flow stress and strain rate. This relationship is usually expressed by the Arrhenius-type constitutive equation. The exponent-type Zener–Hollomon parameter (Z) is mostly employed to characterise the correlation between hot compression temperature and strain rate on the flow behaviour of most metals. This parameter is expressed as follows:

$$Z = \dot{\varepsilon} \exp\left(\frac{Q}{RT}\right) \quad (1)$$

In the above equation, $\dot{\varepsilon}$ is the strain rate (s^{-1}), Q is the DRX activation energy (kJ/mol^{-1}), R is the universal gas constant (8.314 J/mol K), and T is the hot compression temperature (K).

Then the Arrhenius-type equation is used to show the influence of main deformation parameters on the flow stress behaviour. This equation is expressed as follows:

$$\dot{\varepsilon} = \begin{cases} A_1 \sigma^{n_1} & \alpha\sigma < 0.8 \\ A_2 \exp(\beta\sigma) & \alpha\sigma > 1.2 \\ A[\sinh(\alpha\sigma)]^n \exp\left(-\frac{Q}{RT}\right) & \text{for all } \sigma \end{cases} \quad (2)$$

Here, A, n, n_1, α, and β are material constants that are independent of temperature and strain rate, and σ (Mpa) is the peak stress. Accordingly, the material constants n_1 and β are derived from the slopes of $\ln(\dot{\varepsilon}) - \sigma$ and $\ln(\dot{\varepsilon}) - \ln(\sigma)$, respectively, as shown in Figure 5a,b. Here, n_1 and β are calculated as 6.13 and 0.06 Mpa^{-1}, respectively. The adjustable scaling factor α can then be calculated using the following formula $\alpha = \beta/n_1$, which gives an α-value of 0.0098, Table 3.

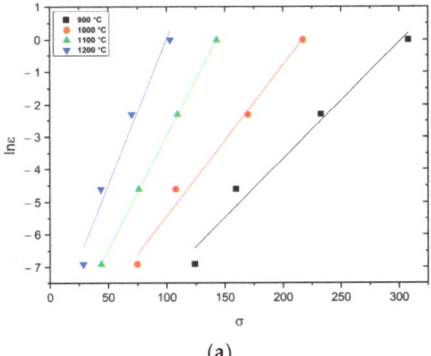
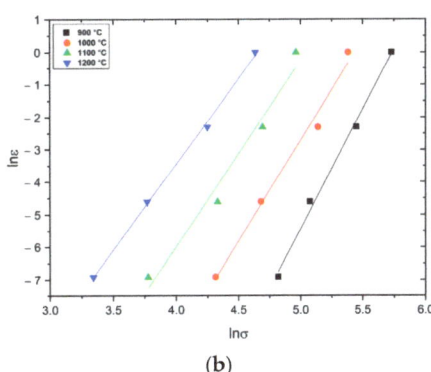

(a) (b)

Figure 5. Relationship between $\ln(\dot{\varepsilon})$ and (**a**) σ, (**b**) $\ln\sigma$.

Table 3. Detailed values of material constants and activation energy for the AM 18Ni-300 maraging steel.

Parameter	Value
α (MPa^{-1})	0.0098
n	4.13
S	11.04
Q (kJ/mol^{-1})	379
A	7.81×10^{12}

The n parameter can be derived from Equation (3), by conducting a linear fitting of $\ln(\dot{\varepsilon})$ against $\ln(\sinh(\alpha\sigma))$, Figure 6a.

$$n = \left[\frac{\partial \ln(\dot{\varepsilon})}{\partial \ln[\sinh(\alpha\sigma)]} \right]_T \quad (3)$$

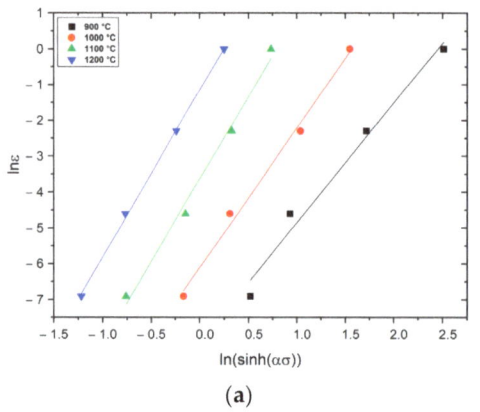 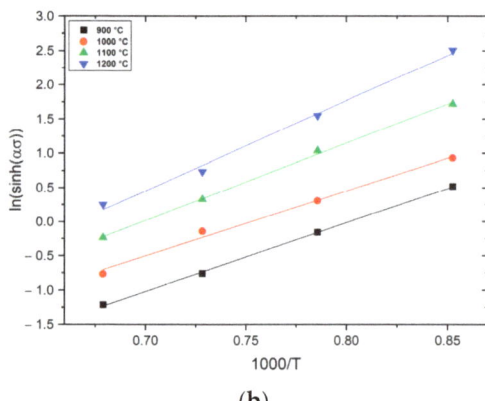

(a)   (b)

Figure 6. Relationship between (**a**) $\ln(\dot{\varepsilon})$ and $\ln\sinh(\alpha\sigma)$ and (**b**) $\ln\sinh(\alpha\sigma)$ and $1000/T$.

Then on the basis of Equation (2), at the constant of $\dot{\varepsilon}$, the activation energy (Q) can be expressed as follows:

$$Q = Rn \frac{\partial \ln[\sinh(\alpha\sigma)]}{\partial(1/T)} \quad (4)$$

In this equation, the $\partial \ln(\sinh(\alpha\sigma))/\partial(1/T)$ value (S) is obtained by plotting the $\ln(\sinh(\alpha\sigma))$ against $(1/T)$ at different $\dot{\varepsilon}$ and conducting a linear fitting, Figure 6b. Substituting the calculated values of n and α into Equation (4), the value of Q can be derived.

Then the parameter A can be calculated on the basis of the following equation:

$$\ln Z = \ln A + n \ln(\sinh(\alpha\sigma)) \quad (5)$$

The A value can be determined on the basis of the intercept of the linearly fitted line in the $\ln Z$ vs. $\ln(\sinh(\alpha\sigma))$ plot, Figure 7.

The values of A, n, and Q are derived using the formulas above, and when they are substituted into Equation (4) they give the following constitutive relationship for the hot working of AM 18Ni-300 maraging steel:

$$\dot{\varepsilon} = 7.81 \cdot 10^{12} \left[\sinh(0.0098\sigma) \right]^{4.13} \exp\left(-\frac{379 \cdot 10^3}{RT} \right) \quad (6)$$

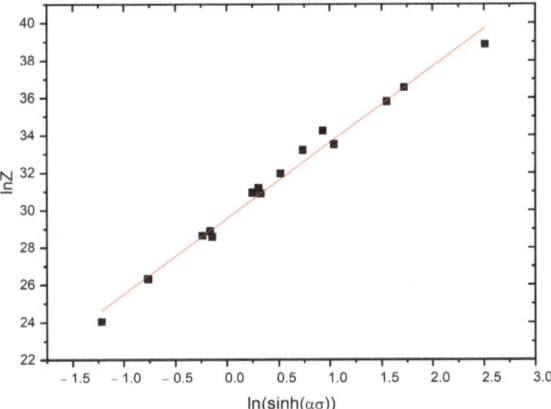

Figure 7. Relationship between ln Z and ln(sinh($\alpha\sigma$)) for the AM 18Ni-300 maraging steel (R2 value of 0.99).

According to the calculations performed in this research article, the deformation activation energy of the additively fabricated maraging steel 18-Ni300 is 379 kJ mol^{-1}. It is apparent that the calculated here activation energy value for AM 18Ni300 steel is similar to that of conventionally manufactured M300 grade (391 kJ/mol) [19], M350 grade (371 kJ/mol) [20] and is much lower than that of CF250 grade (458.8 kJ/mol) [27] in magnitude. Further, the activation energy value for AM 18Ni300 value is much larger than the values for self-diffusion in γ-iron (Q = 280 kJ mol^{-1}), implying that dynamic recovery and dynamic recrystallization are the dominant mechanisms instead of diffusion during hot deformation [28]. It is also worth noting that the calculated Q value for the maraging steel AM 18Ni-300 is similar to that of the conventionally produced grade M300 (Q = 390 kJ mol^{-1}) [19].

3.4. Microstructure after Deformation

The typical optical microscopic microstructures of the additively manufactured 18Ni-300 maraging steel deformed at different deformation conditions are given in Figure 8. For purposes of comparison, all microstructures were captured at the centre of the cross-section of the deformed specimens, and all micrographs were captured at the same magnification. Each microstructure shows the combined effect of temperature and strain rate at the end of deformation.

At the lowest deformation temperature, the microstructure consists mainly of strain-hardened grains (DRV). When the hot deformation temperature increases at a certain strain rate, the dynamically recovered grains are gradually transformed into recrystallized grains as thermally activated processes dominate. When the specimens are deformed in the temperature range of 1100–1200 °C and at a strain rate of 0.001–0.01 s^{-1}, the typical deformed grains are replaced by coarsely recrystallized grains (Figure 8).

Similarly, the effect of strain rate is evident at a given hot compression temperature. The material displays flow-softening behaviour at low strain rates and high temperatures because at high temperatures there is enough time to reduce the system's overall energy [1,20]. whereas in the temperature range of 1100–1200 °C and the strain rate range of 0.1–10 s^{-1}, DRX microstructures can be clearly distinguished. It is worth noting that similar behaviour was reported for M350 grade maraging steel [29].

To obtain a more comprehensive characterization of microstructural evolution and changes in grain orientation, EBSD studies were conducted. Figure 9 shows the EBSD maps, where the red lines represent interfaces with a low angle of misorientation (2°–15°), while the green lines represent interfaces with a higher angle of misorientation (>15°).

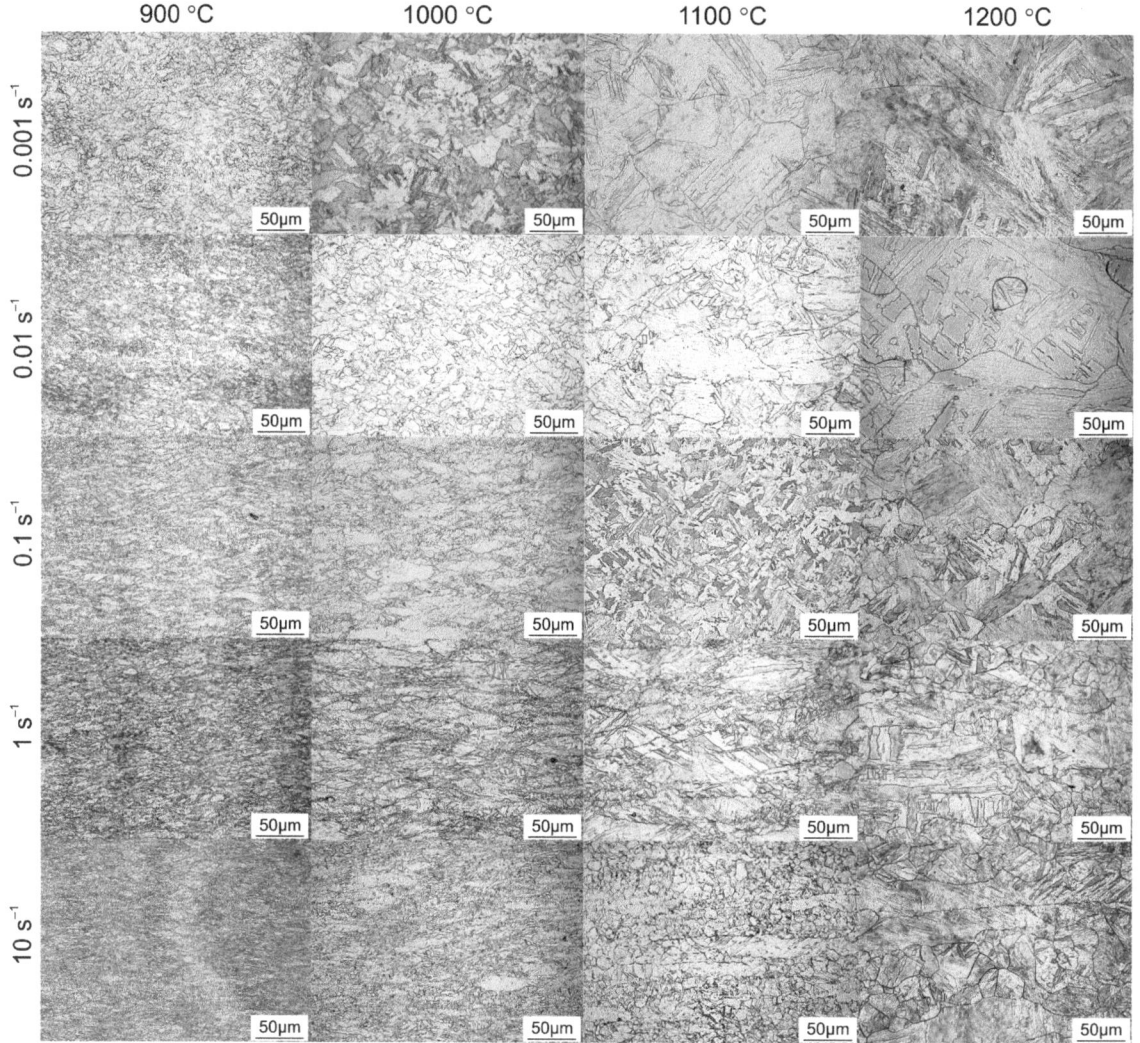

Figure 8. Light microscopy microstructures of hot compression-tested samples of AM 18Ni-300 maraging steel.

At a low deformation temperature of 900 °C and a moderate strain rate of 0.1 s^{-1}, finer grains with a significant fraction (51.8%) of interfaces with a low angle of misorientation are seen in a microstructure. This relatively high number of LAGBs can be associated with a higher accumulation of dislocations. As shown in Figure 9 and Table 4, when the strain rate is 0.1 s^{-1} and the hot compression temperature is increased to 1200 °C, the percentage of HGABs also increases to 65.1%. This means that the DRX degree is larger at higher deformation temperatures and moderate strain rates than for the specimens deformed at lower temperatures and lower strain rates. The increase in deformation temperature is also accompanied by significant grain growth. As can be seen from Table 4, the grain size increases from 1.8 to 6.5 μm.

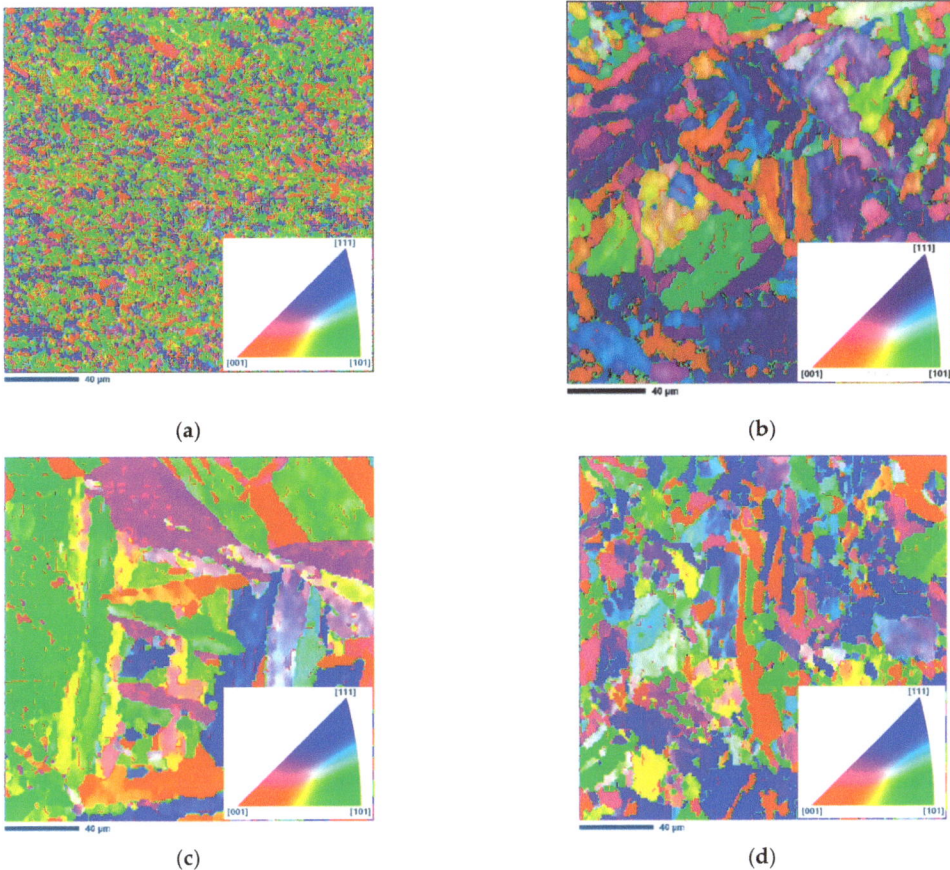

Figure 9. IPF-Z images of 18Ni-300 maraging steel samples (**a**) 900 °C 0.1 s^{-1}, (**b**) 1200 °C 0.1 s^{-1}, (**c**) 1200 °C 0.001 s^{-1}, and (**d**) 1200 °C 10 s^{-1}. IPF coloring triangles are shown in the lower right corners of figures, red [001], blue [111] and green [101].

Table 4. Main microstructural parameters derived from electron backscatter diffraction.

Sample	f_{HAGBs}	f_{LAGBs}	Grain Size, μm	Average KAM (°)
900 °C 0.1 s^{-1}	48.2	51.8	1.8	1.12
1200 °C 0.1 s^{-1}	75.6	24.4	6.5	0.68
1200 °C 0.001 s^{-1}	65.1	34.9	8.1	0.54
1200 °C 10 s^{-1}	76.1	23.9	5.2	0.98

At any given deformation temperature, the effect of strain rate is also clearly seen. At the lowest strain rate considered in this study of 0.001 s^{-1} and a hot compression temperature of 1200 °C, the measured average grain size is 8.1 μm. When the strain rate is increased to 10 s^{-1}, the grain size decreases to 5.2 μm. Meanwhile, the percentage of HAGBs also increases with increasing strain rate. For the samples with a strain rate of 0.001 s^{-1} at 1200 °C and 10 s^{-1} at 1200 °C, it increased from 65.1% to 76.1%, confirming some occurrence of DRX.

From the inverse pole figures of the specimens deformed at 1200 °C, it can be concluded that the thickness of the martensite plates is also affected by the strain rate. It can be seen that the width of the martensite plates increases with decreasing strain rate. At the highest strain rate of 10 s^{-1}, the competition between dynamic recovery and dynamic recrystallization is also evident, so the microstructure consists of relatively thin martensite plates/lamellae with a considerable amount of ultrafine DRX grains (Figure 9d).

Figure 10 shows the corresponding Kernel Average Misorientation (KAM) maps of selected AM 18Ni-300 maraging steel samples. The high KAM values represent strain accumulation (high stored energy), while the lower KAM values represent (dislocation free) recrystallized grains [30]. As can be deduced from Table 4 and Figure 10a, the highest KAM value of 1.12° shows a specimen compressed at 900 °C and a moderate strain rate of 0.1 s^{-1} with no evidence of dynamic recrystallization. This indicates the presence of a network of accumulated dislocations and confirms the dislocation slip activities [31,32]. It is also worth noting that KAM evolves inhomogeneously in this sample. The highest KAM values are found in the smallest grains and close to the boundaries, which are therefore thought to store the largest amount of dislocations.

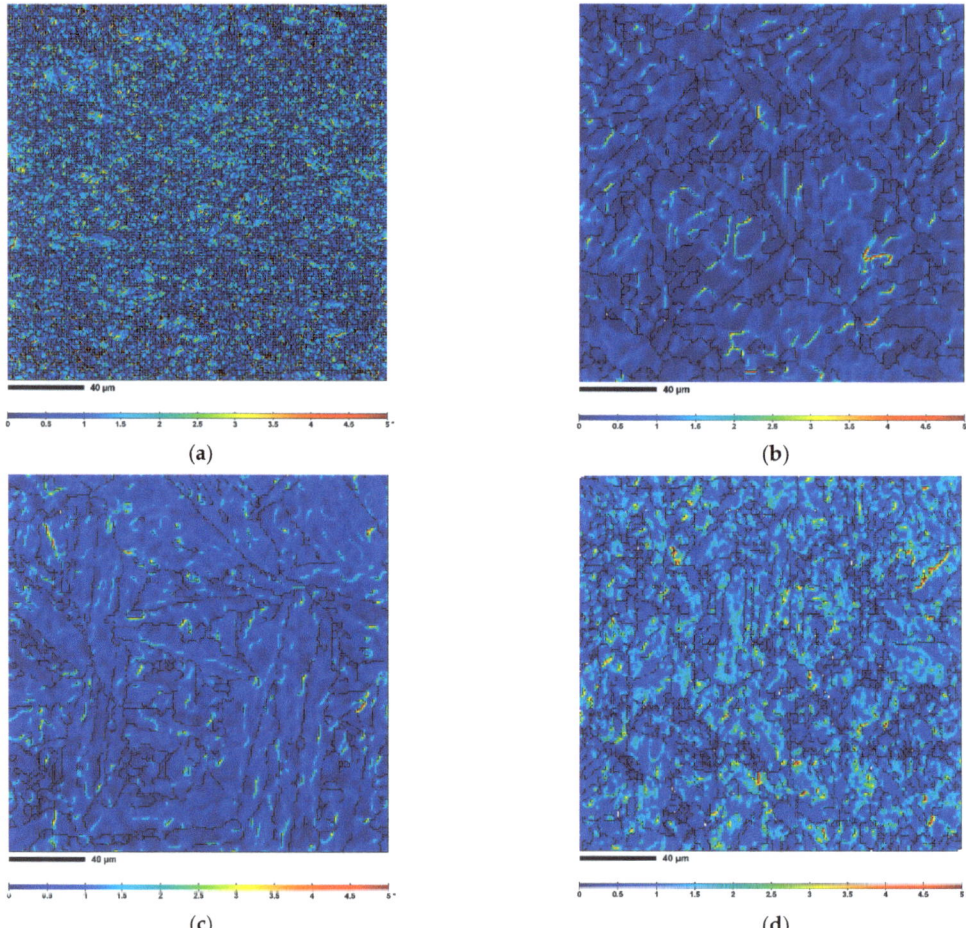

Figure 10. EBSD KAM maps of AM 18Ni-300 maraging steel samples (**a**) 900 °C 0.1 s^{-1}, (**b**) 1200 °C 0.1 s^{-1}, (**c**) 1200 °C 0.001 s^{-1}, (**d**) 1200 °C 10 s^{-1} (black lines correspond to the grain boundaries).

The 1200 °C specimen deformed at the same strain rate of 0.1 s^{-1} has a relatively low KAM value of 0.68°, indicating that the higher deformation temperature allows the release of stored deformation energy. It is also worth noting that the effect of strain rate is also clearly visible in the KAM maps. As the strain rate decreases (Figure 10c), the value of KAM also decreases slightly to 0.54°, indicating the lowest dislocation density and the lowest slip activity. On the other hand, the highest value of KAM of 0.98° at a strain rate of 10 s^{-1} indicates the highest dislocation accumulation, which means that the driving force for dynamic recrystallization (DRX) is also highest (the insufficient dislocation density cannot activate dynamic recrystallization process). This is can explain why more dynamically recrystallized grains developed at this strain rate (see Figure 9d).

4. Conclusions

Additive manufactured 18Ni-300 maraging steel samples were hot compressed to investigate intrinsic workability, hot working properties and microstructure development. The following conclusions are made based on the above analysis:

- The hot deformation experiments were performed at strains of 0.7 at 900, 1000, 1100, and 1200 °C with initial strain rates of 0.001, 0.01, 0.1, 1, and 10 s^{-1}. Typically, the flow stress of the additive manufactured 18Ni-300 maraging steel decreased with increasing temperature and increased with increasing strain rate.
- Dynamic recrystallization is the softening mechanism occurring at high temperatures and at intermediate/high strain rates as confirmed by the EBSD study.
- The constitutive relation corresponding to the peak flow stress is:

$$\dot{\varepsilon} = 7.81 \cdot 10^{12} \left[\sinh\left(0.0098\sigma\right) \right]^{4.13} \exp\left(-\frac{379 \cdot 10^3}{RT}\right)$$

in which the activation energy, Q = 379 kJ/mol.

- With increasing deformation temperature, the grain size increases, which is accompanied by an increase in the width of the martensite laths.

Author Contributions: Conceptualization, B.T. and W.B.; methodology, W.B. and M.K.; validation, A.R.G.; investigation, B.T., W.B., M.K., G.M. and P.S.; resources, M.K.; data curation, B.T., W.B., M.K., G.M. and P.S.; writing—original draft preparation, B.T., W.B., M.K., G.M. and P.S.; writing—review and editing, A.R.G.; project administration, B.T.; funding acquisition, B.T. All authors have read and agreed to the published version of the manuscript.

Funding: The research was carried out as part of project no LIDER/49/0196/L-9/17/NCBR/2018 financed by the National Center for Research and Development of Poland.

Institutional Review Board Statement: Not applicable.

Informed Consent Statement: Not applicable.

Data Availability Statement: All the raw data supporting the conclusion of this paper were provided by the authors.

Conflicts of Interest: The authors declare no conflict of interest.

References

1. Ansari, S.S.; Mukhopadhyay, J.; Murty, S.V.S.N. Analysis of Stress-Strain Curves to Predict Dynamic Recrystallization During Hot Deformation of M300 Grade Maraging Steel. *J. Mater. Eng. Perform.* **2021**, *30*, 5557–5567. [CrossRef]
2. Dehgahi, S.; Pirgazi, H.; Sanjari, M.; Seraj, P.; Odeshi, A.; Kestens, L.; Mohammadi, M. High strain rate torsional response of maraging steel parts produced by laser powder bed fusion techniques: Deformation behavior and constitutive model. *Mech. Mater.* **2022**, *168*, 104296. [CrossRef]
3. Chadha, K.; Tian, Y.; Bocher, P.; Spray, J.G.; Aranas, J.C. Microstructure Evolution, Mechanical Properties and Deformation Behavior of an Additively Manufactured Maraging Steel. *Materials* **2020**, *13*, 2380. [CrossRef] [PubMed]
4. Ong, J.K.; Tan, Q.Y.; Silva, A.; Tan, C.C.; Chew, L.T.; Wang, S.; Stanley, C.; Vastola, G.; Tan, U.-X. Effect of process parameters and build orientations on the mechanical properties of maraging steel (18Ni-300) parts printed by selective laser melting. *Mater. Today Proc.* **2022**, *70*, 438–442. [CrossRef]
5. Sanjari, M.; Mahmoudiniya, M.; Pirgazi, H.; Tamimi, S.; Ghoncheh, M.H.; Shahriairi, A.; Hadadzadeh, A.; Amirkhiz, B.S.; Purdy, M.; de Araujo, E.G.; et al. Microstructure, texture, and anisotropic mechanical behavior of selective laser melted maraging stainless steels. *Mater. Charact.* **2022**, *192*, 112185. [CrossRef]

6. Król, M.; Snopiński, P.; Hajnyš, J.; Pagáč, M.; Łukowiec, D. Selective Laser Melting of 18NI-300 Maraging Steel. *Materials* **2020**, *13*, 4268. [CrossRef]
7. Cruces, A.; Branco, R.; Borrego, L.; Lopez-Crespo, P. Energy-based critical plane fatigue methods applied to additively manufactured 18Ni300 steel. *Int. J. Fatigue* **2023**, *170*, 107548. [CrossRef]
8. Snopiński, P.; Appiah, A.N.S.; Hilšer, O.; Kotoul, M. Investigation of Microstructure and Mechanical Properties of SLM-Fabricated AlSi10Mg Alloy Post-Processed Using Equal Channel Angular Pressing (ECAP). *Materials* **2022**, *15*, 7940. [CrossRef] [PubMed]
9. Snopiński, P.; Matus, K.; Tatiček, F.; Rusz, S. Overcoming the strength-ductility trade-off in additively manufactured AlSi10Mg alloy by ECAP processing. *J. Alloy. Compd.* **2022**, *918*, 9165817. [CrossRef]
10. Habibiyan, A.; Hanzaki, A.Z.; Abedi, H.R. An investigation into microstructure and high-temperature mechanical properties of selective laser-melted 316L stainless steel toward the development of hybrid Ampliforge process. *Int. J. Adv. Manuf. Technol.* **2020**, *110*, 383–394. [CrossRef]
11. Fette, M.; Sander, P.; Wulfsberg, J.; Zierk, H.; Herrmann, A.; Stoess, N. Optimized and Cost-efficient Compression Molds Manufactured by Selective Laser Melting for the Production of Thermoset Fiber Reinforced Plastic Aircraft Components. *Procedia CIRP* **2015**, *35*, 25–30. [CrossRef]
12. Bambach, M.; Sizova, I.; Emdadi, A. Development of a processing route for Ti-6Al-4V forgings based on preforms made by selective laser melting. *J. Manuf. Process.* **2019**, *37*, 150–158. [CrossRef]
13. Duda, T.; Raghavan, L.V. 3D metal printing technology: The need to re-invent design practice. *AI Soc.* **2018**, *33*, 241–252. [CrossRef]
14. Jiang, J.; Hooper, P.; Li, N.; Luan, Q.; Hopper, C.; Ganapathy, M.; Lin, J. An integrated method for net-shape manufacturing components combining 3D additive manufacturing and compressive forming processes. *Procedia Eng.* **2017**, *207*, 1182–1187. [CrossRef]
15. Shakil, S.I.; Smith, N.R.; Yoder, S.P.; Ross, B.E.; Alvarado, D.J.; Hadadzadeh, A.; Haghshenas, M. Post fabrication thermomechanical processing of additive manufactured metals: A review. *J. Manuf. Process.* **2022**, *73*, 757–790. [CrossRef]
16. Bambach, M.; Sizova, I.; Sydow, B.; Hemes, S.; Meiners, F. Hybrid manufacturing of components from Ti-6Al-4V by metal forming and wire-arc additive manufacturing. *J. Mater. Process. Technol.* **2020**, *282*, 116689. [CrossRef]
17. Jackson, M.A.; Van Asten, A.; Morrow, J.D.; Min, S.; Pfefferkorn, F.E. Energy Consumption Model for Additive-Subtractive Manufacturing Processes with Case Study. *Int. J. Precis. Eng. Manuf. Technol.* **2018**, *5*, 459–466. [CrossRef]
18. Chakravarthi, K.V.A.; Koundinya, N.T.B.N.; Sarkar, A.; Murty, S.V.S.N.; Rao, B.N. Optimization of Hot Workability and Control of Microstructure in 18Ni (M250 Grade) Maraging Steel Using Processing Maps. *Mater. Perform. Charact.* **2018**, *7*, 20180082. [CrossRef]
19. Chakravarthi, K.; Koundinya, N.; Murty, S.N.; Rao, B.N. Microstructure, properties and hot workability of M300 grade maraging steel. *Def. Technol.* **2018**, *14*, 51–58. [CrossRef]
20. Chakravarthi, K.V.A.; Koundinya, N.T.B.N.; Murty, S.V.S.N.; Rao, B.N. Microstructural Evolution and Constitutive Relationship of M350 Grade Maraging Steel During Hot Deformation. *J. Mater. Eng. Perform.* **2017**, *26*, 1174–1185. [CrossRef]
21. Kučerová, L.; Burdová, K.; Jeníček, S.; Chena, I. Effect of solution annealing and precipitation hardening at 250 °C–550 °C on microstructure and mechanical properties of additively manufactured 1.2709 maraging steel. *Mater. Sci. Eng. A* **2021**, *814*, 141195. [CrossRef]
22. Tan, C.; Ma, W.; Deng, C.; Zhang, D.; Zhou, K. Additive manufacturing SiC-reinforced maraging steel: Parameter optimisation, microstructure and properties. *Adv. Powder Mater.* **2023**, *2*, 100076. [CrossRef]
23. Shamsdini, S.; Pirgazi, H.; Ghoncheh, M.; Sanjari, M.; Amirkhiz, B.S.; Kestens, L.; Mohammadi, M. A relationship between the build and texture orientation in tensile loading of the additively manufactured maraging steels. *Addit. Manuf.* **2021**, *41*, 101954. [CrossRef]
24. Shamsdini, S.; Ghoncheh, M.H.; Sanjari, M.; Pirgazi, H.; Amirkhiz, B.S.; Kestens, L.; Mohammadi, M. Plastic deformation throughout strain-induced phase transformation in additively manufactured maraging steels. *Mater. Des.* **2021**, *198*, 109289. [CrossRef]
25. Casati, R.; Lemke, J.N.; Tuissi, A.; Vedani, M. Aging Behaviour and Mechanical Performance of 18-Ni 300 Steel Processed by Selective Laser Melting. *Metals* **2016**, *6*, 218. [CrossRef]
26. Król, M.; Snopiński, P.; Czech, A. The phase transitions in selective laser-melted 18-NI (300-grade) maraging steel. *J. Therm. Anal. Calorim.* **2020**, *142*, 1011–1018. [CrossRef]
27. Chakravarthi, K.V.A.; Koundinya, N.T.B.N.; Murty, S.V.S.N.; Sivakumar, D.; Rao, B.N. Optimization of Hot Workability and Control of Microstructure in CF250 Grade Cobalt-Free Maraging Steel: An Approach Using Processing Maps. *Met. Microstruct. Anal.* **2018**, *7*, 35–47. [CrossRef]
28. Zhang, L.; Wang, W.; Shahzad, M.B.; Shan, Y.-Y.; Yang, K. Hot Deformation Behavior of an Ultra-High-Strength Fe–Ni–Co-Based Maraging Steel. *Acta Met. Sin.* **2019**, *32*, 1161–1172. [CrossRef]
29. Ansari, S.S.; Chakravarthi, K.V.A.; Murty, S.V.S.N.; Rao, B.N.; Mukhopadhyay, J. Hot Workability and Microstructure Control through the Analysis of Stress–Strain Curves during Hot Deformation of M350 Grade Maraging Steel. *Mater. Perform. Charact.* **2019**, *8*, 20190030. [CrossRef]
30. Hadadzadeh, A.; Amirkhiz, B.S.; Li, J.; Odeshi, A.; Mohammadi, M. Deformation mechanism during dynamic loading of an additively manufactured AlSi10Mg_200C. *Mater. Sci. Eng. A* **2018**, *722*, 263–268. [CrossRef]

31. Dehgahi, S.; Alaghmandfard, R.; Tallon, J.; Odeshi, A.; Mohammadi, M. Microstructural evolution and high strain rate compressive behavior of as-built and heat-treated additively manufactured maraging steels. *Mater. Sci. Eng. A* **2021**, *815*, 141183. [CrossRef]
32. Dehgahi, S.; Pirgazi, H.; Sanjari, M.; Alaghmandfard, R.; Tallon, J.; Odeshi, A.; Kestens, L.; Mohammadi, M. Texture evolution during high strain-rate compressive loading of maraging steels produced by laser powder bed fusion. *Mater. Charact.* **2021**, *178*, 111266. [CrossRef]

Disclaimer/Publisher's Note: The statements, opinions and data contained in all publications are solely those of the individual author(s) and contributor(s) and not of MDPI and/or the editor(s). MDPI and/or the editor(s) disclaim responsibility for any injury to people or property resulting from any ideas, methods, instructions or products referred to in the content.

Article

Development of Temperature-Controlled Shear Tests to Reproduce White-Etching-Layer Formation in Pearlitic Rail Steel

Léo Thiercelin [1,*], Sophie Cazottes [2,*], Aurélien Saulot [3], Frédéric Lebon [4], Florian Mercier [2], Christophe Le Bourlot [2], Sylvain Dancette [2] and Damien Fabrègue [2]

1 Arts et Métiers Institute of Technology, CNRS, Université de Lorraine, LEM3-UMR 7239, F-57078 Metz, France
2 Université de Lyon, INSA Lyon, CNRS UMR 5510, MATEIS, F-69621 Villeurbanne, France
3 Université de Lyon, INSA Lyon, CNRS UMR 5259 LaMCoS, F-69621 Villeurbanne, France
4 Aix Marseille Université, CNRS, Centrale Marseille, LMA UMR 7031, F-13453 Marseille, France
* Correspondence: leo.thiercelin@ensam.eu (L.T.); sophie.cazottes@insa-lyon.fr (S.C.)

Citation: Thiercelin, L.; Cazottes, S.; Saulot, A.; Lebon, F.; Mercier, F.; Le Bourlot, C.; Dancette, S.; Fabrègue, D. Development of Temperature-Controlled Shear Tests to Reproduce White-Etching-Layer Formation in Pearlitic Rail Steel. *Materials* **2022**, *15*, 6590. https://doi.org/10.3390/ma15196590

Academic Editor: Wojciech Borek

Received: 1 September 2022
Accepted: 19 September 2022
Published: 22 September 2022

Publisher's Note: MDPI stays neutral with regard to jurisdictional claims in published maps and institutional affiliations.

Copyright: © 2022 by the authors. Licensee MDPI, Basel, Switzerland. This article is an open access article distributed under the terms and conditions of the Creative Commons Attribution (CC BY) license (https://creativecommons.org/licenses/by/4.0/).

Abstract: The formation of a white etching layer (WEL), a very hard and brittle phase on the rail surface, is associated with a progressive transformation of the pearlitic grain to very fragmented grains due to the cumulative passage of trains. Its formation is associated with a complex thermomechanical coupling. To predict the exact conditions of WEL formation, a thermomechanical model previously proposed by the authors needs to be validated. In this study, monotonic and cyclic shear tests using hat-shaped specimens were conducted in the temperature range of 20 °C to 400 °C to reproduce the WEL formation. The tests showed a strong sensitivity of the material to temperature, which does not necessarily favor WEL formation. For the monotonic tests, no WELs were produced; however, a localization of the plastic deformation was observed for tests performed at 200 °C and 300 °C. In this temperature range, the material was less ductile than at room temperature, leading to failure before WEL formation. At 400 °C, the material exhibited a much more ductile behavior, and nanograins close to WEL stages were visible. For the cyclic tests, a WEL zone was successfully reproduced at room temperature only and confirmed the effect of shear in WEL formation. The same cyclic tests conducted at 200 °C and 300 °C yielded results consistent with those of the monotonic tests; the deformation was much more localized and did not lead to WEL formation.

Keywords: hat-shaped specimen; shear stress; pearlitic steel; thermomechanical test; dynamic recovery; white etching layer

1. Introduction

The increase in rail traffic and cumulative tonnage over the past decades has led to an increase in rolling-contact fatigue defects on the rail surface [1]. The emergence of this increasing number of defects is generally associated with microstructural changes on the rail surface, such as severe plastic deformation and the formation of new, harder, and more brittle phases commonly known as brown and white etching layers (BEL and WEL, respectively) [2].

Understanding the kinetics of WEL formation is complex, as it depends on contact conditions such as thermomechanical, cyclic, multiaxial, and dynamic loading. Therefore, the thermomechanical path leading to WEL formation is not unique [3,4]. Nevertheless, microstructural characterization of the rail surface has provided evidence that the temperature and/or mechanical-stress field are the driving forces in WEL formation.

On the one hand, a purely thermal mechanism of WEL formation has been proposed by several authors [5–10]. Indeed, under high-braking conditions with high wheel sliding [3], the rail surface undergoes thermal cycles of heating with temperatures above 700 °C followed by rapid cooling that could progressively produce WEL spots. In addition, Nakkalil [11] explained that dynamic loading applied to the rail surface induces localized plasticity and would lead to the formation of multiple adiabatic shear bands in the rail surface. These shear bands would then accumulate to form a homogeneous WEL, as also explained by Baumann [12]. This process would explain the presence of martensite and residual austenite within the thermal WEL [13].

On the other hand, under conventional traffic conditions on aligned tracks, the relative sliding ratio between the wheel and rail is generally limited to 2% by anti-skid devices [14], which limits the contact temperature to only several hundred degrees [15,16]. Therefore, wheel–rail contact tests performed at a sliding ratio above 2% are not representative of the wheel–rail contact and highlight the thermal WEL formation [17,18]. WELs observed in areas subjected to low temperature rise are composed of highly deformed grains associated with an accumulation of plastic deformation [3,4,19,20]. The severe plastic deformation (SPD) progressively transforms pearlitic grains of several dozen of micrometers into nanograins of dislocated ferrite supersaturated in carbon [2,21]. In fact, the increase of the dislocation density would facilitate the fragmentation of the grains [22–28], increase the kinetics of cementite dissolution [29,30], and favor the carbon-atom mobility [31]. In addition, such a level of deformation would be linked to the intrinsic conditions of the wheel–rail contact that combines high compressive stresses with high shear stresses. Under these severe conditions, the material could locally reach a level of severe deformation that would be analogous to the severely deformed specimens in the case of cold drawing [32–34] or SPD experiments [23,35,36]. Indeed, high-pressure torsion tests [36] performed for pearlitic steels at room temperature confirmed the mechanically induced WEL formation under hydrostatic pressure and shear coupling.

These two scenarios cannot really be distinguished, and the hypothesis of a thermomechanical coupling is the most probable to explain and predict WEL formation [2,37]. A macroscopic model developed by Antoni et al. [38] was then proposed to simulate the mechanical-stress effect on WEL formation. In this model, a coupling between the hydrostatic pressure and temperature was proposed. Nevertheless, this model does not take into account the shearing of the grains at the extreme surface of the rails. To overcome these limits, Thiercelin et al. [39] improved this model by adding a shear contribution to the WEL formation criterion. The WEL formation would result from a coupling between the hydrostatic pressure and the shear stress, which would be enhanced by the temperature. The tendencies of WEL formation kinetics have already been confirmed in the literature [36,40–42]; however, the model needs to be more accurately experimentally identified. For this purpose, this coupling has been de-correlated to separately quantify the effects of temperature, hydrostatic pressure, and shear.

First, Merino [43], Lafilé [44], and Thiercelin [45] successfully reproduced the kinetics of WEL formation under representative wheel–rail contact conditions at 1/15th scale. These tests demonstrated the predominantly mechanical formation of the WEL, as the slip level was less than 2%. Moreover, the combined effect of contact pressure and sliding was confirmed, validating one part of the model proposed by Thiercelin et al. [39], i.e., the pressure–shear coupling.

Second, temperature-controlled shear tests have demonstrated the formation of adiabatic shear bands in steels [46,47]. In addition, tests performed by Lins et al. [48] on low-carbon interstitial-free steels (IF steel) revealed shear-band formation kinetics, called "progressive subgrain misorientation", which is similar to the mechanically induced WEL kinetics. Such tests conducted on pearlitic steels are still lacking and would constitute a major breakthrough for the identification of the model proposed by Thiercelin et al. and more generally for the formation of shear bands in pearlitic steels.

The objective of the current study was to reproduce shear tests under controlled temperatures representative of wheel–rail contact. Monotonic and cyclic shear tests were conducted for different temperature ranges (from 20 °C to 400 °C). Optical and scanning microscopy observations were performed to characterize each test condition. First, static monotonic tests with microstructure analysis are presented for each temperature tested (20 °C, 200 °C, 300 °C, and 400 °C). Next, the results of the cyclic tests are presented for each tested temperature. Finally, the effect of temperature and loading path are discussed with respect to the microstructural evolution eventually leading to WEL formation.

2. Materials and Methods

2.1. Material

The material studied was a pearlitic rail steel grade R260 used in most of the French railway network. The chemical composition of the material is given in Table 1.

Table 1. Chemical composition of R260 pearlitic steel (weight%) [49].

C	Si	Mn	P	S	Cr	Al
0.62	0.80	0.15–0.58	0.70–1.20	<0.025	<0.15	<0.004

In its as-received state, the microstructure consists of pearlitic grains with a diameter measured to be approximately 9.6 ± 8.1 μm. The grains initially have a rather globular shape with an aspect ratio of 2.0 ± 0.9 and are weakly disoriented. There are initially 30% low-angle grain boundaries (LAGBs), a majority of medium-angle grain boundaries (MAGBs) (40%), and 30% high-angle grain boundaries (HAGBs) (Figure 1a). The group of grains having a similar crystallographic orientation constitutes a pearlitic colony whose average size is generally several tens of micrometers [50] (Figure 1b).

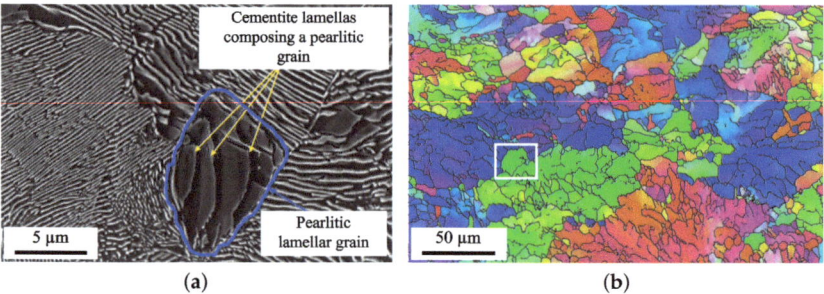

Figure 1. (a) SEM microstructure of the as-received structure and (b) EBSD IPFZ map of the as-received microstructure.

2.2. Thermomechanical Test Bench

Thermomechanical tests were performed using the Gleeble 3800 device, a test bench that can simulate a complex thermomechanical loading path by performing mechanical cycles and thermal cycles simultaneously (Figure 2). The Gleeble thermomechanical test bench can perform compression tests at a rate up to 2000 mm/s. The measurement and control of the strain was performed by the displacement of the mobile die previously calibrated. In addition, the specimen temperature was measured with type K thermocouple welded on the sample. The specimen was heated by conduction, and the temperature was continuously monitored with a control loop adapting the amount of current in the sample in order to get the desired thermal cycle. For more information, please refer to the Gleeble company website [51].

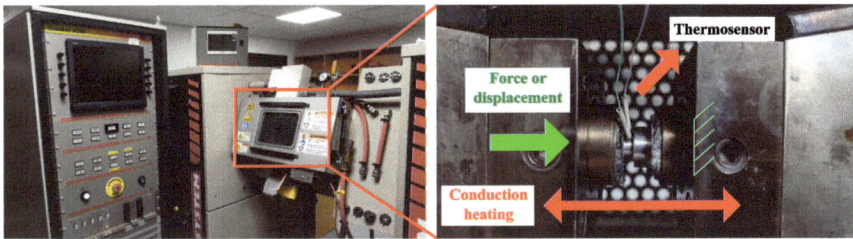

Figure 2. Gleeble 3800 test bench with the positioning of the hat-test specimen: the force (or displacement) is applied by a mobile die, and the temperature is induced by thermal conduction and is controlled with a type K thermocouple. For more information please see [51].

2.3. Hat-Shaped Specimen

The shear tests under controlled temperature were performed with the help of a hat-shaped specimen, which, under the effect of a compression, generated a localized shear in a crown of width $L = 2$ mm and thickness $d = 0.2$ mm (Figure 3). The thickness d corresponded to the size of the deformation gradient observed at the rail surface. The other dimensions were chosen from previous work on similar alloys [52,53] and are summarized in Table 2.

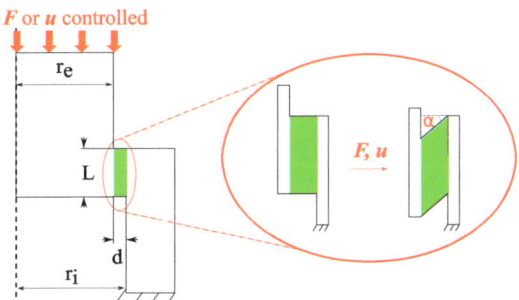

Figure 3. Scheme of an axisymmetric section of a hat-shaped specimen subjected to a compressive load with r_i and r_e representing the internal and external diameters of the specimen, respectively, and $d = r_i - r_e$ and L representing the thickness and width of the sheared zone, respectively. The sheared zone is then a crown of length L and thickness d.

Table 2. Hat-shaped specimen dimensions.

r_i (mm)	r_e (mm)	$d = r_i - r_e$ (mm)	L (mm)
4	3.8	0.2	2

To analyze the experimental data, some theoretical assumptions about the stress and strain field in the sheared zone are required. The hat specimens are assumed to undergo pure shear, which is confined to the theoretical shear zone (green area in Figure 3). The stress and strain tensors are then expressed as follows:

$$\sigma = \begin{pmatrix} 0 & \tau & 0 \\ \tau & 0 & 0 \\ 0 & 0 & 0 \end{pmatrix} \quad \text{with} \quad \tau = \frac{F}{\pi L (r_i + r_e)} \quad (1)$$

$$\varepsilon = \begin{pmatrix} 0 & \gamma/2 & 0 \\ \gamma/2 & 0 & 0 \\ 0 & 0 & 0 \end{pmatrix} \quad \text{with} \quad \gamma = \tan(\alpha) = \frac{u}{d} \quad (2)$$

with τ representing the shear stress induced in the sheared zone, which depends on F, the applied compressive load; L, the length of the sheared zone; and r_i and r_e, the inner and outer radius of the specimen, respectively (Figure 3). In addition, γ is the shear strain, which depends on u, the displacement in the compression direction, and d, the theoretical thickness of the sheared zone.

It must be pointed out that the heterogeneous shape of the specimen induces temperature heterogeneity in the specimen. Preliminary heat conduction tests showed that there was a difference of approximately 10 °C between the lower and upper part of the specimen. Nevertheless, the current study was focused on the effect of the temperature in the shear zone; therefore, only one thermosensor was welded in the vicinity of the shear stress (Figure 4), and the temperature was considered uniform in the specimen during all the mechanical tests.

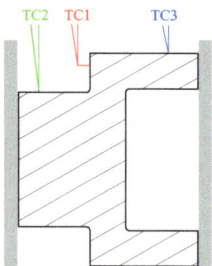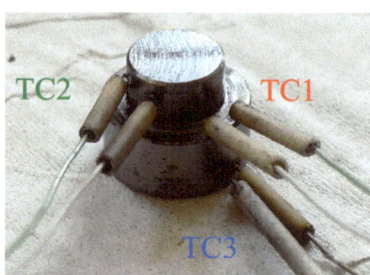

Figure 4. Welding of the thermosensor. The control of the temperature was achieved with the sensor labelled "TC1".

2.4. Microstructural Characterization

Optical microscopy characterization was performed after mechanical polishing down to 1 µm and subsequent Nital etching to observe cementite lamellae and the possible WEL.

The samples were also prepared for observation in secondary electron (SE) mode and electron backscatter diffraction (EBSD) mode in the scanning electron microscope. For the EBSD mode, the samples were mechanically polished down to 1 µm using a conventional grinding machine. The final preparation step consisted of vibratory polishing for approximately 1 h using a colloidal silica suspension (Struers OP-S) with a grain size of 0.05 µm. For some cross-section samples, it was possible to perform an additional ionic polishing step with a GATAN ILLION 2 device with an accelerating voltage of 4 kV. The observations were performed using a FEG Zeiss Supra 55VP microscope equipped with an Oxford EBSD Symmetry detector. The accelerating voltage was set to 12 or 15 kV, depending on the size of the microstructural elements to be analyzed. The EBSD data treatment was performed using Atex software [54]. For the data treatment of the EBSD maps, the disorientation angle selected for grain detection was set to 5°, and only grains with at least 10 pixels were considered, as the detection is not reliable below this level. The statistical distributions of the grain size and the aspect ratio in each EBSD map were characterized from mean values followed by the the standard deviation noted "±". Figure 5 indicates the color code used to define the crystallographic textures of all the inverse pole figures (IPF) presented in this study.

The sheared areas of each specimen were then analyzed after having been axially cut and prepared as previously described. It is worth noting that some sheared zones were analyzed after the failure of the specimen and some were analyzed before. Depending on the final state of the specimen, an associated scheme is included to facilitate understanding of the microstructural analyses (Figure 6).

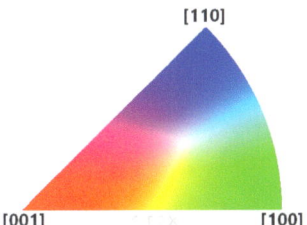

Figure 5. Inverse pole figure color coding of orientation maps presented in this study.

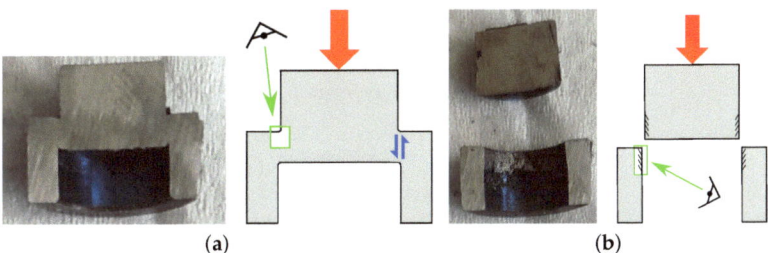

Figure 6. Axial cut of hat-shaped specimen: (**a**) case without failure and (**b**) case after failure.

The sheared zones were systematically compared to the stages of microstructural evolution leading to the WEL formation observed on the rail surface. Based on a mechanism proposed by several authors [25,28] and the formalism initially proposed by Thiercelin et al. [45], three microstructural indicators were used to quantitatively describe the progressive evolution of the microstructure: mean grain size, aspect ratio, and grain-boundary disorientation. For the latter, three intervals of grain disorientation were considered:

- LAGBs for angles between 5° and 15°,
- MAGBs for angles from 15° to 40°,
- HAGBs for angles above 40°.

Figure 7 is a longitudinal cross-section of worn pearlitic rail with evidence of a microstructural gradient from the pearlitic stage to the WEL stage over approximately 60 μm from the contact surface. The microstructural gradient analysis provides insights into the kinetics of WEL transformation. This gradient can be divided into several successive stages of transformation located at different depths. First, the grains begin to fiber and remain mainly unfragmented (stage 2) from 40-μm depth. The average grain size starts to decrease, the aspect ratio is very high, and the grain disorientation remains quite low. Next, between 20- and 40-μm depth, the grains fragment but still keep a rather elongated shape (stage 3). At 10-μm depth, there is a fine area of some micrometers composed of strongly disoriented nanograins. This is the non-fibrous and nanostructured stage (stage 4). Finally, in the first 10 μm from the contact surface, the grains appear white after Nital etching, which corresponds to a WEL spot. At this stage, the nanograins are spherical and much more disoriented than in the previous stages. This stage is considered the final stage in WEL formation (stage 5). The stages of transformation leading to WEL formation and the associated microstructural indicators proposed by [45] are presented in Table 3.

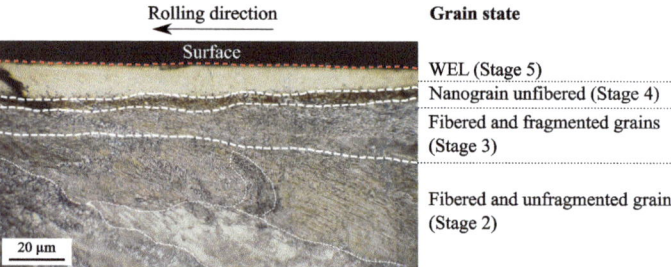

Figure 7. Longitudinal cross section of a rail extracted from region with a high WEL frequency [55]; the microstructural gradient consists of four areas representing the differents evolution stages leading to WEL formation. Stage 2 is the fibrous state of the grains without fragmentation; stage 3 is a state where the grains are still fibrous but fragmented; stage 4 is a state where the grains have no particular orientation and are very fragmented; and stage 5 is the state that appears white in optical microscopy after Nital etching.

Table 3. Evolution of indicators and the stages of evolution from the pearlitic state to the final WEL state according to [45].

Stage	Grain State	Grain Size (μm)	Aspect Ratio (−)	LAGB-MAGB-HAGB (%)
1	As-received pearlitic	9.6 ± 8.1	2.0 ± 0.9	30-41-29
2	Fibered and unfragmented	0.6 ± 0.3	4.1 ± 2.7	20-48-32
3	Fibered and highly fragmented	0.6 ± 0.4	2.8 ± 1.4	19-50-31
4	Unfibered and nanostructured	0.2 ± 0.1	1.7 ± 0.7	9-48-43
5	WEL	0.2 ± 0.1	1.6 ± 0.5	20-18-61

3. Results

3.1. Monotonic Tests

As the temperature in the wheel–rail contact does not exceed several hundred degrees, testing temperatures of less than 400 °C were selected. First, monotonic shear tests were performed at four temperatures until failure (20 °C, 200 °C, 300 °C, and 400 °C) at a quasi-static loading rate ($\dot{\gamma} = 0.5\ \text{s}^{-1}$). The conditions of the monotonic tests are given in Table 4. A macroscopic analysis of the experimental curves combined with microstructural characterization of each condition tested was conducted.

Table 4. Monotonic test conditions until failure.

Temperature (°C)	$\dot{\gamma}$ (s^{-1})	Number of Tests
20	0.5	2
200	0.5	2
300	0.5	1
400	0.5	2

3.1.1. Macroscopic Analysis

The macroscopic behavior of the material was studied by plotting the stress–strain curve (Figure 8) obtained with the stress and the associated shear strain as theoretically observed by the material in the sheared zone (Equations (1) and (2)). For each temperature, at least two experiments were performed and showed sim confirming the repeatability of the experiments.

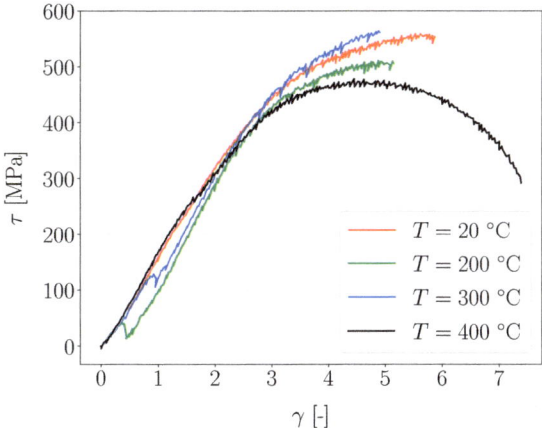

Figure 8. Shear stress τ vs. shear strain γ curve using (1) and (2).

For temperatures below 300 °C, the ductility of the material decreased with temperature. The failure strain was on the order of 5.8 at room temperature, whereas it was 5.1 at 200° and 4.9 at 300 °C. Moreover, in this temperature range, the stress–strain curves at 20 °C and 300 °C followed the same trend, differing from that at 200 °C. Indeed, for the latter, the material became significantly less resistant than at 20 °C and 300 °C as its apparent yield stress (stress beyond which the behavior becomes non-linear) and its critical shear stress before failure were much lower. The observations at these three temperatures reveal three different behaviors, which do not allow conclusions to be drawn on the trend of behavior in this temperature range.

In addition, the curves at 200 °C and 300 °C exhibit serrations characterized by microstress drops during monotonic loading. This mechanism can likely be attributed to the Portevin–Le Chatelier effect (PLC) already encountered for pearlitic steels in this temperature range [56–58]. This mechanism, also known as dynamic strain aging (DSA), corresponds to an instability of the plastic flow in metals when dislocations interact with atoms in solid solutions. During plastic deformation, the dislocations are blocked by the atoms in solid solutions until a critical force is reached. Subsequently, the stress falls until the next obstacle, explaining the serrations.

It is, however, quite surprising that similar serrations were observed regardless of the temperature. Indeed, previous studies have shown that the PLC is favored at higher temperatures in pearlitic steels. Therefore, we consider that the serrations observed could also be experimental artifacts.

For a temperature of 400 °C, the material response differed completely from that at the previous temperatures tested. It became softer and more ductile compared with the material at the other temperatures, with a failure strain above 7. It must be pointed out that for this experiment, the specimen did not fail. Moreover, the apparent yield strength estimated at 300 MPa drastically decreased compared with that at 20 °C (approximately 450 MPa).

The following section will focus on the microstructural characterization of the shear zone performed at the end of the monotonic test for each temperature.

3.1.2. Microstructural Characterization

Figure 9 presents optical and EBSD micrographs of the sheared area obtained at the end of the test for each temperature. Similar to the macroscopic analyses, the microstructural behavior in the range between 20 °C and 300 °C differed slightly from that at 400 °C. The observations are thus given separately.

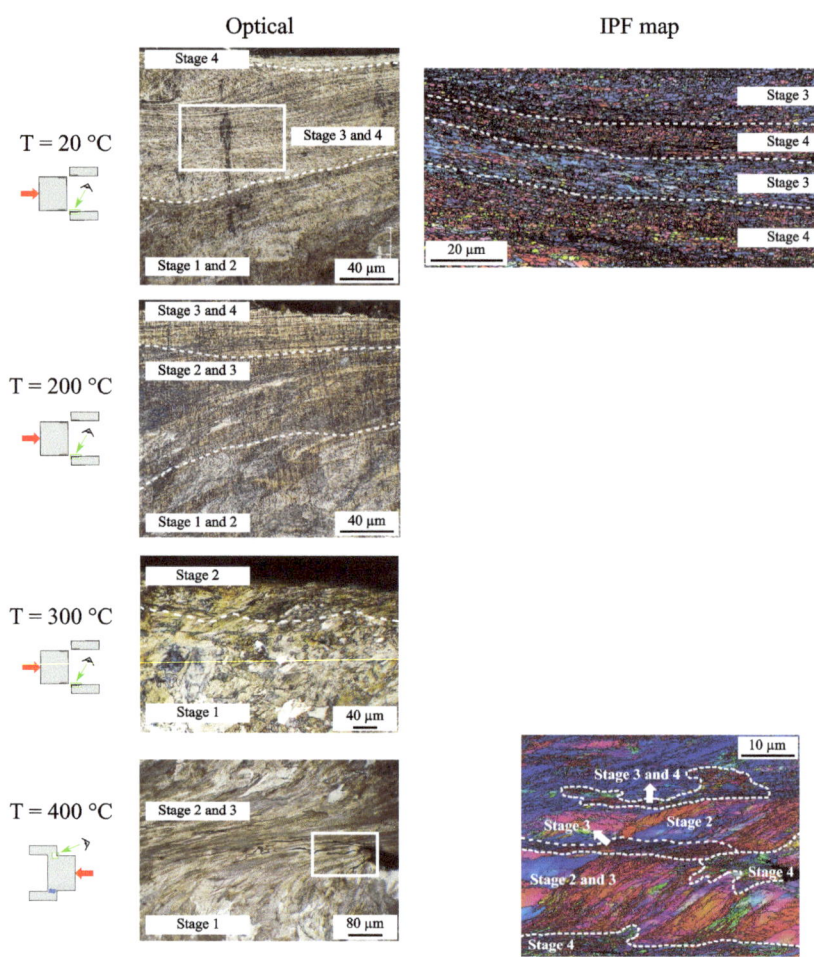

Figure 9. Effect of temperature on the microstructural gradient in the sheared zone of the specimen at the end of the monotonic tests. The specimen failed at the tests performed at T = 20 °C, 200 °C, and 300 °C but did not fail at 400 °C. IPF maps were constructed for the tests performed at room temperature (step size = 0.08 μm) and 400 °C (step size = 0.1 μm).

First, between 20 °C and 300 °C, a microstructural gradient similar to that of the rail surface was observed. At room temperature, the microstructural gradient extended over a hundred micrometers with grains that gradually transformed from the broken surface, in a fragmented and unfibered state (stage 4), to a fibrous stage at a depth of 100 μm (stage 2). In the intermediate area, between 30 and 60 μm, a mixture of grains making the transition between stage 3 and 4 was observed. These optical observations are confirmed by the EBSD map within this area, where the grain have a relatively globular shape with an aspect ratio of 2 ± 1 and are submicrometric in size (0.5 ± 0.2 μm). In addition, their disorientation is rather high as the proportion of MAGBs and HAGBs is predominant at 39% and 33%, respectively. The evaluation of all these indicators lead to the conclusion that the grains in the area studied by EBSD are in a transition stage between stages 3 and 4 (Table 3) as seen in Table 5.

Table 5. Evaluation of microstructural indicators for the EBSD map at room temperature.

Grain Size (µm)	Aspect Ratio	LAGB-MAGB-HAGB (%)	Stage Estimated
0.5 ± 0.2	1.9 ± 0.7	28-39-33	3 and 4

The same test performed at 200 °C revealed a similar microstructural gradient as that at room temperature. The sheared zone consisted of sheared grains that gradually became less fragmented and fibered to a depth of 60 µm. This sheared zone appears more confined than that for the test performed at 20 °C, which was approximately 100 µm thick. The effect of temperature on the strain localization appears to confirm the results at 300 °C, where the grains were slowly sheared. The analysis of the stress–strain curves combined with the microstructural gradient indicate that the plastic deformation was much less severe at 200 °C and 300 °C, where it is very confined.

As noted above, the material was much more ductile at 400 °C and did not fail. The micrograph shows the microstructural gradient of the specimen strained until a shear strain of 6.9. No WEL spots were observed; however, the grains were highly fibered over a large thickness of 200 µm, which corresponds to the theoretical sheared zone d (Figure 3). A main crack that initiated at the corner of the specimen propagated along the interface between the fibered structure and the as-received pearlitic grains. We also denote the presence of many secondary cracks within the sheared zone.

The EBSD map of the area close to the crack tip reveals a strong grain heterogeneity in the sheared zone. The grains on either side of the crack were deformed and reoriented in the shear direction without being fragmented (stage 2). Close to the most sheared zone, the grains were fibered and fragmented (stage 3). At the tip of the crack, the grains were not fibered and were very fragmented and randomly oriented, corresponding to a critical grain state close to the WEL state (stage 4). In addition, a network of secondary cracks in the presence of grains at the same critical stage (stage 4) nearby were observed.

3.2. Cyclic Tests

3.2.1. Macroscopic Analysis

Cyclic shear tests were then conducted for different temperatures to simulate the thermomechanical loading path undergone by the rail. Because WEL formation is related to the cementite dissolution by mechanical stresses and, in particular, by dislocations, a maximal cyclic stress slightly higher than the apparent elastic shear stress determined with the monotonic stress–strain curves was applied (Figure 8). In addition, to maintain contact and ensure good thermal conduction, a minimum cyclic shear stress of 100 MPa was imposed, which corresponds to the minimum contact force of 5 kN.

A maximum stress of 560 MPa was considered for 20 °C and 300 °C. The specimen failure occurred after 600 cycles compared to 1400 cycles at 300 °C.

For a temperature of 200 °C, only a few cycles would have led to the failure of the specimen. At this temperature, a lower maximum stress of 500 MPa was thus considered. A test of 10,000 cycles was conducted without breaking the specimen. The thermomechanical conditions for the cyclic tests are summarized in Table 6.

Table 6. Cyclic tests conditions.

Temperature (°C)	τ_{min} (MPa)	τ_{max} (MPa)	Number of Cycles
20	100	560	500 and 600 (failure)
200	100	500	10,000
300	100	560	1000 and 1400 (failure)

These preliminary results highlight the significant effect of temperature on the fatigue limit of the material. In a similar manner as for the monotonic case, a different behavior was observed for each temperature. Indeed, between 20 °C and 200 °C, the fatigue limit of the material decreased as the temperature increased. In contrast, at 300 °C, the fatigue limit unexpectedly became greater than at 20 °C. Cyclic tests before specimen failure were conducted for all three temperatures. The following section will present the microstructural observations of the three tests before failure.

3.2.2. Microstructural Characterization

Figure 10 presents optical micrographs after Nital etching of the shear zone at the end of each cyclic test stopped before the specimen broke. Regardless of the test temperature, a crack usually started in the corner of the cap and propagated over a few hundred micrometers.

For the test at 20 °C, the grains downstream of the crack were sheared but appeared slightly fragmented (stage 2). Moreover, on both sides of the crack, a transformation gradient up to the WEL was visible (see the optical zoom in this area). It can be assumed that cracks might have propagated due to incompatibility of the microstructure.

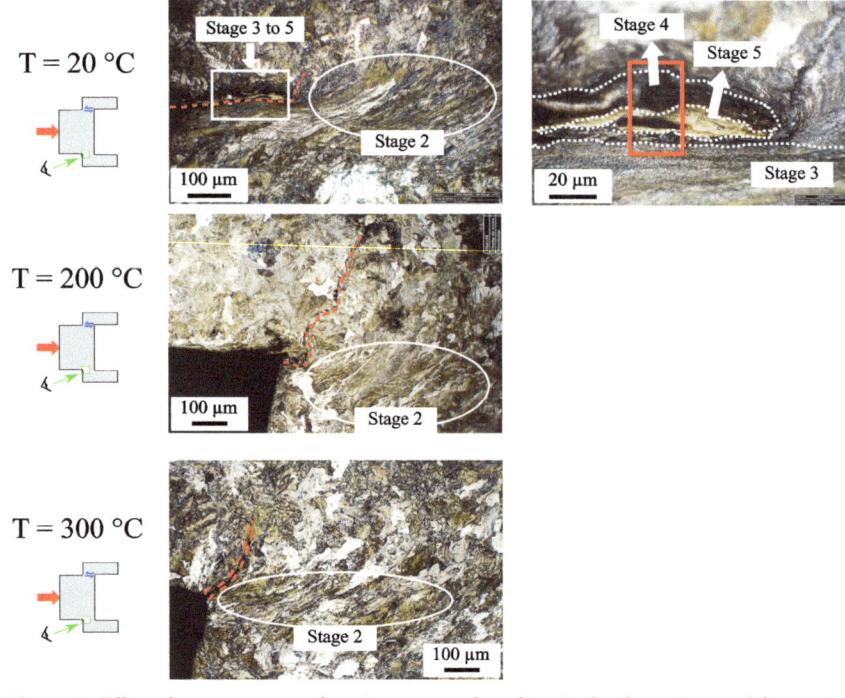

Figure 10. Effect of temperature on the microstructural gradient in the sheared zone of the specimen at the end of the cyclic tests. The mechanical loading and the number of cycles at each temperature are shown in Table 6. The red dotted line denotes the start of cracking in the corner of the specimen. The red box corresponds to the area investigated by EBSD in Figure 11.

An EBSD map of the region around the crack (red box in the micrograph at room temperature of Figure 10) is presented to measure the microstructural indicators in this area (Figure 11). This map is divided into four distinct areas as follows:

- A fibered and fragmented zone in the lower part of the image (zone A);
- A non-indexed zone in the upper part of the crack, which corresponds to the white zone observed optically (zone B);
- A very fragmented zone without fibration above the very poorly indexed zone (zone C);
- A transition zone where the material fibered and flowed until it fragmented (zone D).

Region A, the lower part of the crack, was composed of fibered and elongated grains with an average size of 0.4 ± 0.2 μm and an aspect ratio of 2 ± 1, on average. In this region, there is a majority proportion of low disoriented grains (45%) and fairly close proportions of weakly and very disoriented grains (38% and 18%, respectively). All of these indicators are consistent with the identification of stage 3 in zone A.

Region B, corresponding to a WEL optically, was unindexed in EBSD, suggesting a significant grain-size reduction that is smaller than the indexation stepsize of 0.06 μm and a high level of grain disorientation.

Region C, which was optically dark (Figure 10, T = 20 °C), did not show evidence of any apparent grain fibering. This zone consisted of spherically shaped nanograins (aspect ratio of 1.6 ± 0.5), with a fairly uniform size (0.3 ± 0.1 μm). The grains had no particular crystallographic orientation and were much more disoriented than those in region A. Indeed, weakly disoriented grains occupied only 12% compared to 45% in zone A. There was then a large majority of large disorientation angles equally distributed between the moderately disoriented and highly disoriented grains (43% and 45%, respectively). All these observations are indicative of stage 4 in this region.

The set of indicator values in zones A and C of Figure 11 are summarized in Table 7.

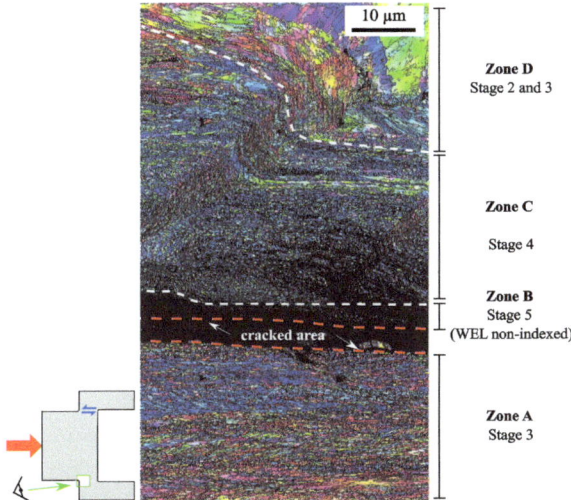

Figure 11. T = 20 °C − N_{cycle} = 500 with τ_{min}/τ_{max} = 100/560 MPa, γ_{final} = 3.5. IPF X map of the microstructural gradient of the cracked zone that corresponds to the white box in Figure 10 for the case at room temperature, stepsize = 0.06 μm. This region is divided into four areas: zone A is composed of fibered and fragmented grains (stage 3). Zone B is a WEL area with very small grains and disoriented grains. Zone C is composed of very small and disoriented grains (stage 4), whereas zone D is a transition area with grains that begin to fiber and fragment themselves (stage 2 and 3).

Table 7. Summary of measurements of indicators for the EBSD map in Figure 11.

Zone	Grain Size (μm)	Aspect Ratio	LAGB-MAGB-HAGB (%)
A	0.4 ± 0.2	2.0 ± 1.0	45-37-18
C	0.3 ± 0.1	1.6 ± 0.5	12-43-45

The micrographs at 200 °C and 300 °C exhibit the same microstructure (Figure 10). Contrary to the test performed at room temperature, the crack, being initiated at the corner, propagated normally to the shear direction. The grains were slowly deformed downstream to the crack, and no WEL was observed.

4. Discussion

4.1. Effect of the Thermomechanical Path on WEL Formation Kinetics

The monotonic tests conducted at the four temperatures from 20 °C to 400 °C did not lead to WEL formation. The monotonic tests revealed a strongly non-linear behavior of the material with temperature. Indeed, at 200 °C, the material was softer and more brittle than at room temperature. At 300 °C, behavior similar to that at 20 °C was observed, with the material being more resistant and ductile. At 400 °C, the stress–strain curve indicated a much more ductile behavior, which can be explained by the higher ductility of the grains [56]. WEL formation was never observed after monotonic tests. However, the analysis of their microstructure indicates that the temperature does not seem to favor the deformation/fragmentation required for WEL formation. Indeed, an analysis of the microstructures using the WEL formation indicators defined in Table 3 reveals that the most advanced stages were found for the RT sample. Table 8 summarizes the different microstructures obtained at the end of the monotonic tests.

Table 8. Synthesis for the monotonic tests.

Temperature (°C)	$\gamma_{failure}$ (—)	Presence of WEL	Final Stage
20	5.3	no	4
200	4.2	no	4
300	4.4	no	2
400	6.9	no	4

Regarding the cyclic tests, a WEL was observed at 20 °C. However, at 200 °C and 300 °C, little deformation or transformation of the grains was observed. Table 9 summarizes the different microstructures obtained at the end of the cyclic tests.

It appears that WEL formation is possible under the effect of shear at room temperature via a mechanism of grain fragmentation and transformation for a sufficient level of deformation, as presented by Thiercelin [45]. Indeed, the tests performed at higher temperatures (200 °C and 300 °C) resulted in less deformation of the microstructure for both the static and cyclic tests compared with that at room temperature. Localization of the stress was observed, which is likely related to the geometry of the specimens, which would crack before forming a WEL.

In addition, in the 200–300 °C temperature range, the kinetics of WEL formation could be inhibited by other microstructural mechanisms, such as dynamic recovery. The latter would enhance the annihilation of dislocations and therefore enhance the deformability of the microstructure.

Another physical mechanism could concern the cementite precipitation beginning at 300 °C [36], which would counterbalance its dissolution under the effect of mechanical stresses. Finally, the more ductile behavior at 400 °C can be attributed to a mechanism of deformation of each phase of the grains (ferrite and cementite) without fragmentation [57].

Table 9. Synthesis for the cyclic shear tests. The same maximal stress of 560 MPa was considered for 20 °C and 300 °C. For the test at 200 °C, such a maximum stress would have been too close to the monotonic failure stress (Figure 8). A few cycles would have led to the failure of the specimen, which explains the choice of a lower maximum shear stress (500 MPa rather than 560 MPa).

Temperature (°C)	τ_{min}/τ_{max} (MPa)	Number of Cycles	γ_{final} [—]	Final Stage
20	100/560	500, 600 (failure)	3.5	5
200	100/560	1 (monotonic case)	no data	no data
200	100/500	10,000	1.8	2
300	100/560	1000, 1400 (failure)	2.1	2

4.2. Effect of Temperature on Fatigue Strength

In this study, the effect of contact temperature on the fatigue strength of the rails was studied in order to anticipate the failure of the rails. At room temperature, the material was much more ductile, and failure occurred after WEL formation. For higher temperatures (200 °C and 300 °C), the material was less ductile and the failure occurred after a more localized deformation. Finally, the monotonic test conducted at 400 °C revealed crack initiation in highly nanostructured zones.

The failure mechanism clearly differs with temperature, and the deformation of the microstructure at stages less advanced than WEL formation can already be critical for the material depending on the contact temperature.

Nevertheless, the stage of transformation of the microstructure combined with the contact temperature is a good indicator of the probability of material failure. The WEL stage is the most critical. This assumption can be related to the studies of Saxena et al. [59], where the toughness of WELs and work-hardened zones of rail steels were studied. The authors concluded that in the case of wheel–rail contact, the toughness of the material is inversely proportional to its hardness. As hardness is linked to the dislocation density, grain size, and carbon content in solid solution [32,60,61], the toughness would then be directly dependent on the evolution stages (Table 3).

4.3. Thermomechanical Model and Wheel-Rail Contact Conditions

The role of temperature in the kinetics of WEL formation is contestable, as temperatures above 200 °C will activate other microstructural mechanisms that inhibit grain fragmentation–transformation kinetics. The model proposed by Thiercelin [39] would then be invalidated for temperatures of 200 °C, 300 °C, and 400 °C. Future tests at lower temperatures (below 200 °C) will have to be conducted to determine whether low-temperature elevations could still favor the WEL formation mechanism, as noted by Newcombs et al. [19]. In addition, for temperatures close to the austenitization temperature, the stresses could facilitate WEL formation [1,62,63].

The second point concerns a noteworthy difference in the thermomechanical path of the tests and the reality of the wheel–rail contact. Indeed, contrary to the contact conditions, the temperature was kept constant during the entire duration of the applied cyclic mechanical loading. In reality, the cyclic thermal load is more complex than in those tests and could modify the kinetics of the microstructural transformations. Therefore, the WEL formation criteria would then depend explicitly on the temperature evolutions with time (heating and cooling rates) and on the accumulation process of deformation/temperature/shear rather than on the temperature itself.

Finally, the interactions of the different scales of the wheel–rail contact (train, wheel–rail interaction, and the effect of wear particles) lead to a variability of the thermomechanical field applied to the rail surface. The tests performed in this study showed a strongly non-linear behavior of the material, which was explained by the activation of various microstructural mechanisms, such as dynamic recovery, cementite precipitation and or dissolution, grain fragmentation, phase transformation, and/or deformation. Depending on the temperature, the material will have different cyclic responses, which may lead to the development of rolling-contact fatigue defects or contribute to the wear rate if the contact temperature is sufficiently high, as observed during bi-disc tests for temperatures above 300 °C [64].

5. Conclusions

Monotonic and cyclic shear tests under controlled temperatures were performed using hat-shaped specimens and the Gleeble thermomechanical simulator in an attempt to reproduce the WEL formation induced at the rail surface.

The monotonic tests conducted at four temperatures (20 °C, 200 °C, 300 °C, and 400 °C) did not lead to WEL formation but resulted in grain-transformation stages close to WEL formation. The cyclic tests conducted at 20 °C, 200 °C, and 300 °C in a plastic regime confirmed the monotonic trends and resulted in WEL formation at room temperature only. The effect of medium temperature appears to be unfavorable for WEL formation by fragmentation-transformation of grains as it activates other mechanisms, such as dynamic recovery or cementite precipitation.

The thermal coupling part of the model proposed by Thiercelin et al. [39] was then invalidated for temperatures between 200 °C and 400 °C; however, the formation of WEL by pure shear was confirmed. Further shear tests at temperatures between 20 °C and 200 °C must be considered to validate the model at low temperature.

The residual stresses in the sheared zone will be further estimated to improve the constitutive model of WEL formation. Moreover, these measurements will be compared to the residual stresses measured on the real rail surface in presence of WEL [6,7].

The tests confirmed that the evolution stages leading to WEL formation could reflect a probabilistic criterion of crack initiation. The probability of cracking of the material increases inversely with decreasing grain size, which is related to the stage of evolution of the microstructure. In the railroad, the crack initiation could then be explained by an advanced stage of evolution of the microstructure (stage 4 or 5).

These tests showed limitations due to premature cracking, which could be overcome by performing identical tests under hydrostatic pressure to limit the damage [65,66] and to be more representative of the multiaxial stresses undergone by the rail. Moreover, the achievement of thermal cycles simultaneously with the mechanical cycles constitutes future perspectives to simulate more faithfully the thermomechanical path undergone by the rail.

Author Contributions: L.T.: conceptualization, methodology, software, investigation, writing—original draft. S.C.: conceptualization, methodology, investigation, resources, writing—original draft, supervision. A.S.: conceptualization, methodology, resources, writing—review and editing, supervision, project administration. F.L.: conceptualization, methodology, software, writing—review and editing, supervision, project administration. F.M.: investigation, resources, writing—review and editing. S.D.:investigation, resources, writing—review and editing. C.L.B.: investigation, resources, writing—review and editing. D.F.: investigation, resources, writing—review and editing. All authors have read and agreed to the published version of the manuscript.

Funding: This research received no external funding.

Institutional Review Board Statement: Not applicable.

Informed Consent Statement: Not applicable.

Data Availability Statement: Not applicable.

Acknowledgments: This work is part of the multi-disciplinary project MOPHAB, which aims to improve our knowledge and understanding of the mechanisms leading to the formation of the white etching layer in the materials used to construct railways and to develop corresponding numerical models. This project was supported by IRT Railenium and other industrial partners (RATP: Régie Autonome des Transports Parisiens, France, SNCF: Société Nationale des Chemins de Fer Francais, France, SAARSTAHL rail).

Conflicts of Interest: The authors declare no conflict of interest.

References

1. Zhu, H.; Li, H.; Al-Juboori, A.; Wexler, D.; Lu, C.; McCusker, A.; McLeod, J.; Pannila, S.; Barnes, J. Understanding and Treatment of Squat Defects in a Railway Network. *Wear* **2020**, *442–443*, 203139. [CrossRef]
2. Kumar, A.; Agarwal, G.; Petrov, R.; Goto, S.; Sietsma, J.; Herbig, M. Microstructural Evolution of White and Brown Etching Layers in Pearlitic Rail Steels. *Acta Mater.* **2019**, *171*, 48–64. [CrossRef]
3. Al-Juboori, A.; Wexler, D.; Li, H.; Zhu, H.; Lu, C.; McCusker, A.; McLeod, J.; Pannil, S.; Wang, Z. Squat Formation and the Occurrence of Two Distinct Classes of White Etching Layer on the Surface of Rail Steel. *Int. J. Fatigue* **2017**, *104*, 52–60. [CrossRef]
4. Al-Juboori, A.; Zhu, H.; Wexler, D.; Li, H.; Lu, C.; McCusker, A.; McLeod, J.; Pannila, S.; Barnes, J. Characterisation of White Etching Layers Formed on Rails Subjected to Different Traffic Conditions. *Wear* **2019**, *436–437*, 202998. [CrossRef]

5. Wang, L.; Pyzalla, A.; Stadlbauer, W.; Werner, E. Microstructure Features on Rolling Surfaces of Railway Rails Subjected to Heavy Loading. *Mater. Sci. Eng. A* **2003**, *359*, 31–43. [CrossRef]
6. Wild, E.; Wang, L.; Hasse, B.; Wroblewski, T.; Goerigk, G.; Pyzalla, A. Microstructure Alterations at the Surface of a Heavily Corrugated Rail with Strong Ripple Formation. *Wear* **2003**, *254*, 876–883. [CrossRef]
7. Pyzalla, A.; Wang, L.; Wild, E.; Wroblewski, T. Changes in Microstructure, Texture and Residual Stresses on the Surface of a Rail Resulting from Friction and Wear. *Wear* **2001**, *251*, 901–907. [CrossRef]
8. Österle, W.; Rooch, H.; Pyzalla, A.; Wang, L. Investigation of White Etching Layers on Rails by Optical Microscopy, Electron Microscopy, X-ray and Synchrotron X-ray Diffraction. *Mater. Sci. Eng. A* **2001**, *303*, 150–157. [CrossRef]
9. Takahashi, J.; Kawakami, K.; Ueda, M. Atom Probe Tomography Analysis of the White Etching Layer in a Rail Track Surface. *Acta Mater.* **2010**, *58*, 3602–3612. [CrossRef]
10. Daniel, W. *Final Report on the Rail Squat Project R3-105*; CRC for Rail Innovation: Brisbane, Australia, 2013.
11. Nakkalil, R. Formation of Adiabatic Shear Bands in Eutectoid Steels in High Strain Rate Compression. *Acta Metall. Mater.* **1991**, *39*, 2553–2563. [CrossRef]
12. Baumann, G.; Fecht, H.; Liebelt, S. Formation of White-Etching Layers on Rail Treads. *Wear* **1996**, *191*, 133–140. [CrossRef]
13. Zhang, H.; Ohsaki, S.; Mitao, S.; Ohnuma, M.; Hono, K. Microstructural Investigation of White Etching Layer on Pearlite Steel Rail. *Mater. Sci. Eng. A* **2006**, *421*, 191–199. [CrossRef]
14. Simon, S. De la Dynamique Ferroviaire à L'Accommodation Microstructurale du Rail—Contribution des TTS à La réponse Tribologique des Aciers—Cas du Défaut de Squat. Ph.D. Thesis, INSA Lyon, Villeurbanne, France, 2014.
15. Knothe, K.; Liebelt, S. Determination of Temperatures for Sliding Contact with Applications for Wheel-Rail Systems. *Wear* **1995**, *189*, 91–99. [CrossRef]
16. Lian, Q.; Deng, G.; Tieu, A.K.; Li, H.; Liu, Z.; Wang, X.; Zhu, H. Thermo-Mechanical Coupled Finite Element Analysis of Rolling Contact Fatigue and Wear Properties of a Rail Steel under Different Slip Ratios. *Tribol. Int.* **2020**, *141*, 105943. [CrossRef]
17. Zhou, Y.; Peng, J.; Luo, Z.; Cao, B.; Jin, X.; Zhu, M. Phase and Microstructural Evolution in White Etching Layer of a Pearlitic Steel during Rolling–Sliding Friction. *Wear* **2016**, *362–363*, 8–17. [CrossRef]
18. Bernsteiner, C.; Müller, G.; Meierhofer, A.; Six, K.; Künstner, D.; Dietmaier, P. Development of White Etching Layers on Rails: Simulations and Experiments. *Wear* **2016**, *366–367*, 116–122. [CrossRef]
19. Newcomb, S.; Stobbs, W. A Transmission Electron Microscopy Study of the White-Etching Layer on a Rail Head. *Mater. Sci. Eng.* **1984**, *66*, 195–204. [CrossRef]
20. Lojkowski, W.; Millman, Y.; Chugunova, S.; Goncharova, I.; Djahanbakhsh, M.; Bürkle, G.; Fecht, H.J. The Mechanical Properties of the Nanocrystalline Layer on the Surface of Railway Tracks. *Mater. Sci. Eng. A* **2001**, *303*, 209–215. [CrossRef]
21. Takahashi, J.; Kobayashi, Y.; Ueda, M.; Miyazaki, T.; Kawakami, K. Nanoscale Characterisation of Rolling Contact Wear Surface of Pearlitic Steel. *Mater. Sci. Technol.* **2013**, *29*, 1212–1218. [CrossRef]
22. Tao, N.; Wang, Z.; Tong, W.; Sui, M.; Lu, J.; Lu, K. An Investigation of Surface Nanocrystallization Mechanism in Fe Induced by Surface Mechanical Attrition Treatment. *Acta Mater.* **2002**, *50*, 4603–4616. [CrossRef]
23. Sauvage, X.; Ivanisenko, Y. The Role of Carbon Segregation on Nanocrystallisation of Pearlitic Steels Processed by Severe Plastic Deformation. *J. Mater. Sci.* **2007**, *42*, 1615–1621. [CrossRef]
24. Pan, R.; Ren, R.; Chen, C.; Zhao, X. The Microstructure Analysis of White Etching Layer on Treads of Rails. *Eng. Fail. Anal.* **2017**, *82*, 39–46. [CrossRef]
25. Pan, R.; Ren, R.; Chen, C.; Zhao, X. Formation of Nanocrystalline Structure in Pearlitic Steels by Dry Sliding Wear. *Mater. Charact.* **2017**, *132*, 397–404. [CrossRef]
26. Li, S.; Wu, J.; Petrov, R.H.; Li, Z.; Dollevoet, R.; Sietsma, J. "Brown Etching Layer": A Possible New Insight into the Crack Initiation of Rolling Contact Fatigue in Rail Steels? *Eng. Fail. Anal.* **2016**, *66*, 8–18. [CrossRef]
27. He, C.; Ding, H.; Shi, L.; Guo, J.; Meli, E.; Liu, Q.; Rindi, A.; Zhou, Z.; Wang, W. On the Microstructure Evolution and Nanocrystalline Formation of Pearlitic Wheel Material in a Rolling-Sliding Contact. *Mater. Charact.* **2020**, *164*, 110333. [CrossRef]
28. Pan, R.; Chen, Y.; Lan, H.; E, S.; Ren, R. Investigation into the Evolution of Tribological White Etching Layers. *Mater. Charact.* **2022**, *190*, 112076. [CrossRef]
29. Languillaume, J.; Kapelski, G.; Baudelet, B. Cementite Dissolution in Heavily Cold Drawn Pearlitic Steel Wires. *Acta Mater.* **1997**, *45*, 1201–1212. [CrossRef]
30. Sauvage, X.; Copreaux, J.; Danoix, F.; Blavette, D. Atomic-Scale Observation and Modelling of Cementite Dissolution in Heavily Deformed Pearlitic Steels. *Philos. Mag. A* **2000**, *80*, 781–796. [CrossRef]
31. Gavriljuk, V. Decomposition of Cementite in Pearlitic Steel Due to Plastic Deformation. *Mater. Sci. Eng. A* **2003**, *345*, 81–89. [CrossRef]
32. Zhang, X.; Godfrey, A.; Huang, X.; Hansen, N.; Liu, Q. Microstructure and Strengthening Mechanisms in Cold-Drawn Pearlitic Steel Wire. *Acta Mater.* **2011**, *59*, 3422–3430. [CrossRef]
33. Lamontagne, A.; Massardier, V.; Kléber, X.; Sauvage, X.; Mari, D. Comparative Study and Quantification of Cementite Decomposition in Heavily Drawn Pearlitic Steel Wires. *Mater. Sci. Eng. A* **2015**, *644*, 105–113. [CrossRef]
34. Djaziri, S.; Li, Y.; Nematollahi, G.A.; Grabowski, B.; Goto, S.; Kirchlechner, C.; Kostka, A.; Doyle, S.; Neugebauer, J.; Raabe, D.; et al. Deformation-Induced Martensite: A New Paradigm for Exceptional Steels. *Adv. Mater.* **2016**, *28*, 7753–7757. [CrossRef] [PubMed]

35. Lojkowski, W.; Djahanbakhsh, M.; Bürkle, G.; Gierlotka, S.; Zielinski, W.; Fecht, H.J. Nanostructure Formation on the Surface of Railway Tracks. *Mater. Sci. Eng. A* **2001**, *303*, 197–208. [CrossRef]
36. Ivanisenko, Y.; Lojkowski, W.; Valiev, R.; Fecht, H.J. The Mechanism of Formation of Nanostructure and Dissolution of Cementite in a Pearlitic Steel during High Pressure Torsion. *Acta Mater.* **2003**, *51*, 5555–5570. [CrossRef]
37. Steenbergen, M.; Dollevoet, R. On the Mechanism of Squat Formation on Train Rails – Part I: Origination. *Int. J. Fatigue* **2013**, *47*, 361–372. [CrossRef]
38. Antoni, G.; Désoyer, T.; Lebon, F. A Combined Thermo-Mechanical Model for Tribological Surface Transformations. *Mech. Mater.* **2012**, *49*, 92–99. [CrossRef]
39. Thiercelin, L.; Saint-Aimé, L.; Lebon, F.; Saulot, A. Thermomechanical Modelling of the Tribological Surface Transformations in the Railroad Network (White Etching Layer). *Mech. Mater.* **2020**, *151*, 103636. [CrossRef]
40. Hilliard, J.E. Iron-Carbon Phase Diagram Isobaric Sections of the Eutectoid Region at 35, 50 and 65 Kilobars. *Trans. Metall. Soc. AIME* **1963**, *227*, 429–438.
41. Wu, J.; Petrov, R.H.; Naeimi, M.; Li, Z.; Dollevoet, R.; Sietsma, J. Laboratory Simulation of Martensite Formation of White Etching Layer in Rail Steel. *Int. J. Fatigue* **2016**, *91*, 11–20. [CrossRef]
42. Vargolici, O.; Merino, P.; Saulot, A.; Cavoret, J.; Simon, S.; Ville, F.; Berthier, Y. Influence of the Initial Surface State of Bodies in Contact on the Formation of White Etching Layers under Dry Sliding Conditions. *Wear* **2016**, *366–367*, 209–216. [CrossRef]
43. Merino, P.; Cazottes, S.; Lafilé, V.; Risbet, M.; Saulot, A.; Bouvier, S.; Marteau, J.; Berthier, Y. An Attempt to Generate Mechanical White Etching Layer on Rail Surface on a New Rolling Contact Test Bench. *Wear* **2021**, *482–483*, 203945. [CrossRef]
44. Lafilé, V.; Marteau, J.; Risbet, M.; Bouvier, S.; Merino, P.; Saulot, A. Characterization of the Microstructure Changes Induced by a Rolling Contact Bench Reproducing Wheel/Rail Contact on a Pearlitic Steel. *Metals* **2022**, *12*, 745. [CrossRef]
45. Thiercelin, L. Modélisation Multi-Physique Des Mécanismes de Formation de La Phase Blanche. Ph.D. Thesis, Aix-Marseille Université, LMA, Marseille and LaMCoS, INSA Lyon, Marseille, France, 2021. Available online: https://www.theses.fr/2021AIXM0050 (accessed on 31 August 2022).
46. Habak, M. Etude de L'Influence de la Microstructure et des Paramètres de Coupe sur le Comportement en Tournage dur de L'Acier à Roulement 100Cr6. Ph.D. Thesis, Arts et Métiers Institute of Technology, Angers, France, 2006. Available online: https://www.theses.fr/2006ENAM0057 (accessed on 31 August 2022).
47. Dougherty, L.; Cerreta, E.; Gray, G.; Trujillo, C.; Lopez, M.; Vecchio, K.; Kusinski, G. Mechanical Behavior and Microstructural Development of Low-Carbon Steel and Microcomposite Steel Reinforcement Bars Deformed under Quasi-Static and Dynamic Shear Loading. *Metall. Mater. Trans. A* **2009**, *40*, 1835–1850. [CrossRef]
48. Lins, J.; Sandim, H.; Kestenbach, H.J.; Raabe, D.; Vecchio, K. A Microstructural Investigation of Adiabatic Shear Bands in an Interstitial Free Steel. *Mater. Sci. Eng. A* **2007**, *457*, 205–218. [CrossRef]
49. Rail Steel Grades—Steel Compositions and Properties. Available online: https://britishsteel.co.uk/media/40810/steel-grade-dimensions-and-properties.pdf (accessed on 31 August 2022).
50. Dylewski, B.; Risbet, M.; Bouvier, S. The Tridimensional Gradient of Microstructure in Worn Rails – Experimental Characterization of Plastic Deformation Accumulated by RCF. *Wear* **2017**, *392–393*, 50–59. [CrossRef]
51. Gleeble Systems. Available online: https://www.gleeble.com/products/gleeble-systems/gleeble-3800.html (accessed on 31 August 2022).
52. Hor, A. Simulation Physique des Conditions Thermomécaniques de Forgeage et d'Usinage: Caractérisation et Modélisation de la Rhéologie et de l'Endommagement. Ph.D. Thesis, Arts et Métiers Institute of Technology, Angers, France, 2011.
53. Hor, A.; Morel, F.; Lebrun, J.L.; Germain, G. An Experimental Investigation of the Behaviour of Steels over Large Temperature and Strain Rate Ranges. *Int. J. Mech. Sci.* **2013**, *67*, 108–122. [CrossRef]
54. Beausir, B.; Fundenberger, J.J. ATEX; Analysis Tools for Electron and X-ray Diffraction, 2017. Available online: www.atex-software.eu (accessed on 31 August 2022).
55. Simon, S.; Saulot, A.; Dayot, C.; Quost, X.; Berthier, Y. Tribological Characterization of Rail Squat Defects. *Wear* **2013**, *297*, 926–942. [CrossRef]
56. Inoue, A.; Ogura, T.; Masumoto, T. Microstructures of Deformation and Fracture of Cementite in Pearlitic Carbon Steels Strained at Various Temperatures. *Metall. Trans. A* **1977**, *8*, 1689–1695. [CrossRef]
57. Tsuzaki, K.; Matsuzaki, Y.; Maki, T.; Tamura, I. Fatigue Deformation Accompanying Dynamic Strain Aging in a Pearlitic Eutectoid Steel. *Mater. Sci. Eng. A* **1991**, *142*, 63–70. [CrossRef]
58. Gonzalez, B.; Marchi, L.; Fonseca, E.; Modenesi, P.; Buono, V. Measurement of Dynamic Strain Aging in Pearlitic Steels by Tensile Test. *ISIJ Int.* **2003**, *43*, 428–432. [CrossRef]
59. Saxena, A.K.; Kumar, A.; Herbig, M.; Brinckmann, S.; Dehm, G.; Kirchlechner, C. Micro Fracture Investigations of White Etching Layers. *Mater. Des.* **2019**, *180*, 107892. [CrossRef]
60. Carroll, R.I. Surface Metallurgy and Rolling Contact Fatigue of Rail. Ph.D. Thesis, University of Sheffield, Sheffield, UK, 2005.
61. Park, J. Quantitative Measurement of Cementite Dissociation in Drawn Pearlitic Steel. *Mater. Sci. Eng. A* **2011**, *528*, 4947–4952. [CrossRef]
62. Murugan, H. Study on White Etching Layer in R260Mn Rail Steel by Thermo-Mechanical Simulation. Master's Thesis, Technische Universiteit Delft, Delft, The Netherlands, 2018. Available online: http://resolver.tudelft.nl/uuid:1b330f31-4ec7-4ff7-aff7-486174395ab4 (accessed on 31 August 2022).

63. Al-Juboori, A. Thermomechanical Simulation of White Etching Layer Formation on Rail Steel. *Mater. Forum* **2018**, *51*, 50–55.
64. Lewis, R.; Olofsson, U. Mapping Rail Wear Regimes and Transitions. *Wear* **2004**, *257*, 721–729. [CrossRef]
65. Eleöd, A.; Baillet, L.; Berthier, Y.; Törköly, T. *Deformability of the Near Surface Layer of the First Body*; Tribology Series; Elsevier: Amsterdam, The Netherlands, 2003; Volume 41, pp. 123–132. [CrossRef]
66. Eleöd, A.; Berthier, Y.; Baillet, L.; Törköly, T. *Transient and Stationary Changes of the Mechanical Properties of the First Body Governed by the Hydrostatic Pressure Component of the Local Stress State during Dry Friction*; Tribology Series; Elsevier: Amsterdam, The Netherlands, 2003; Volume 43, pp. 553–561. [CrossRef]

MDPI AG
Grosspeteranlage 5
4052 Basel
Switzerland
Tel.: +41 61 683 77 34

Materials Editorial Office
E-mail: materials@mdpi.com
www.mdpi.com/journal/materials

Disclaimer/Publisher's Note: The title and front matter of this reprint are at the discretion of the Guest Editor. The publisher is not responsible for their content or any associated concerns. The statements, opinions and data contained in all individual articles are solely those of the individual Editor and contributors and not of MDPI. MDPI disclaims responsibility for any injury to people or property resulting from any ideas, methods, instructions or products referred to in the content.

www.ingramcontent.com/pod-product-compliance
Lightning Source LLC
LaVergne TN
LVHW072348090526
838202LV00019B/2503